Lecture Notes in Physics

The Lecture Notes in Physics

The series Lecture Notes in Physics (LNP), founded in 1969, reports new developments in physics research and teaching – quickly and informally, but with a high quality and the explicit aim to summarize and communicate current knowledge in an accessible way. Books published in this series are conceived as bridging material between advanced graduate textbooks and the forefront of research and to serve three purposes:

- to be a compact and modern up-to-date source of reference on a well-defined topic

- to serve as an accessible introduction to the field to postgraduate students and nonspecialist researchers from related areas

- to be a source of advanced teaching material for specialized seminars, courses and schools

Both monographs and multi-author volumes will be considered for publication. Edited volumes should, however, consist of a very limited number of contributions only. Proceedings will not be considered for LNP.

Volumes published in LNP are disseminated both in print and in electronic formats, the electronic archive being available at springerlink.com. The series content is indexed, abstracted and referenced by many abstracting and information services, bibliographic networks, subscription agencies, library networks, and consortia.

Proposals should be sent to a member of the Editorial Board, or directly to the managing editor at Springer:

Christian Caron
Springer Heidelberg
Physics Editorial Department I
Tiergartenstrasse 17
69121 Heidelberg / Germany
christian.caron@springer.com

D. Husemöller
M. Joachim
B. Jurčo
M. Schottenloher

Basic Bundle Theory and K-Cohomology Invariants

With contributions by Siegfried Echterhoff, Stefan
Fredenhagen and Bernhard Krötz

 Springer

Authors

D. Husemöller
MPI für Mathematik
Vivatsgasse 7
53111 Bonn, Germany
dale@mpim-bonn.mpg.de

B. Jurčo
MPI für Mathematik
Vivatsgasse 7
53111 Bonn, Germany
jurco@mathematik.uni-muenchen.de

M. Joachim
Universität Münster
Mathematisches Institut
Einsteinstr. 62
48149 Münster, Germany
joachim@math.uni-muenster.de

M. Schottenloher
Universität München
Mathematisches Institut
Theresienstr. 39
80333 München, Germany
Martin.Schottenloher@Mathematik.
Uni-Muenchen.de

Contributors

Siegfried Echterhoff (*Chap. 17, Appendix*)
Mathematisches Institut
Westfälische Wilhelms-Universität
Münster, Germany

Bernhard Krötz (*Chap. 16, Appendix*)
MPI für Mathematik, Bonn, Germany

Stefan Fredenhagen (*Physical Background to the K-Theory Classification of D-Branes*), MPI für Gravitationsphysik, Potsdam, Germany

D. Husemöller et al., *Basic Bundle Theory and K-Cohomology Invariants*, Lect. Notes Phys. 726 (Springer, Berlin Heidelberg 2008), DOI 10.1007/978-3-540-74956-1

ISSN 0075-8450
ISBN 978-3-642-09436-1   e-ISBN 978-3-540-74956-1

Springer is a part of Springer Science+Business Media
springer.com
© Springer-Verlag Berlin Heidelberg 2008
Softcover reprint of the hardcover 1st edition 2008

Cover design: eStudio Calamar S.L., F. Steinen-Broo, Pau/Girona, Spain

Dedicated to the memory of Julius Wess

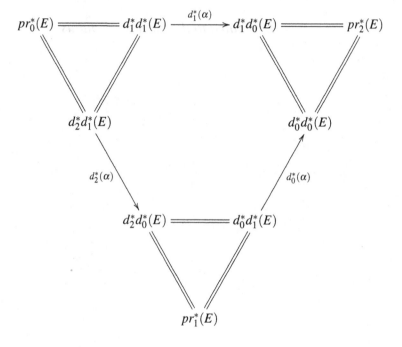

Preface

This lecture notes volume has its origins in a course by Husemöller on fibre bundles and twisted K-theory organized by Brano Jurčo for physics students at the LMU in München, summer term 2003. The fact that K-theory invariants, and in particular twisted K-theory invariants, were being used in the geometric aspects of mathematical physics created the need for an accessible treatment of the subject. The course surveyed the book *Fibre Bundles*, 3rd. Ed. 1994, Springer-Verlag by Husemöller, and covered topics used in mathematical physics related to K-theory invariants. This book is referred to just by its title throughout the text.

The idea of lecture notes came up by J. Wess in 2003 in order to serve several purposes. Firstly, they were to be a supplement to the book *Fibre Bundles* providing companion reading and alternative approaches to certain topics; secondly, they were to survey some of the basic results of background to K-theory, for example operator algebra K-theory, not covered in the *Fibre Bundles*; and finally the notes would contain information on the relation to physics. This we have done in the survey following this introduction "Physical background to the K-theory classification of D-branes: Introduction and references" tracing the papers how and where K-theory invariants started to play a role in string theory. The basic references to physics are given at the end of this survey, while the mathematics references are at the end of the volume.

Other lectures of Husemöller had contributed to the text of the notes. During 2001/2002 in Münster resp. during Summer 2002 in München, Husemöller gave Graduate College courses on the topics in the notes, organized by Joachim Cuntz resp. Martin Schottenloher, and in the Summer 2001, he had a regular course on C*-algebras and K-theory in München. The general question of algebra bundles was studied with the support of Professor Cuntz in Münster during short periods from 2003 to 2005. Finally, Husemöller lectured on these topics during a workshop at IPM, Tehran, Iran, September 2005. It is with a great feeling of gratitude that these lecture opportunities are remembered here.

The notes are organized into five parts. The first part on basic bundle theory emphasizes the concept of bundle as one treats the concepts of set, space, homotopy, group, or ring in basic mathematics. A bundle is just a map called the projection from the total space to its base space. As with commutative groups, topological groups, transformation groups, and Lie groups, the concept of bundle is enhanced or enriched with additional axioms and structures leading to étale bundles, principal bundles, fibre bundles, vector bundles, and algebra bundles. A topic discussed in the first part, which is not taken up in the *Fibre Bundles*, is the Serre–Swan theorem which relates vector bundles on a compact space X with finitely generated projective modules over the ring $C(X)$ of continuous complex valued functions on the space X. This is one of the points where topological, algebraic, and operator K-theory come together.

The second part of the notes takes up the homotopy classification of principal bundles and fibre bundles. Applications to the case of vector bundles are considered and the role of homotopy theory in K-theory is developed. This is related to the fact that K-theory is a representable functor on the homotopy category. The theory of characteristic classes in describing orientation and spin structures on vector bundles is carried out in detail, also leading to the notion of a string structure on a bundle and on a manifold.

There are various versions of topological K-theory, and their relation to Bott periodicity is considered in the third part of the notes. An advanced version of operator K-theory, called KK-theory which integrates K-cohomology and K-homology, is introduced, and various features are sketched.

The fourth part of the notes begins with algebra bundles with fibres that are either matrix algebras or algebras of bounded operators on a separable Hilbert space. The infinite dimensional algebra bundles are classified by only one characteristic class in the integral third cohomology group of the base space along the lines of the classification of complex line bundles with its first Chern class in the integral second cohomology group of the base space. The twisting of twisted K-theory is given by an infinite dimensional algebra bundle, and the twisted K-theory is defined in terms of cross sections of Fredholm bundles related to the algebra bundle describing the twist under consideration.

A fundamental theme in bundle theory centers around the gluing of local bundle data related to bundles into a global object. In the fifth part we return to this theme and study gluing on open sets in a topological space of not just simple bundle data but also data in a more general category where the gluing data may satisfy transitivity conditions only up to an isomorphism. The resulting objects are gerbes or stacks.

August 2007   Dale Husemöller

Contents

Physical Background to the K-Theory Classification of D-Branes: Introduction and References

by S. Fredenhagen

Topological D-Brane Charges

In quantum field theories, we are often confronted with the situation that there are extended field configurations that are topologically different, for example, instantons and monopoles. They can carry charges that are topological invariants and so are conserved under small fluctuations. Similarly in string theory, D-branes carry topological charges. In the semiclassical geometric description, these charges can be understood as sources of Ramond–Ramond (RR) fields, higher form fields that couple electrically and magnetically to the D-branes. These charges have to be quantized, similar to the Dirac quantization of electric and magnetic charges in electrodynamics.

The classification of D-branes and their charges was a topic of great importance since their discovery. Minasian and Moore (1997) suggested that D-brane charges are classified by K-theory and not just by homology as was proposed first. See Chap. 4 and 9 for K-theory and cohomology as well as their connection in Part 3.

Drawbacks of the Homological Classification

To discuss the homological classification and its drawbacks, we consider D-branes in type II string theory on a spacetime with the topology $\mathbb{R} \times M$. Here, $\mathbb{R}$ represents the time direction and M is a compact 9-manifold. Dp-branes are extended along the time direction and also wrapped around a p-dimensional submanifold of M.

The strategy to study D-brane charges is always the same, firstly, identify the set of all possible, static D-branes, and then quotient this set out by all dynamical transformations between different D-brane configurations. Viewing the D-brane as an object with tension, a static D-brane cannot have a boundary, and the wrapped p-dimensional submanifold is a p-cycle (of minimal volume). Smooth deformations

D. Husemöller et al.: *Physical Background to the K-Theory Classification of D-Branes*, Lect. Notes Phys. **726**, 1–6
(2008)
DOI 10.1007/978-3-540-74956-1_1

of the submanifold should not change the charge, so one could think that the submanifolds wrapped by D-branes are classified by homotopy classes of p-cycles in M. This is not completely true, because there is one more dynamical input, namely branes that are boundaries of $(p+1)$-dimensional submanifolds are unstable, and they can be removed (not smoothly, but with splitting and joining) from M along the $(p+1)$-dimensional submanifold. So we conclude that D-brane charges are classified by the homology group $H_p(M,\mathbb{Z})$ (see Chap. 9; if M is noncompact, the compactly supported homology has to be used, see Chap. 11). This has an interesting consequence as the group $H_p(M,\mathbb{Z})$ can contain torsion elements, so it opens the possibility of a process in which a stack of n coincident Dp-branes are annihilated. Such configurations can never preserve any supersymmetry. A superposition of supersymmetric D-branes that satisfy a Bogomolnyi–Prasad–Sommerfield (BPS) bound can never decay. The homological classification thus captures stable D-branes that are missed by supersymmetry-based classifications.

The homological classification—despite its successes—is not the correct one. It contains charges that are not conserved, and it contains charges that even cannot be realized by any physical brane. An example of the latter point is the Freed–Witten anomaly (Freed and Witten) (1999) that forbids branes to wrap some nontrivial homology cycles. To obtain the group of conserved charges, one has to subtract the unphysical ones and quotient out the unstable ones—this leads naturally to the K-theory classification of D-brane charges (see Sect. 5 in (Diaconescu et al. 2003)). In the case of a non-trivial H-field, it leads to twisted K-theory, where the three-form field H defines a class $[H]$ in $H^3(M,\mathbb{Z})$ (see Chap. 20).

K-Theory Invariants (Minasian and Moore)

As it was already mentioned, the first proposal for the K-theory classification of D-brane charges in type II string theory was due to Minasian and Moore (1997). Besides sourcing RR fields, D-branes also support ordinary gauge fields, the so-called Chan–Paton bundles. This is related to the fact that D-branes are submanifolds where open strings end and these endpoints of open strings carry charges, the so-called Chan–Paton factors. Minasian and Moore have found a formula for the D-brane charge Q with respect to all RR fields

$$Q = \mathrm{ch}(f_! E) \cup \sqrt{\hat{A}(M)} \in H^{\mathrm{even}}(M,\mathbb{Q}).$$

Here, E is a gauge bundle on the D-brane worldvolume N, $f: N \to M$ describes the embedding of the D-brane into M, $f_!$ is the corresponding push-forward, ch is the Chern character, and $\hat{A}(M)$ is the Atiyah–Hirzebruch class of M (see Chap. 10 for characteristic classes).

To explain the construction $f_! E$, we follow Grothendieck and Atiyah–Hirzebruch by introducing the group $K(M)$ whose elements are formal differences $E' - E''$ of bundles E', E'' on M up to isomorphism. The following relations are to be satisfied in $K(M)$:

$$E' - E'' = E' \oplus E - E'' \oplus E$$

and

$$E = E' + E'' \text{ in } K(M) \text{ for } E = E' \oplus E'' \text{ on } M.$$

The operation $f_!$ on a bundle is not in general a bundle but an element of the K-group $K(M)$ for $f : N \to M$.

Minasian and Moore noticed that the charge Q can be interpreted as a modified Chern isomorphism between K-theory and cohomology, $Q : K(M) \to H^{\text{even}}(M, \mathbb{Q})$ (see 10(5.4), which is Sect. 5.4 in Chap. 10). This can be made into an isometry if one equips the groups $K(M)$ and $H^{\text{even}}(M, \mathbb{Q})$ with suitable pairings, namely the K-theory pairing which is defined as an index of a properly chosen Dirac operator of the tensored K-classes and the cohomology pairing which is obtained as Poincaré duality applied to the cohomology classes modified by the square root of the Atiyah–Hirzebruch class. Their suggestion was that the K-theory classes might contain more information about the D-branes as the corresponding charge in the cohomology.

Physical Explanation due to Witten for a Vector Bundle Description

While the work of Minasian and Moore gave a hint that K-theory is the more natural description for D-brane charges, Witten (1998) gave a physical explanation why K-theory should be the correct framework. The formal difference of bundles $E' - E''$ appearing in K-theory is interpreted (in type IIB string theory) as a configuration of space-filling branes with gauge bundle E' and space-filling anti-D-branes with gauge bundle E''. Sen's (1998) conjecture states that—when the H-field is trivial— all brane configurations can be obtained from such stacks of space-filling branes and anti-branes via tachyon condensation. The annihilation of branes and anti-branes with isomorphic bundles E is the physical interpretation of the K-theory relation

$$E' - E'' = E' \oplus E - E'' \oplus E .$$

This argument was extended by Hořava (1999) to type IIA D-brane configurations, which are classified by $K^1(M)$.

In the presence of a nontrivial H-field, the picture based on Sen's conjecture has to be modified, and one is led to consider twisted K-theory. This was suggested by Witten (1998) in the case of a torsion H-field and by Bouwknegt and Mathai (2000) in the nontorsion case.

Role of the Freed–Witten Anomaly and the Twisted K-Theory

A different approach to the K-theory classification of D-branes based on the Freed–Witten anomaly was pioneered by Maldacena et al. (2001). The Freed–Witten anomaly is a global world-sheet anomaly of string theory in the presence of D-branes and a nontrivial H-field. The strategy of Maldacena, Moore, and Seiberg was again to start with all allowed (here, anomaly-free) D-branes and then identify configurations that can be dynamically transformed into each other. This leads to the following two conditions:

1. A D-brane can wrap a cycle $N \subset M$ only if

$$W_3(N) + [H]|_N = 0$$

   in $H^3(M, \mathbb{Z})$. Here, $W_3(N)$ is the third integral Stiefel–Whitney class of TN. If the H-field is trivial, then the first condition just says that the D-brane must be spin$^\mathbb{C}$.
2. Branes wrapping homologically nontrivial N can nevertheless be unstable if, for some $N' \subset M$ containing N,

$$PD(N \subset N') = W_3(N') + [H]|_{N'},$$

   where $PD(N \subset N')$ stands for Poincaré dual of N in N'.

These two conditions lead naturally to the twisted K-theory classification of D-brane charges. The twisted K-theory class on the spacetime comes as a (twisted K-theory, see Part 4) push-forward of an ordinary untwisted K-theory class on the D-brane. Finally, both the unphysical and unstable branes are nicely interpreted within the Atiyah–Hirzebruch spectral sequence (see Chap. 23).

D-Branes as Boundary Conditions for Open Strings

As already mentioned, open strings end on D-branes, which in turn means that D-branes can be characterized by (conformally invariant) boundary conditions on the open string worldsheet. From this worldsheet point of view, the D-brane charge group is again obtained along a similar strategy as before. Firstly, we classify all boundary conditions for a given closed string background and then identify those which are connected by renormalization group (RG) flows on the world-sheet boundary. Unfortunately, in most situations, one is neither able to classify all boundary conditions nor to classify all RG flows between them. In some cases, however, this strategy was nevertheless successfully pursued as will be discussed in the following paragraph. One should stress that this classification is in a way complementary to the ones already discussed, because it does not rely on any geometrical structures of the target space.

For a two-dimensional topological field theory, where the whole content of the theory is encoded in a finite-dimensional Frobenius algebra, this problem was addressed by Moore and Segal (2004). By using sewing constraints, they obtained the complete classification of D-branes in these theories in terms of K-theory.

WZW Models and the Freed, Hopkins, Teleman Theorem

Wess–Zumino–Witten (WZW) models provide an important class of backgrounds, where a lot is known on the conformal field theory description of D-branes. Alekseev and Schomerus (2007) worked out the charge group from the worldsheet approach for the $SU(2)$ WZW model, which led to precise agreement with the twisted K-theory. A more structural connection between twisted K-theory and conformal field theory data was shown by Freed et al. (2005) in the following form. Consider a simple, simply connected, compact Lie group G of dimension d. Central extensions of its smooth loop group LG by the circle group T are classified by their level k. Positive energy representations of LG at fixed level are the ones which are important in string theory. The free abelian group $R^k(LG)$ of irreducible isomorphism classes with the multiplication given by fusion rules of conformal field theory is called the Verlinde ring of G at level k (see Chap. 24). If h is the dual Coxeter number and $k + h > 0$ is interpreted as an equivariant twisting class in $H_G^3(G, \mathbb{Z})$, then the ring $R^k(LG)$ is isomorphic to the d mod 2-shifted equivariant twisted K-theory $K_G^{k+h,+d}(G)$. Here, G acts on itself by the adjoint action and the ring structure on K-theory is the convolution (Pontryagin) product. An explicit realization of this isomorphism given by a Dirac operator—the gauge coupled supercharge of the level $k + h$ supersymmetric WZW model—was proposed by Mickelsson (2004). Physically, the equivariant twisted K-theory describes D-branes in the coset model G/G, which is a topological field theory whose Frobenius algebra is precisely the Verlinde algebra.

Suggested Reading

A good point to start studying the use of K-theory in string theory is the original paper by Witten (1998). There is also a good review by Olsen and Szabo (1999). For the classification of D-brane charges in the presence of a nontrivial H-field in terms of twisted K-theory, one might consult the paper by Maldacena, Moore and Seiberg (2001) and the review by Moore (2004). A more recent review on K-theory in string theory that also discusses the limitations of the K-theory classification is the one by Evslin (2006).

References

Alekseev, A., Schomerus, V.: RR charges of D2-branes in the WZW model (arXiv:hep-th/0007096)

Bouwknegt, P., Mathai, V.: D-branes, B-fields and twisted K-theory. JHEP **0003**:007 (2000) (arXiv:hep-th/0002023)

Diaconescu, D.E., Moore, G.W., Witten, E.: E(8) gauge theory, and a derivation of K-theory from M-theory. Adv. Theor. Math. Phys. **6**:1031 (2003) (arXiv:hep-th/0005090)

Evslin, J.: What does(n't) K-theory classify? (arXiv:hep-th/0610328)

Freed, D.S., Witten, E.: Anomalies in string theory with D-branes, Asian J. Math., Vol. 3, **4**:819–851 (1999) (arXiv:hep-th/9907189)

Freed, D.S., Hopkins, M.J., Teleman, C.: Twisted K-theory and Loop Group Representations (arXiv:math/0312155)

Freed, D.S., Hopkins, M.J., Teleman, C.: Loop Groups and Twisted K-Theory II (arXiv:math/0511232)

Hořava, P.: Type IIA D-branes, K-theory, and matrix theory. Adv. Theor. Math. Phys. **2**:1373 (1999) (arXiv:hep-th/9812135)

Maldacena, J.M., Moore, G.W., Seiberg, N.: D-brane instantons and K-theory charges. JHEP **0111**:062 (2001) (arXiv:hep-th/0108100)

Mickelsson, J.: Gerbes, (twisted) K-theory, and the supersymmetric WZW model, Infinite dimensional groups and manifolds, IRMA Lect. Math. Theor. Phys. Vol. 5, 93–107; de Gruyter; Berlin (2004) (arXiv:hep-th/0206139)

Minasian, R. and Moore, G.W.: K-theory and Ramond-Ramond charge. JHEP **9711**:002 (1997) (arXiv:hep-th/9710230)

Moore, G.W., Segal, G.: D-branes and K-theory in 2D topological field theory (arXiv:hep-th/0609042)

Moore, G.W.: K-theory from a physical perspective; Topology, geometry and quantum field theory, London Math. Soc. Lecture Note Ser., Vol. 308, 194–234, Cambridge Univ. Press, Cambridge, (2004) (arXiv:hep-th/0304018)

Olsen, K., Szabo, R.J.: Constructing D-branes from K-theory. Adv. Theor. Math. Phys. **3**:889 (1999) (arXiv:hep-th/9907140)

Sen, A.: Tachyon condensation on the brane antibrane system. JHEP **9808**:012 (1998) (arXiv:hep-th/9805170)

Witten, E.: D-branes and K-theory. JHEP **9812**:019 (1998) (arXiv:hep-th/9810188)

Part I
Bundles over a Space and Modules over an Algebra

In his notes, "A General Theory of Fibre Spaces with Structure Sheaf" published by Kansas University (1955 resp. 1958), Grothendieck found it useful for general questions to think of a map $p : E \to B$ as an object which he called a fibre space. The fibres $E_b = p^{-1}(b)$ are parametrized by elements $b \in B$ and assembled in the set $E = \bigcup_{b \in B} E_b$ in terms of the topology of E.

Since *fibre space* had already a precise meaning in the thesis of Serre, we followed the Grothendieck idea but changed the terminology to *bundle* as in *Fibre Bundles*. The term bundle fits with the idea that a bundle with additional structure would be compatible with the usual terminology for such notions as vector bundle, principal bundle, and fibre bundle as explained in Chaps. 2 and 5.

Later Grothendieck extended the idea of associating to a category $\mathscr{C}$, the category of morphisms or bundles $p : E \to B$. The idea of the category of morphisms is found in many contexts in mathematics. For example, a variety or scheme $p : X \to S$ over a scheme S is considered as a family of varieties or schemes $X_s = p^{-1}(s)$ parametrized by $s \in S$.

In another direction, we can associate to a space X the C^*-algebra $C(X)$ of all bounded continuous functions $f : X \to \mathbb{C}$ with the uniform norm $||f|| = \sup_{x \in X} |f(x)|$. In the case of a compact X, we can recover X from the algebra $C(X)$ as the space of maximal ideals. This is the Gelfand–Naimark theorem. For compact B, we can describe vector bundles over B as finitely generated projective modules over $C(B)$. This is the Serre–Swan theorem which is proved in Chap. 3. In this same chapter, we go further and show how finitely generated projective modules M over a ring R and idempotents $e = e^2$ in the matrix algebras $M_n(R)$ are related. Isomorphism classes of projective modules correspond to algebraic equivalence classes of idempotents in all the possible matrix algebras.

In Chap. 4, we relate stable classes of vector bundles over X to stable classes of finitely generated projective $C(X)$-modules. Then, we relate stable classes of finitely generated projective R-modules to stable classes of idempotents $e \in M_n(R)$ for all n. These are the three versions of K-theory discussed in the first part.

As preparation for the homotopy version of K-theory considered in Part II, we introduce the basic properties of principal fibre bundles in Chap. 5. Every vector bundle is the fibre bundle of a principal $GL(n)$-bundle, and frequently, the special extra features of a vector bundle can be described in terms of the related principal bundle. Local triviality is considered, and the related notion of descent which plays a basic role in algebraic geometry is introduced for principal bundles. The principal bundle is also the natural domain for the study of the gauge group of a bundle.

Grothendieck, A.: A General Theory of Fibre Spaces with Structure Sheaf, University of Kansas, Lawrence (1955)

Grothendieck, A.: Le groupe de Brauer. Part I. Séminaire Bourbaki, Vol. 9, Exp. 290: 199–219 University of Kansas, Lawrence, 1955

Chapter 1
Generalities on Bundles and Categories

This is a preliminary chapter introducing much of the general terminology in the topology of bundles with the general language of category theory. We use the term bundle in the most general context, and then in the next chapters, we define the main concepts of our study, that is, vector bundles, principal bundles, and fibre bundles, as bundles with additional structure.

In Grothendieck's notes "A General Theory of Fibre Spaces with Structure Sheaf," University of Kansas, Lawrence, 1955 (resp. 1958) he mentions that "the functor aspect of the notions dealt with has been stressed through, and as it now appears should have been stressed even more."

Hopefully, we are carrying this out to the appropriate extent in this approach to bundles mixed with a general introduction to category theory where examples are drawn from the theory of bundles. The reader with a background in category theory and topology will see only a slightly different approach from the usual one.

We will introduce several notations used through the book, for example, (set), (top), (gr), and (k) denote, respectively, the categories of sets, spaces, groups, and k-modules for a commutative ring k. These are explained in the context of the definition of a category in Sect. 4 and in the notations for categories at the end of the book.

Chap. 2 of *Fibre Bundles* (Husemöller 1994) is a reference for this chapter.

1 Bundles Over a Space

The following section, as mentioned in the introduction, is motivated by Grothendieck's Kansas notes where we use the term "bundle" instead of "fibre space." A space is a topological space, and a map or mapping is a continuous function.

1.1. Definition Let B be a space. A bundle E over B is a map $p : E \to B$. The space E is called the total space of the bundle, the space B is called the base space of the bundle, the map p is called the projection of the bundle, and for each $b \in B$, the subspace $p^{-1}(b)$, denoted often by E_b, is called the fibre of the bundle over $b \in B$.

D. Husemöller et al.: *Generalities on Bundles and Categories*, Lect. Notes Phys. **726**, 9–22 (2008)
DOI 10.1007/978-3-540-74956-1_2

1.2. Example The product bundle $B \times Y$ over B with fibre Y is the product space with the projection $pr_B = p : B \times Y \to B$. Observe that the fibre $(B \times Y)_b = \{b\} \times Y$ which is isomorphic to Y for all $b \in B$. Another general example comes by restricting a bundle $p : E \to B$ to subspaces $p|E' : E' \to B'$, where E' is a subspace of E and B' is a subspace of B with the property that $p(E') \subset B'$. Many bundles arise as restrictions of a product bundle.

1.3. Definition A morphism from the bundle $p : E \to B$ to the bundle $p' : E' \to B$ over B is a map $u : E \to E'$ such that $p'u = p$.

This condition $p'u = p$ is equivalent to the condition $u(E_b) \subset E_b'$ for all $b \in B$, that is, a morphism u is a map which is fibre preserving. The identity $E \to E$ is a morphism, and if $u' : E' \to E''$ is a second morphism of bundles over B, then $u'u : E \to E''$ is a morphism of bundles.

In Sect. 4, we give the formal definition of a category, but for now, we can speak of this data of bundles and morphisms of bundles with composition as a category. It is the category (bun/B) of bundles over B.

In fact, we can relate bundles over distinct base spaces with morphisms generalizing (1.3).

1.4. Definition A morphism from the bundle $p : E \to B$ to the bundle $p' : E' \to B'$ is a pair of maps (u, f), where $u : E \to E'$ and $f : B \to B'$ such that $p'u = fp$. In particular, the following diagram is commutative

$$
\begin{array}{ccc}
E & \xrightarrow{\ u\ } & E' \\
\downarrow{\scriptstyle p} & & \downarrow{\scriptstyle p'} \\
B & \xrightarrow{\ f\ } & B' .
\end{array}
$$

This condition $p'u = fp$ is equivalent to the condition $u(E_b) \subset E_{f(b)}'$ for all $b \in B$, that is, a morphism u is a map which carries the fibre over b to the fibre over $f(b)$. The pair of identities $(\mathrm{id}_E, \mathrm{id}_B)$ is a morphism $E \to E$, and if $(u', f') : E' \to E''$ is a second morphism of bundles, then $(u'u, f'f) : E \to E''$ is a morphism of bundles defining composition $(u', f')(u, f)$. With these definitions, we have a new category (bun) which contains (top) the category of all spaces as the full subcategory of bundles $\mathrm{id}_B : B \to B$. Each category (bun/B) is contained in (bun) as a subcategory but not as a full subcategory except in the case where B reduces to a point $*$ or the empty set. In this case, $(\mathrm{bun}/*)$ is equivalent to the category (top) viewed as the category of total spaces, and it gives another inclusion of the category (top) into (bun).

We have an important functor associated with bundles over a space which we introduce directly. The formal definition is in a later section.

1.5. Definition Let $p : E \to B$ be a bundle over B, and let U be an open subset of B. The set $\Gamma(U, E)$ of sections (or cross sections) of E over U is the set of continuous maps $\sigma : U \to E$ with $p(\sigma(b)) = b$ for all $b \in U$. For a morphism $u : E \to E'$ in (bun/B), we define a function $\Gamma(U, u) : \Gamma(U, E) \to \Gamma(U, E')$ by

$$\Gamma(U,u)(\sigma) = u\sigma \in \Gamma(U,E')$$

for $\sigma \in \Gamma(U,E)$.

The formal algebraic properties of the set of cross sections are explained in two remarks (5.5) and (5.6).

1.6. Example Let $p : E = B \times Y \to B$ be the product bundle with $p(b,y) = b$ as in (1.2). A cross section σ over an open subset $U \subset X$ is a map $\sigma : U \to E = B \times Y$ of the form

$$\sigma(b) = (b, f_\sigma(b))$$

for $b \in U$. This means that $\Gamma(U,E)$ is just equivalent to the set of maps $f_\sigma : U \to Y$ to the fibre Y.

Now, we consider the special case of an induced bundle where the general concept is considered in Sect. 3.

1.7. Definition Let A be a subspace of a space B and let $p : E \to B$ be in (bun/B). Then, the restriction $E|A$ is the restriction $q = p|p^{-1}(A) : p^{-1}(A) \to A$ in (bun/A) so that as a space $E|A = p^{-1}(A)$. For a morphism $u : E \to E'$, the morphism $u|A : E|A \to E'|A$ is the subspace restriction.

For a pair of subspaces $C \subset A \subset B$, we have $E|C = (E|A)|C$, and the restriction is a functor $(\text{bun}/B) \to (\text{bun}/A)$ as will be further discussed in Sect. 5. With the restriction, we can introduce an important concept.

1.8. Definition A bundle $p : E \to B$ is locally trivial with fibre Y provided each $b \in B$ has an open neighborhood U with $E|U$ isomorphic to the product bundle $pr_1 : U \times Y \to U$. A bundle $p : E \to B$ is trivial with fibre Y provided it is isomorphic to the product bundle $pr_1 : B \times Y \to B$.

2 Examples of Bundles

Many examples come under the following construction which consists of an arbitrary base space and finite dimensional vector spaces in the fibre. In general, the definitions and examples extend to the case of an infinite dimensional vector space with a given topology as a fibre, for example, an infinite dimensional Hilbert space. See Chaps. 20–22.

2.1. Example Let $pr_X : X \times V \to X$ be the product vector bundle over a space X with fibre a real vector space V with its natural topology. A subvector bundle $p : E \to X$ is given by a subspace $E \subset X \times V$ such that for $p(x,v) = x$, each fibre $p^{-1}(x) = E_x \subset \{x\} \times V$ is a subvector space of V. A cross section $s \in \Gamma(U,E)$ is given by $s(x) = (x, v(x))$, where v is a map $v : U \to V$ such that $(x, v(x)) \in E \subset X \times V$ for each $x \in X$ in X the base space. Such a cross section is called a vector field, and most vector fields in mathematics and physics are of this simple form. The constraint on the values of the vector field is contained in the nature of the bundle $p : E \to X$.

For the next example, we use the inner product

$$(x|y) = x_0y_0 + \ldots + x_ny_n$$

for $x, y \in \mathbb{R}^{n+1}$ in Euclidean space.

2.2. Example Tangent and normal bundle to the sphere. The sphere S^n of dimension n is the closed subspace of $\mathbb{R}^{n+1}$ consisting of all $(x_0, \ldots, x_n)$ satisfying $(x|x) = x_0^2 + \ldots + x_n^2 = 1$. As in (2.1), the tangent bundle $T(S^n)$ is the subbundle of the product bundle $p_1 : S^n \times \mathbb{R}^{n+1} \to S^n$ consisting of all $(b, x) \in S^n \times \mathbb{R}^{n+1}$ satisfying $(b|x) = 0$, and the normal bundle $N(S^n)$ is the subbundle of the product bundle $p_1 : S^n \times \mathbb{R}^{n+1} \to S^n$ consisting of all $(b, tb) \in S^n \times \mathbb{R}^{n+1}$, where $t \in \mathbb{R}$.

2.3. Remark Any $(b, y) \in S^n \times \mathbb{R}^{n+1}$ can be decomposed as a perpendicular sum in the fibre $\{b\} \times \mathbb{R}^{n+1}$ by the formula

$$y = (y - (y|b)b) + (y|b)b,$$

where $y - (y|b)b \in T(S^n)$ and $(y|b)b \in N(S^n)$. This is an example of a fibre product considered in (3.7). Classically, the fibre product of two bundles of vector bundles is called the Whitney sum and denoted by

$$T(S^n) \oplus N(S^n) = S^n \times \mathbb{R}^{n+1}.$$

When a space B is a quotient space of a space X, it is in many cases possible to extend the quotient operation to the product bundle $X \times Y \to X$ giving a bundle $E \to B$ with fibre Y. We consider one basic example.

2.4. Example The real projective space $P_n(\mathbb{R})$ of lines in $\mathbb{R}^{n+1}$ through the origin can be viewed as the quotient of S^n by the relation $x = -x$, since for such a line L we have $L \cap S^n = \{x, -x\}$. For any real vector space V, we can take the quotient of the product bundle $S^n \times V \to S^n$ by the relation $(x, v) = (-x, -v)$ giving a bundle $q : E(V) \to P_n(\mathbb{R})$, where the fibre $q^{-1}(\{x, -x\}) = \{x\} \times V$ is identified with $\{-x\} \times V$ under $(x, v) = (-x, -v)$. Again, the fibre is isomorphic to the vector space V. If we were to identify $(x, v) = (-x, v)$, the quotient bundle would again be the product bundle

$$P_n(\mathbb{R}) \times V \to P_n(\mathbb{R})$$

up to isomorphism. On the other hand, $E(V)$ is very far from being the product bundle as we shall see as the theory develops in the first two parts.

2.5. Remark The Whitney sum in (2.3) of the product bundle over S^n denoted by $T(S^n) \oplus N(S^n) = S^n \times \mathbb{R}^{n+1}$ has a quotient version on the real projective space of a Whitney sum denoted by

$$T(P_n(\mathbb{R})) \oplus (P_n(\mathbb{R}) \times \mathbb{R}) = E(\mathbb{R}^{n+1}).$$

Here, $T(P_n(\mathbb{R}))$ is the tangent bundle to the real projective space $P_n(\mathbb{R})$, and the relation $(b,y) = (-b,-y)$ becomes under the sum

$$y = (y - (y|b)b) + (y|b)b$$

the relation

$$(b, y - (y|b)b) \oplus (b, (y|b)b) = (-b, -(y - (y| - b)(-b)) \oplus (-b, (y|b)(-b))$$

The second term is a product since $(b|y) = (-b| - y)$ is independent of the representative of $\{b, -b\}$.

2.6. Remark We have a natural double winding giving $P_1(\mathbb{R}) = S^1$, the circle. The space $S^1 \times \mathbb{R}$ is a band, and $E(\mathbb{R})$ over S^1 or the real projective line $P_1(\mathbb{R})$ is the Möbius band.

3 Two Operations on Bundles

The first operation on bundles transfers a bundle from one space to another under a continuous map, but before doing the general case, let us introduce the following definition.

3.1. Definition Let $f : B' \to B$ be a continuous map, and let $p : E \to B$ be in (bun/B). Then the induced bundle $f^{-1}(E)$ is the subspace $f^{-1}(E) \subset B' \times E$ consisting of all $(b',x) \in B' \times E$ such that $f(b') = p(x)$ together with the restriction of the first projection $q : B' \times E \to B'$ to the subspace $f^{-1}(E)$. For a morphism $u : E \to E'$ over B, the morphism $f^{-1}(u) : f^{-1}(E) \to f^{-1}(E')$ is the map defined by $f^{-1}(u)(b',x) = (b', u(x))$ for $(b',x) \in f^{-1}(E)$.

We also use the notation $f^*(E) \to B'$ for the induced bundle $f^{-1}(E) \to B'$.

3.2. Remark The bundles $f^{-1}(E) \to B$ and the bundle morphisms $f^{-1}(u)$ define a functor from the category (bun/B) to the category (bun/B') in the sense that for the identity id_E, the induced map $f^{-1}(\mathrm{id}_E)$ is the identity on $f^{-1}(E)$ and for a composite of two bundle morphisms $u : E \to E'$ and $u' : E' \to E''$, we have a composition of two bundle morphisms $f^{-1}(u'u) = f^{-1}(u')f^{-1}(u)$.

3.3. Remark For two continuous maps $g : B'' \to B'$ and $f : B' \to B$ and for a bundle $p : E \to B$ in (bun/B), we have a natural isomorphism given by a projection $g^{-1}f^{-1}(E) \to (fg)^{-1}(E)$ which carries $(b'',(b',x))$ to (b'',x). Since $b' = g(b'')$ is determined by b'', this projection is an isomorphism under which for a morphism $u : E \to E'$ we have a commutative diagram as in (1.7)

$$g^{-1}f^{-1}(E) \xrightarrow{\ g^{-1}f^{-1}(u)\ } g^{-1}f^{-1}(E')$$

$$\downarrow \qquad\qquad\qquad\qquad \downarrow$$

$$(fg)^{-1}(E) \xrightarrow{\ (gf)^{-1}(u)\ } (fg)^{-1}(E').$$

In Sect. 5, this kind of diagram is described in general as a morphism or natural transformation of functors which is in fact more, namely a natural equivalence between the functors

$$g^{-1}f^{-1}(E) \longrightarrow (fg)^{-1}(E).$$

Finally, we consider products in the categories (bun) and (bun/B).

3.4. Definition Let $(p_i : E_i \to B_i)_{i \in I}$ be a family of bundles. Then the product bundle of this family in (bun) is the product of the maps p_i given by $\Pi_{i \in I} p_i : \Pi_{i \in I} E_i \to \Pi_{i \in I} B_i$.

Let $(p_i : E_i \to B)_{i \in I}$ be a family of bundles over B, that is, in the category (bun/B).

3.5. Definition The product of this family in (bun/B), also called the fibre product, is the subspace $E = \Pi_{B, i \in I} E_i \subset \Pi_{i \in I} E_i$ consisting of I-tuples $(x_i)_{i \in I}$ in the product space such that for $i, j \in I$ we have $p_i(x_i) = p_j(x_j)$ which we define as $[(x_i)_{i \in I}] \in E$. The projection is $p((x_i)_{i \in I}) = x_j$ for any $j \in I$.

For $p' : E' \to B$ and $p'' : E'' \to B$, the fibre product is denoted by $p : E' \times_B E'' \to B$, where $E' \times_B E''$ is the subspace of $E' \times E''$ consisting of (x', x'') with $p'(x') = p''(x'') = p(x', x'')$.

Observe that the space $f^{-1}(E) = B' \times_B E$ is the total space of the fibre product of $f : B' \to B$ and $p : E \to B$ in (1.9). As another example, the fibre product of the product bundle $B \times Y' \to B$ with the product bundle $B \times Y'' \to B$ is "in a natural way" isomorphic to the product bundle $B \times (Y' \times Y'') \to B$. In Sect. 5, we explain the notion of isomorphism of functors.

3.6. Example Returning to the n-sphere S^n and the examples of (2.2) and (2.3), we have the fibre product or Whitney sum

$$T(S^n) \oplus N(S^n) = T(S^n) \times_{S^n} N(S^n) = S^n \times \mathbb{R}^{n+1}.$$

Returning to the n-dimensional real projective space $P_n(\mathbb{R})$ and the example of (2.5), we have the fibre product or Whitney sum

$$T(P_n(\mathbb{R})) \oplus (P_n(\mathbb{R}) \times \mathbb{R}) = T(P_n(\mathbb{R})) \times_{P_n(\mathbb{R})} (P_n(\mathbb{R}) \times \mathbb{R}) = E(\mathbb{R}^{n+1}).$$

4 Category Constructions Related to Bundles

In (1.3), we began speaking of the category of bundles (bun/B) over a space B, and now we give some of the basic definitions in category theory for the reader unfamiliar with these concepts.

4.1. Definition A category $\mathscr{C}$ consists of three sets of data.

(I) A class whose elements are called objects of $\mathscr{C}$. When we say X is in $\mathscr{C}$ we mean X is an object in $\mathscr{C}$.
(II) For each pair of objects X, Y in $\mathscr{C}$ a set $\mathrm{Hom}(X,Y)$ [also denoted by $\mathrm{Hom}_{\mathscr{C}}(X,Y)$] whose elements f are called morphisms from X to Y, this is denoted by $f : X \to Y$ or $X \xrightarrow{f} Y$.
(III) For each triple of objects X, Y, Z in $\mathscr{C}$ a function

$$c : \mathrm{Hom}(X,Y) \times \mathrm{Hom}(Y,Z) \longrightarrow \mathrm{Hom}(X,Z)$$

called composition and denoted by $c(f,g) = gf : X \to Z$ for $f : X \to Y$ and $g : Y \to Z$ in $\mathscr{C}$.

For a category, the following axioms are satisfied:

(1) If $f : X \to Y$, $g : Y \to Z$, and $h : Z \to W$ are three morphisms, then associativity of composition $h(gf) = (hg)f$ holds.
(2) For each X in $\mathscr{C}$, there exists a morphism $1_X : X \to X$ such that for all $f : Y \to X$ and $g : X \to Y$ we have $f = 1_X f$ and $g 1_X = g$.

Observe that $1_X : X \to X$ satisfying the axiom (2) is unique because if $1'_X$ were a second one, then $1'_X = 1'_X 1_X = 1_X$. We also denote $1_X : X \to X$ by simply $X : X \to X$.

Before considering examples of categories, we introduce the notion of isomorphism in a category. It is a general concept in category theory, and in a specific category it is something which has to be identified, not defined, in special examples.

4.2. Definition Let $\mathscr{C}$ be a category. A morphism $f : X \to Y$ is called an isomorphism provided there exists a morphism $f' : Y \to X$ with $f'f = X$ and $ff' = Y$. Two objects X and Y in a category $\mathscr{C}$ are isomorphic provided there exists an isomorphism $f : X \to Y$.

4.3. Remark For a morphism $f : X \to Y$, if there are two morphisms $f', f'' : Y \to X$ with $f'f = X$ and $ff'' = Y$, then we have

$$f' = f'(ff'') = (f'f)f'' = f'',$$

and f is an isomorphism, and the unique morphism $f' = f''$ is called its inverse. The composition of two isomorphisms and the inverse of an isomorphism are isomorphisms. The relation two objects are isomorphic is an equivalence relation. Let $\mathrm{Aut}(X)$ denote the group of automorphisms of X, that is, the set of isomorphisms $X \to X$ with composition as group operation.

4.4. Example Let (set) denote the category of sets, that is, the class of objects is the class of sets, $\mathrm{Hom}(A,B)$ is the set of all functions $A \to B$, and composition is composition of functions. Let (top) denote the category of spaces, that is, the class of objects is the class of topological spaces, $\mathrm{Hom}(X,Y)$ is the set of all continuous functions $X \to Y$, and composition is composition of continuous functions.

Let (gr) denote the category of groups, that is, the class of objects is the class of groups, $\mathrm{Hom}(G,H)$ is the set of all group morphisms $G \to H$, and composition is composition of group morphisms. For a commutative ring k with unit, we denote by (k) the category of k-modules, that is, the class of objects is the class of (unitary) k-modules, $\mathrm{Hom}(M,N)$ is the set of all k-linear functions $M \to N$, and composition is composition of k-linear functions.

4.5. Remark The isomorphisms in (set) are the bijective functions, the isomorphisms in (top) are the bijective continuous maps whose inverse function is continuous, the isomorphisms in (gr) are the bijective group morphisms, and the isomorphisms in (k) for a commutative ring k are the bijective k-linear maps.

4.6. Definition A subcategory $\mathscr{C}'$ of a category $\mathscr{C}$ is a category where the objects of $\mathscr{C}'$ form a subclass of the objects in $\mathscr{C}$ and whose morphism set $\mathrm{Hom}_{\mathscr{C}'}(X',Y')$ is a subset of $\mathrm{Hom}_{\mathscr{C}}(X',Y')$ such that the composition function for $\mathscr{C}'$ is the restriction of the composition function for $\mathscr{C}$. A subcategory $\mathscr{C}'$ of a category $\mathscr{C}$ is called full provided $\mathrm{Hom}_{\mathscr{C}'}(X',Y') = \mathrm{Hom}_{\mathscr{C}}(X',Y')$ for X',Y' in $\mathscr{C}'$.

Observe that a full subcategory $\mathscr{C}'$ of a category $\mathscr{C}$ is determined by its subclass of objects. For the ring of integers $\mathbb{Z}$, a $\mathbb{Z}$-module is just an abelian group, and hence the category of abelian groups (ab) $= (\mathbb{Z})$ is a full subcategory of (gr) determined by the groups satisfying the commutative law.

Using the previous section as a model, we are able to define bundles in any category $\mathscr{C}$.

The category (bun) becomes the category $\mathrm{Mor}(\mathscr{C})$ of morphisms in $\mathscr{C}$, and the category (bun/B) becomes the category $\mathscr{C}/B$ of objects E over B, that is, morphisms $p : E \to B$.

4.7. Definition Let $\mathscr{C}$ be a category. Let $\mathrm{Mor}(\mathscr{C})$ be the category whose objects are morphisms $p : E \to B$ in $\mathscr{C}$, and a morphism in $\mathrm{Mor}(\mathscr{C})$ from $p : E \to B$ to $p' : E' \to B'$ is a pair of morphisms (u,f), where $u : E \to E'$ and $f : B \to B'$ such that $p'u = fp$. Composition of (u,f) and (u',f') from $p' : E' \to B'$ to $p'' : E'' \to B''$ is

$$(u',f')(u,f) = (u'u, f'f) : (p : E \to B) \longrightarrow (p'' : E'' \to B'').$$

The category (bun) is just the category $\mathrm{Mor}((\mathrm{top}))$.

4.8. Definition For B in a category $\mathscr{C}$, the subcategory $\mathscr{C}/B$ of $\mathrm{Mor}(\mathscr{C})$ consists of all objects $p : E \to B$ with base B, and a morphism in $\mathscr{C}/B$ from $p : E \to B$ to $p' : E' \to B$ is a morphism $u : E \to E'$ such that $p'u = p$. Composition of u and u' from $p' : E' \to B$ to $p'' : E'' \to B$ is $u'u : (p : E \to B) \to (p'' : E'' \to B)$.

The category (bun/B) is just the category (top)/B.

5 Functors Between Categories

5.1. Definition Let $\mathscr{C}$ and $\mathscr{C}'$ be two categories. A functor $T : \mathscr{C} \to \mathscr{C}'$ consists of two sets of data

(I) A function T from the objects in $\mathscr{C}$ to the objects in $\mathscr{C}'$, hence the function type of notation for a functor.

(II) For each pair of objects X, Y in $\mathscr{C}$ we have a function $T : \mathrm{Hom}_{\mathscr{C}}(X, Y) \to \mathrm{Hom}_{\mathscr{C}'}(T(X), T(Y))$, that is, if $u : X \to Y$ is a morphism in $\mathscr{C}$, then $T(u) : T(X) \to T(Y)$ is a morphism in $\mathscr{C}'$.

For a functor, the following axioms are satisfied.

(1) If $u : X \to Y$ and $v : Y \to Z$ are two morphisms in $\mathscr{C}$, then for the composites $T(vu) = T(v)T(u) : T(X) \to T(Z)$ in $\mathscr{C}'$.

(2) For an object X in $\mathscr{C}$, we have $T(\mathrm{id}_X) = \mathrm{id}_{T(X)}$ in $\mathscr{C}'$. In particular the notation $X : X \to X$ under T becomes $T(X) : T(X) \to T(X)$.

5.2. Proposition *If $F : \mathscr{C} \to \mathscr{C}'$ is a functor and if $u : X \to Y$ is an isomorphism in $\mathscr{C}$ with inverse $u^{-1} : Y \to X$, then $T(u) : T(X) \to T(Y)$ is an isomorphism in $\mathscr{C}'$ with inverse*

$$T(u)^{-1} = T(u^{-1}) : T(Y) \longrightarrow T(X).$$

Proof. This follows directly from the axioms for a functor.

5.3. Elementary Operations If $F : \mathscr{C} \to \mathscr{C}'$ and $F' : \mathscr{C}' \to \mathscr{C}''$ are two functors, then the composite on objects and morphisms $F'F : \mathscr{C} \to \mathscr{C}''$ is a functor, for it is immediate to check axioms (1) and (2). The identity functor $\mathrm{id}_{\mathscr{C}} : \mathscr{C} \to \mathscr{C}$ is the identity on objects and morphisms of $\mathscr{C}$. If we think of categories as objects and functors as morphisms between categories, then we have the first idea about the category of categories. Unfortunately, the concept has to be modified, and this we do in later chapters.

5.4. *Examples* The functor $F : (\mathrm{top}) \to (\mathrm{set})$ which deletes the system of open sets of a space X leaving the underlying set is an elementary example. The functors assigning to a bundle $p : E \to B$ the total space $T(p : E \to B) = E$ or the base space $B(p : E \to B) = B$ are defined as functors $T : (\mathrm{bun}/B) \to (\mathrm{top})$ or $B : (\mathrm{bun}) \to (\mathrm{top})$.

5.5. *Example* The sets $\Gamma(U, E)$ and the functions $\Gamma(U, u)$ define a functor from the category (bun/B) to the category (set) of sets in the sense that for the identity $\Gamma(U, \mathrm{id}_E)$ is the identity on the set of section $\Gamma(U, E)$, and for a composite of two bundle morphisms $u : E \to E'$ and $u' : E' \to E''$, we have a composition of two functions $\Gamma(U, u'u) = \Gamma(U, u')\Gamma(U, u)$. These are the two axiomatic properties of a functor in the definition (5.1).

5.6. *Remark* Now, we can change from an open set U to a smaller open set V in B, that is, for open sets $V \subset U \subset B$, we can restrict $\rho(V, U) : \Gamma(U, E) \to \Gamma(V, E)$ continuous functions

$$\rho(V, U)(s) = s|V.$$

There is a compatibility condition between this restriction and function defined by composition by an B-bundle morphism $u : E \to E'$. For open sets $V \subset U \subset B$, we have a commutative square, which in the next section we will interpret in two ways as a morphism of functors.

$$\begin{array}{ccc} \Gamma(U,E) & \xrightarrow{\;\;\Gamma(U,u)\;\;} & \Gamma(U,E') \\ {\scriptstyle\rho(V,U)}\Big\downarrow & & \Big\downarrow{\scriptstyle\rho(V,U)} \\ \Gamma(V,E) & \xrightarrow{\;\;\Gamma(V,u)\;\;} & \Gamma(V,E') \end{array}$$

Now, we formulate a definition where this kind of diagram is described in general as a morphism or natural transformation of functors in the next section.

6 Morphisms of Functors or Natural Transformations

6.1. Definition Let $S, T : \mathscr{C} \to \mathscr{C}'$ be two functors between two categories. A morphism or natural transformation $\theta : S \to T$ is given by a morphism $\theta(X) : S(X) \to T(X)$ in $\mathscr{C}'$ for each object X in $\mathscr{C}$ such that for each morphism $u : X' \to X''$ in $\mathscr{C}$ we have

$$T(u)\theta(X') = \theta(X'')S(u),$$

or equivalently, the following diagram is commutative

$$\begin{array}{ccc} S(X') & \xrightarrow{\;\;\theta(X')\;\;} & T(X') \\ {\scriptstyle S(u)}\Big\downarrow & & \Big\downarrow{\scriptstyle T(u)} \\ S(X'') & \xrightarrow{\;\;\theta(X'')\;\;} & T(X''). \end{array}$$

6.2. Definition Let $R, S, T : \mathscr{C} \to \mathscr{C}'$ be three functors between two categories, and let $\phi : R \to S$ and $\psi : S \to T$ be two morphisms of functors. For each X consider the morphism

$$\psi(X)\phi(X) = (\psi\phi)(X) : R(X) \to T(X)$$

which is called the composition $\psi\phi$ of ϕ and ψ.

Then, $\psi\phi : R \to T$ is a morphism of functors, for if $u : X' \to X''$ is a morphism in $\mathscr{C}$, then we have

$$T(u)(\psi\phi)(X') = \psi(X'')S(u)\phi(X') = \psi\phi(X'')R(u).$$

The identity morphism $1 : T \to T$ is given by $1(X) =$ identity on $T(X)$ for each X in $\mathscr{C}$.

6.3. Definition An isomorphism of functors or natural equivalence $\theta : S \to T$ between two functors $S, T : \mathscr{C} \to \mathscr{C}'$ is a morphism of functors θ such that each $\theta(X)$ is an isomorphism

$$\theta(X) : S(X) \longrightarrow T(X)$$

in $\mathscr{C}'$.

If $\theta(X)^{-1}$ is inverse of each $\theta(X)$, then the family of

$$\theta(X)^{-1} : T(X) \longrightarrow S(X)$$

defines a morphism $\theta^{-1} : T \to S$ of functors which is inverse to $\theta : S \to T$ in the sense of the composition defined in (6.2).

6.4. Definition A functor $T : \mathscr{C}' \to \mathscr{C}''$ is an equivalence of categories provided there exists a functor $S : \mathscr{C}'' \to \mathscr{C}'$ such that the composite ST is isomorphic to $\mathrm{id}_{\mathscr{C}'}$, the identity functor on $\mathscr{C}'$, and TS is isomorphic to $\mathrm{id}_{\mathscr{C}''}$ the identity functor on $\mathscr{C}''$.

The composition of two equivalences of categories is again an equivalence of categories.

6.5. Definition Two categories $\mathscr{C}'$ and $\mathscr{C}''$ are equivalent provided there exists an equivalence of categories $\mathscr{C}' \to \mathscr{C}''$.

Observe that equivalence of categories is a weaker relation than isomorphism of categories. It almost never happens that two categories are isomorphic, but equivalent categories will have a bijection between isomorphism classes and related morphism sets. In fact, this is the way of recognizing that a functor is an equivalence by starting with its mapping properties on the morphism sets.

6.6. Definition A functor $T : \mathscr{X} \to \mathscr{Y}$ is faithful (resp. fully faithful, full) provided for every pair of objects X, X' in $\mathscr{X}$ the function $T : \mathrm{Hom}_{\mathscr{X}}(X, X') \to \mathrm{Hom}_{\mathscr{Y}}(T(X), T(X'))$ is injective (resp. bijective, surjective).

The functors $F : (\mathrm{top}) \to (\mathrm{set})$ and $F : (\mathrm{gr}) \to (\mathrm{set})$ which assign to a topological space or a group its underlying set are faithful functors. In the next proposition, we have a useful criterion for a functor to be an equivalence from one category to another. In Chap. 3, we apply this criterion to show that the category of vector bundles on X is equivalent to the category of finitely generated projective $C(X)$-modules.

6.7. Proposition *A functor* $T : \mathscr{X} \to \mathscr{Y}$ *is an equivalence of categories if and only if* T *is fully faithful and for each object* Y *in* $\mathscr{Y}$, *there exists an object* X *in* $\mathscr{X}$ *with* $T(X)$ *isomorphic to* Y *in* $\mathscr{Y}$.

Proof. If $S : \mathscr{Y} \to \mathscr{X}$ is an inverse up to isomorphisms with the identity functor, then $T : \mathrm{Hom}_{\mathscr{X}}(X, X') \to \mathrm{Hom}_{\mathscr{Y}}(T(X), T(X'))$ is a bijection with inverse constructed by composing

$$S : \mathrm{Hom}_{\mathscr{Y}}(T(X), T(X')) \to \mathrm{Hom}_{\mathscr{X}}(ST(X), ST(X'))$$

with the isomorphism $\mathrm{Hom}_{\mathscr{X}}(ST(X), ST(X')) \to \mathrm{Hom}_{\mathscr{X}}(X, X')$. Hence, T is fully faithful. Since for each Y in $\mathscr{Y}$, the object Y in $\mathscr{Y}$ is isomorphic to $TS(Y)$, the second condition is satisfied for an equivalence of categories.

Conversely, we wish to construct $S : \mathscr{Y} \to \mathscr{X}$ by defining S on objects. For each Y, we choose an object $S(Y) = X$ in $\mathscr{X}$ together with an isomorphism $\theta(Y) : Y \to T(X) = TS(Y)$. This is possible by the hypothesis, and we proceed to define

$$S : \mathrm{Hom}_{\mathscr{Y}}(Y,Y') \longrightarrow \mathrm{Hom}_{\mathscr{X}}(S(Y),S(Y'))$$

as the composite of $\theta(Y',Y) : \mathrm{Hom}_{\mathscr{Y}}(Y,Y') \to \mathrm{Hom}_{\mathscr{Y}}(TS(Y),TS(Y'))$ given by $\theta(Y',Y)f = \theta(Y')f\theta(Y)^{-1}$ and the inverse of the bijection T defined as $T^{-1} : \mathrm{Hom}_{\mathscr{Y}}(TS(Y),TS(Y')) \to \mathrm{Hom}_{\mathscr{X}}(S(Y),S(Y'))$. The two properties of a functor follow for S from the observation that T is a functor and the following relations hold $\theta(Y'',Y)(f'f) = \theta(Y'',Y')(f')\theta(Y',Y)(f)$ and $\theta(Y,Y)(\mathrm{id}_Y) = \mathrm{id}_{TS(Y)}$.

Finally, the morphisms $\theta(Y)$ define an isomorphism $\theta : \mathscr{Y} \to TS(\mathscr{Y})$ from the identity on the category $\mathscr{Y}$ to TS, and ST was defined so that it is the identity on $\mathscr{X}$. This proves the proposition.

7 Étale Maps and Coverings

7.1. Definition A map $f : Y \to X$ is open (resp. closed) provided for each open (resp. closed) subset $V \subset Y$, the direct image $f(V)$ is open (resp. closed) in X.

7.2. Example An inclusion map $j : Y \to X$ of a subspace $Y \subset X$ is open (resp. closed) if and only if Y is an open (resp. closed) subspace of X. Any projection $p_j : \Pi_{i \in I} X_i \to X_j$ from a product to a factor space is open. A projection $p_Y : X \times Y \to Y$ is closed for all Y if and only if X is a quasicompact space.

7.3. Remark A space Y is separated (or Hausdorff) provided the following equivalent conditions are satisfied.

(1) For $y,y' \in Y$ with $y \neq y'$, there exists open neighborhoods V of y and V' of y' in Y with $V \cap V'$ empty.
(2) The intersection of all closed neighborhoods V of any $y \in Y$ is just $\{y\}$.
(3) The diagonal map $\Delta_Y : Y \to Y \times Y$ is a closed map.

This is a basic definition and assertion in general topology. Observe that condition (3) can be formulated in terms of the diagonal subset $\Delta \subset Y \times Y$ being a closed subset.

7.4. Remark Let $f,g : X \to Y$ be two maps into a separated space. The set of all $x \in X$, where $f(x) = g(x)$ is the closed subset $(f,g)^{-1}(\Delta(Y))$, where $(f,g)(x) = (f(x),g(x))$ defines the continuous map $(f,g) : X \to Y \times Y$.

7.5. Definition A map $f : Y \to X$ is étale provided either of the following equivalent conditions are satisfied.

(1) For each $y \in Y$, there exists open neighborhoods V of y in Y and U of $f(y)$ in X such that $f(V) = U$ and $f|V : V \to U$ is a homeomorphism. This condition is often referred to as the map f is a local homeomorphism which is another term for an étale map.
(2) The map f is open and the diagonal map $\Delta_Y Y : Y \to Y \times_X Y$ is an open map.

A bundle $p : E \to X$ in (bun/X) is an étale (resp. open) bundle provided the map p is an étale (resp. open) map.

For the proof of the equivalence of (1) and (2), we assume firstly (1). To see that f is open, we consider an open subset W of Y and $y \in W$. There exists open neighborhoods V of y in W and U of $f(y)$ in X such that $f(V) = U$ and $f|V : V \to U$ is a homeomorphism. Thus U, is an open neighborhood of $f(y) \in f(W)$ contained in $f(W)$, and f is an open map. To see that the diagonal set $\Delta(Y)$ is open in $Y \times_X Y$, we consider $(y,y) \in \Delta(Y)$. From the local homeomorphism condition (1), there exists open neighborhoods V of y in Y and U of $f(y)$ in X such that $f(V) = U$ and $f|V : V \to U$ is a homeomorphism. Then we have $V \times_U V \subset \Delta(Y)$.

Conversely, we assume (2). Since $\Delta(Y)$ is open in $Y \times_X Y$, there exists an open set $V \subset Y$ with $(V \times V) \cap (Y \times Y) \subset Y \times_X Y$. Then $f(V) = U$ is an open neighborhood of $f(y)$ in X and the open map f restricts to an open map $f|V : V \to U$ which is a bijection since $(V \times V) \cap (Y \times Y) \subset Y \times_X Y$. This establishes the equivalence of (1) and (2).

7.6. Notation The full subcategory of (bun/X) determined by the étale bundles over X is denoted by $(\mathrm{\acute{e}t}/X)$.

7.7. Proposition *Let $p : E \to X$ be an étale (resp. open) bundle. For each map $f : Y \to X$, the induced bundle $q : f^{-1}(E) \to Y$ is an étale (resp. open) bundle.*

Proof. To show that q is open, we consider an open set W in $f^{-1}(E)$ and $(y,x) \in W \subset f^{-1}(E)$. Then, there exists open sets V in Y and U in E such that $(y,x) \in (V \times U) \cap f^{-1}(E) \subset W$. Then we have

$$y \in q((V \times U) \cap f^{-1}(E)) \subset V \cap f^{-1}(p(U)) \subset q(W),$$

and this shows that $q(W)$ and hence also q is open.

To show that q is a local homeomorphism if p is, we consider $(y,x) \in f^{-1}(E)$ and an open neighborhood V of $x \in E$ such that $p|V : V \to p(V)$ is a homeomorphism. Then we form the open set $W = (f^{-1}(p(V)) \times V) \cap f^{-1}(E)$ in $f^{-1}(E)$ with $(y,x) \in W$. Observe that $q|W : W \to q(W)$ is a homeomorphism. This proves the proposition.

7.8. Corollary *If $p : E \to X$ is an étale (resp. open) bundle and if $Y \subset X$ is a subspace of X, then $p|p^{-1}(Y) : E|Y = p^{-1}(Y) \to Y$ is an étale (resp. open) bundle.*

7.9. Example Let D be a discrete space. Then the product bundle $p = p_X : X \times D \to X$ is an étale bundle. This is called the product bundle with discrete fibre. In fact, for each $z \in D$, the restriction $p|X \times \{z\} : X \times \{z\} \to X$ is a homeomorphism when restricted to the open set $X \times \{z\} \subset X \times D$.

Now, we are in position to define coverings in the mature context and describe the category of coverings as a full subcategory of the category of bundles.

7.10. Definition A covering $p : E \to X$ is an étale bundle which is locally isomorphic to a product bundle with discrete fibre. A trivial covering $p : E \to X$ is a bundle isomorphic to a product bundle with discrete fibre D. We denote by (cov/X) the full subcategory of $(\mathrm{\acute{e}t}/X)$ and hence of (bun/X) determined by the coverings over X.

References

Grothendieck, A.: A General Theory of Fibre Spaces with Structure Sheaf. University of Kansas, Lawrence (1955) (2nd edition 1958)

Husemöller, D.: *Fibre Bundles*, 3rd ed. Springer-Verlag, New York (1994)

Chapter 2
Vector Bundles

The notion of vector bundle is a basic extension to the geometric domain of the fundamental idea of a vector space. Given a space X, we take a real or complex finite dimensional vector space V and make V the fibre of a bundle over X, where each fibre is isomorphic to this vector space. The simplest way to do this is to form the product $X \times V$ and the projection $pr_X : X \times V \to X$ onto the first factor. This is the product vector bundle with base X and fibre V.

On first sight, the product bundle appears to have no special features, but it contains other vector bundles which often reflect the topology of X in a strong way. This happens, for example, for the tangent bundle and the normal bundle to the spheres which are discussed in 1(2.2) and 1(2.3). All bundles of vector spaces that we will consider will have the local triviality property, namely, they are locally isomorphic to a product bundle. The product bundle is also basic because most of the vector bundles we will be considering will be subvector bundles of a product vector bundle of higher dimension. In some cases, they will be so twisted that they can only live in an infinite dimensional product vector bundle.

We will begin by formulating the concept of bundles of vector spaces over X. These will not be necessarily locally trivial, but they form a well-defined concept and category. Then, a vector bundle is a locally trivial bundle of vector spaces. The point of this distinction is that being a vector bundle is a bundle of vector spaces with an additional axiom and not an additional structure. The local charts which result are *not* new elements of structure but only a property, but to state the property, we need the notion of bundle of vector spaces. After this is done, we will be dealing with just vector bundles.

Chapter 3 of *Fibre Bundles* (Husemöller 1994) is a reference for this chapter.

1 Bundles of Vector Spaces and Vector Bundles

All the vector spaces under consideration are defined either over the real numbers $\mathbb{R}$ or the complex numbers $\mathbb{C}$. We need to use the scalars explicitly, and if either number system applies, we will use the symbol F to denote either the field of real or the field of complex numbers.

D. Husemöller et al.: *Vector Bundles*, Lect. Notes Phys. **726**, 23–34 (2008)
DOI 10.1007/978-3-540-74956-1_3
© Springer-Verlag Berlin Heidelberg 2008

1.1. Definition A bundle of vector spaces over B is a bundle $p : E \to B$ with two additional structures

$$E \times_B E \longrightarrow E \quad \text{and} \quad F \times E \longrightarrow E$$

defined over B called addition and scalar multiplication, respectively. As maps over B, they restrict to each fibre

$$E_b \times E_b = (E \times_B E)_b \longrightarrow E_b \quad \text{and} \quad F \times E_b = (F \times E)_b \longrightarrow E_b,$$

and the basic axiom is that they define a vector space structure over F on each fibre. We require further that the function which assigns to each $b \in B$ the unique zero $0_b \in E_b$ is a continuous section.

Although it is not so necessary, we will usually require that the fibres are finite dimensional, hence the subspace topology on the fibres will be the usual vector space topology. In the infinite dimensional case, the main difference is that one has to preassign a topology on the vector space compatible with addition and scalar multiplication. Bundles of infinite dimensional vector spaces are treated in more detail in Chap. 20.

1.2. Example Let V be a finite dimensional F-vector space with the usual topology, and let $p : B \times V \to B$ be the product bundle. Then, the usual vector space structure on V defines a bundle of vector space structure on the product bundle by

$$(b, v') + (b, v'') = (b, v' + v'') \quad \text{and} \quad k(b, v) = (b, kv), \quad \text{for} \quad k \in F.$$

If $p : E \to B$ is a bundle of vector spaces and if $A \subset B$ is a subspace, then the restriction $q : E|A \to A$ is a bundle of vector spaces with the restriction of the globally defined addition and scalar multiplication. Now, this can be generalized as in $1(2.1)$ to subbundles of vector spaces.

1.3. Definition Let $p : E' \to B$ and $p : E'' \to B$ be two bundles of vector spaces over B. A morphism $u : E' \to E''$ of bundles of vector spaces over B is a morphism of bundles such that the restriction to each fibre $u|E'_b : E'_b \to E''_b$ is a linear map.

1.4. Remark The composition of morphisms of bundles of vector spaces over B is again a morphism of bundles of vector spaces over B. Hence, it defines a category.

1.5. Example Let $p : E \to B$ be a bundle of vector spaces, and let A be a subspace of B. Then, the restriction to a subspace $p|(E|A) : E|A \to A$ is a bundle of vector spaces.

Now, we are in a position to make the main definition of this chapter.

1.6. Definition A vector bundle $p : E \to B$ is a bundle of vector spaces such that every point $b \in B$ has an open neighborhood U with the restriction $p|(E|U) : E|U \to U$ isomorphic to the product bundle of vectors spaces $\mathrm{pr}_1 : U \times V \to U$.

Now, we will always be working with vector bundles and observe that the concept of bundle of vector spaces was only introduced as a means of defining the local triviality property of vector bundles. Of course, this could have been done more directly, but in this way, we try to illustrate the difference between structure and axiom.

1.7. Definition A morphism $u : E' \to E''$ of vector bundles over B is a morphism of the bundles of vector spaces from E' to E''.

1.8. Remark The composition of morphisms of vector bundles over B is again a morphism of vector bundles over B. Hence, it defines a category of vector bundles over B. A trivial vector bundle is one which is isomorphic to a product vector bundle.

1.9. Example Let $E' \to B$ and $E'' \to B$ be two bundles of vector spaces over B. The fibre product $E' \times_B E''$, also denoted by $E' \oplus E''$, is a bundle of vector spaces. If E' and E'' are vector bundles, then $E' \oplus E''$ is also a vector bundle. For the product bundles $B \times V'$ and $B \times V''$, the fibre product or Whitney sum is given by $(B \times V') \oplus (B \times V'') = B \times (V' \oplus V'')$. The terminology of Whitney sum comes from the direct sum and the notation $E' \oplus E''$ from the direct sum of vector spaces.

1.10. Definition Let E', E'' and E be three vector bundles over X. A vector bundle morphism $\beta : E' \oplus E'' = E' \times_X E'' \to E$ is bilinear provided $\beta|(E' \oplus E'')_x : (E' \oplus E'')_x \to E_x$ is a bilinear map of vector spaces. The tensor product $E' \otimes E''$ of E' and E'' is a specific choice of a vector bundle with a bilinear morphism $\theta : E' \oplus E'' \to E' \otimes E''$ which has the universal property that every bilinear morphism $\beta : E' \oplus E'' \to E$ factors uniquely as $u\theta$, where $u : E' \otimes E'' \to E$ is a morphism of vector bundles. Note that this is the usual definition of the tensor product for X a point.

2 Isomorphisms of Vector Bundles and Induced Vector Bundles

In the next result, we use the vector bundle local triviality axiom.

2.1. Proposition *Let $u : E' \to E''$ be a morphism of vector bundles over a space B such that $u_b : E'_b \to E''_b$ is an isomorphism for each $b \in B$. Then, u is an isomorphism.*

Proof. The inverse $v : E'' \to E'$ of u exists fibrewise as a function. The only question is its continuity, and this we check on an open covering of the form $u : E'|U \to E''|U$, where E' and E'' are each trivial bundles. In this case, the restriction of u with inverse v has the form $u : U \times F^n \to U \times F^n$ with formula $u(b, y) = (b, T(b)y)$, where $T : U \to GL_n(F)$ is a continuous map. Then, $v(b, z) = (b, T(b)^{-1}(z))$ is also continuous. This proves the proposition.

2.2. Proposition *Let $p : E \to B$ be a bundle of vector spaces over B, and let $f : B' \to B$ be a continuous map. Then, the induced bundle $f^{-1}E \to B'$ has the structure of a bundle of vector spaces such that the natural f-morphism $w : f^{-1}E \to E$ over each $b' \in B'$ on the fibre*

$$w_{b'} : (f^{-1}E)_{b'} \longrightarrow E_{f(b')}$$

is a vector space isomorphism. If E is a vector bundle, then $f^{-1}E$ is a vector bundle.

Proof. Recall that the induced bundle $f^{-1}E$ is a subspace of $B' \times E$ consisting of all (b',x) with $f(b') = p(x)$. Then, we use $f^{-1}(E \times_B E) = f^{-1}(E) \times_{B'} f^{-1}(E)$, and the sum function on the bundle of vector spaces must be of the form $(b',x) + (b',y) = (b',x+y)$. Then, scalar multiplication must be of the form $a(b',y) = (b',ay)$. In both cases, these functions are continuous, and moreover, on the fibre $w_{b'} : (f^{-1}E)_{b'} \to E_{f(b')}$ is a vector space isomorphism.

Finally, if E is a product bundle, then $f^{-1}E$ is a product bundle, and if E is locally trivial, then $f^{-1}E$ is locally trivial, for it is trivial over open sets $f^{-1}(W)$, where E is trivial over W. This proves the proposition.

2.3. Proposition *Let $(u,f) : (p' : E' \to B') \to (p : E \to B)$ be a morphism of bundles, where p' and p are vector bundles such that over each $b' \in B'$ on the fibre $u_{b'} : E'_{b'} \to E_{f(b')}$ is a morphism of vector spaces. Then, u factors by a morphism $v : E' \to f^{-1}E$ of vector bundles over B' followed by the natural f-morphism $w : f^{-1}E \to E$. Moreover, if over each $b' \in B'$ on the fibre $u_{b'} : E'_{b'} \to E_{f(b')}$ is a isomorphism of vector spaces, then $v : E' \to f^{-1}E$ is an isomorphism of vector bundles over B'.*

Proof. The factorization of u is given by $v(x') = (p'(x'), u(x'))$, and it is a vector bundle morphism since u is linear on each fibre. Moreover, $wv(x') = u(x')$ shows that it is a factorization. If $u_{b'}$ is an isomorphism on the fibre of E' at b', then

$$v_{b'}(x') = (b', u_{b'}(x'))$$

is an isomorphism $E_{f(b')} \to f^{-1}(E)_{b'} = \{b'\} \times E_{f(b')}$. Thus, we can apply (2.1) to obtain the last statement. This proves the proposition.

3 Image and Kernel of Vector Bundle Morphisms

3.1. Remark The vector bundle morphism $w : [0,1] \times F \to [0,1] \times F$ defined by the formula $w(t,z) = (t,tz)$ has a fibrewise kernel and a fibrewise image, but neither is a vector subbundle because the local triviality condition is not satisfied at $0 \in [0,1]$, where the kernel jumps from zero and the image reduces to zero.

In order to study this phenomenon, we recall some elementary conditions on rank of matrices.

3.2. Notation Let $M_{q,n}(F)$ denote the F-vector space of $q \times n$ matrices over F, that is, q rows and n columns of scalars from F. For $q = n$, we abbreviate the notation $M_{n,n}(F) = M_n(F)$ and denote by $GL_n(F)$ the group of invertible $n \times n$ matrices under matrix multiplication. We give these spaces as usual the natural Euclidean topology. We have $M_{q,n}(\mathbb{C}) = M_{q,n}(\mathbb{R}) + M_{q,n}(\mathbb{R})i$, and using blocks of matrices, we have the natural inclusion $M_{q,n}(\mathbb{R}) + M_{q,n}(\mathbb{R})i \subset M_{2q,2n}(\mathbb{R})$, where $A + Bi$ is mapped to

$$\begin{pmatrix} A & 0 \\ 0 & B \end{pmatrix} \in M_{2q,2n}(\mathbb{R}).$$

3.3. Definition The rank filtration $R_k M_{q,n}(F)$ on $M_{q,n}(F)$ is the increasing filtration of all $A \in M_{q,n}(F)$ with rank $(A) \le k$. We denote by $R_{=k}M_{q,n}(F) = R_k M_{q,n}(F) - R_{k-1}M_{q,n}(F)$.

Clearly, we have for the rank filtration

$$\{0\} = R_0 M_{q,n}(F) \subset \ldots \subset R_k M_{q,n}(F) \subset \ldots \subset R_{\min\{q,n\}} M_{q,n}(F) = M_{q,n}(F).$$

For square matrices, we have $R_{=n}M_n(F) = GL_n(F)$, the group of invertible n by n matrices with coefficients in F.

3.4. Remark The sets $R_k M_{q,n}(F)$ in the rank filtration are closed sets which for $F = \mathbb{R}$ or $\mathbb{C}$ can be described as consisting of all matrices A for which all $(k+1) \times (k+1)$ subdeterminants are zero. The group $GL_n(F)$ is the group of all square matrices A with $\det(A) \ne 0$. Since the determinant is a polynomial function, these subsets are algebraic varieties. In particular, the terms of the filtration $R_k M_{q,n}(F)$ are closed subsets of $M_{q,n}(F)$, and $GL_n(F)$ is an open subset of $M_n(F)$. The same is true for $F = \mathbb{H}$, and we will see that it depends on only knowing that $GL_n(\mathbb{H})$ is an open subset of $M_n(\mathbb{H})$ by the subdeterminant characterization of rank.

3.5. Proposition *Let $u : E' \to E''$ be a morphism of vector bundles over a space B. If rank (u_b) is locally constant, then the bundles of vector spaces $\ker(u)$ and $\mathrm{im}(u)$ are vector bundles.*

Proof. We form the subspace $\bigcup_{b \in B} \ker(u_b) = \ker(u) \subset E'$ and the subspace $\bigcup_{b \in B} \mathrm{im}(u_b) = \mathrm{im}(u) \subset E''$. With the restriction of the projections $E' \to B$ and $E'' \to B$, the subbundles $\ker(u)$ and $\mathrm{im}(u)$ are bundles of vector spaces. In order to show that they are vector bundles under the rank hypothesis, we can restrict to an open neighborhood U of any point of B, where both E' and E'' are trivial and u_b has constant rank q.

Choose an isomorphism to the product bundle, and then u can be represented by a vector bundle morphism $w : U \times F^n \to U \times F^m$ which has the form $w(b,y) = (b, T(b)y)$. As a continuous function $T : U \to M_{m,n}$, the matrix $T(b)$ has rank q over U. For each $x \in U$, we can choose a change of coordinates so that $T(b) = \begin{pmatrix} A(b) & B(b) \\ C(b) & D(b) \end{pmatrix}$ such that

$$A(x) \in GL_q(F), \ B(x) = 0, \ C(x) = 0, \ D(x) = 0,$$

and choosing a subopen neighborhood of x in U, called again U, we can assume that $A(b)$ is invertible for all $b \in U$. Here, we use that $GL_q(F) \subset M_q(F)$ is open. Let $\pi : U \times F^n \to U \times F^{n-q}$ denote the projection on the second factor, and note that for $x \in U$, it is the projection onto the $\ker(T(x))$. Since $T(b)$ has constant rank q and

$A(b)$ is invertible, the restriction $\pi|\ker(w) = \pi|U \times \ker(T(b))$ is an isomorphism $U \times \ker(T(b)) \to U \times F^{n-q}$. Hence, we see that $\ker(w)$ is locally trivial and thus a vector bundle.

If $w^t : U \times F^m \to U \times F^n$ is the morphism given by the transpose matrix $T(b)^t$ and the formula $w^t(b,y) = (b, T(b)^t y)$, then $\mathrm{im}(w) = \ker(w^t)$ is also a vector bundle. This proves the proposition.

A very useful case where we know the rank is locally constant and used in the next chapter is in the following proposition.

3.6. Proposition *Let $e = e^2 : E \to E$ be an idempotent endomorphism of a vector bundle $p : E \to B$ over B. Then, the set of $b \in B$ with rank$(e_b) = q$ is open and closed in B. Hence, $\ker(e)$ and $\mathrm{im}(e)$ are vector bundles, and $E = \ker(e) \times_B \mathrm{im}(e) = \ker(e) \oplus \mathrm{im}(e)$.*

Proof. Since on each fibre the identity $1 = e_b + (1 - e_b)$ is the sum of two complementary projections, we have rank (e_b)+rank$(1 - e_b) = n$. The set of all $b \in B$ with rank$(e_b) = q$ is at the same time the set of all $b \in B$ with rank $(e_b) \leq q$ and rank $(1 - e_b) \leq n - q$, which is a closed set, and the set of all $b \in B$ with rank$(e_b) \geq q$ and rank$(1 - e_b) \geq n - q$, which is an open set. Now apply (3.5). This proves the proposition.

4 The Canonical Bundle Over the Grassmannian Varieties

4.1. Definition Let X be a union of an increasing family of subspaces

$$X_0 \subset \ldots \subset X_N \subset X_{N+1} \subset \ldots \subset \lim_{\longrightarrow N} X_N = X$$

with the weak topology, that is a subset $M \subset X$ is closed if and only if $X_m \cap M$ is closed in X_m for all $m \geq 0$.

Another name for the weak topology is the inductive limit topology.

The following inductive limits have the weak topology which start with the inclusions

$$F^N \subset F^{N+1} \subset \ldots \subset \lim_{\longrightarrow N} F^N = F^\infty.$$

4.2. Definition Let $P_n(F^N)$ denote the subspace of $(F^N)^n$ consisting of linearly independent n-tuples of vectors in F^N. Let $Gr_n(F^N)$ denote the quotient space of $P_n(F^N)$ which assigns to a linearly independent n-tuple the subspace F^N of dimension n of which the n-tuple is basis. The space $Gr_n(F^N)$ is called the Grassmann variety of n-dimensional subspaces of F^N, and $P_n(F^N)$ is called the Stiefel variety of linearly independent frames in N-dimensional space.

4.3. Remark The quotient morphism $q : P_n(F^N) \to Gr_n(F^N)$ has the structure of a principal $GL_n(F)$ bundle with right action $P_n(F^N) \times GL_n(F) \to P_n(F^N)$ given by right multiplication of an n-tuple of vectors and by $n \times n$ matrix of scalars

$$(x_1,\ldots,x_n).(a_{i,j}) = (y_1,\ldots,y_n),$$

where $\sum_{i=1}^{n} x_i a_{i,j} = y_j$ for $i,j = 1,\ldots,n$. The subject of principal bundles is taken up in Chap. 5 and we will return to this example.

There are natural inclusions

$$P_n(F^N) \subset P_n(F^{N+1}) \quad \text{and} \quad Gr_n(F^N) \subset Gr_n(F^{N+1})$$

induced by the inclusion of $F^N \subset F^{N+1} = F^N \oplus F$ as zero in the last coordinate.

4.4. Definition The product bundle $Gr_n(F^N) \times F^N$ has two subvector bundles: $E^n(N)$ consisting of all $(W,v) \in Gr_n(F^N) \times F^N$ with $v \in W$ and $^\perp E^{N-n}(N)$ consisting of all $(W,v) \in Gr_n(F^N) \times F^N$ with $v \perp W$, that is, v is orthogonal to all vectors in W. The vector bundle $E^n(N)$ is called the universal vector bundle over $Gr_n(F^N)$.

4.5. Remark If p_W denotes the orthogonal projection of F^N onto W, then we have an isomorphism $\theta : Gr_n(F^N) \times F^N \to E^n(N) \oplus {}^\perp E^{N-n}(N)$ onto the Whitney sum, where $\theta(W,v) = (W, p_W(v)) \oplus (W, v - p_W(v))$. Over the natural inclusion $Gr_n(F^N) \subset Gr_n(F^{N+1})$, there is the natural inclusion of product vector bundles $Gr_n(F^N) \times F^N \subset Gr_n(F^{N+1}) \times F^{N+1}$ which under θ induces natural inclusions of the Whitney sum factors $E^n(N) \subset E^n(N+1)$ and $^\perp E^{N-n}(N) \subset {}^\perp E^{N+1-n}(N+1)$.

Now, we consider the inductive limit spaces and vector bundles as N goes to infinity.

4.6. Remark The inductive limit construction of (4.1) yields the n-dimensional vector bundle

$$E^n = \lim_{\substack{\longrightarrow \\ N \geq n}} E^n(N) \longrightarrow \lim_{\substack{\longrightarrow \\ N \geq n}} Gr_n(F^N) = Gr_n(F^\infty).$$

This vector bundle has the universal property saying that every reasonable vector bundle is induced from it and under certain circumstances, from the subbundles on the finite Grassmann varieties as we see in the next section.

5 Finitely Generated Vector Bundles

5.1. Theorem *The following six properties of a vector bundle E over X of dimension n and given $N \geq n$ are equivalent:*

(1) *There is a continuous $w : E \to F^N$ with $w|E_x : E_x \to F^N$ a linear monomorphism for each $x \in X$.*

(2) *There is a vector bundle morphism $(u,f) : E \to E^n(N)$ which is an isomorphism on each fibre of E.*

(3) *There is an isomorphism $E \to f^*(E^n(N))$ over X for some continuous map $f : X \to Gr_n(F^N)$.*

(4) There is a vector bundle E' over X and an isomorphism $\phi : E \oplus E' \to X \times F^N$ to the trivial N-dimensional bundle over X.

(5) There exists a surjective vector bundle morphism $\psi : X \times F^N \to E$ which is surjective on each fibre.

(6) There exist continuous sections $s_1, \ldots, s_N \in \Gamma(X, E)$ such that the vectors $s_1(x), \ldots, s_N(x)$ generate the vector space E_x for each $x \in X$.

Proof. We begin with a circle of implications.

(1) implies (2): Given $w : E \to F^N$ as in (1), we define the morphism of vector bundles $(u, f) : E \to E^n(N)$ by $f(x) = w(E_x)$ and $u(v) = (E_{\pi(v)}, w(v)) \in E^n(N)$ for $v \in E$ and $\pi : E \to X$ the projection.

(2) implies (3): Given (u, f) as in (2), we factor u by the induced bundle $E \xrightarrow{u'} f^*(E^n(N)) \xrightarrow{u''} E^n(N)$, where $u' : E \to f^*(E^n(N))$ is a fibrewise isomorphism, hence the desired isomorphism in (3).

(3) implies (4): Given the isomorphism $E \to f^*(E^n(N))$ as in (3), we define $E' = f^*(^\perp E^{N-n}(N))$ and take the Whitney sum leading to an isomorphism $E \oplus E' \to f^*(E^n(N)) \oplus f^*(^\perp E^{N-n}(N)) = f^*(E^n(N) \oplus^\perp E^{N-n}(N))$. Since $E^n(N) \oplus^\perp E^{N-n}(N) = Gr_n(F^N) \times F^N$, the product bundle, and a map induces the product bundle to a trivial bundle, we have an isomorphism $E \oplus E' \to X \times F^N$ as in (4).

(4) implies (1): Given an isomorphism $v : E \oplus E' \to X \times F^N$ over X. Using the injection $j : E \to E \oplus E'$ and the projection $pr_2 : X \times F^N \to F^N$, we form the composite $w = (pr_2)vj : E \to F^N$, and this is the desired map in (1), where the restriction $w|E_x : E_x \to F^N$ is a linear monomorphism for each $x \in X$.

(4) implies (5): We use the inverse $X \times F^N \to E \oplus E'$ to the isomorphism in (4) composed with the projection $E \oplus E' \to E$ to obtain a vector bundle morphism $v : X \times F^N \to E$ which is surjective on each fibre.

Conversely, (5) implies (4) by considering the kernel $E' = \ker(\psi)$ of ψ which is a subvector bundle of rank $N - n$ and its orthogonal complement E'' which is the subvector bundle of all (x, v) with $v \perp \ker(\psi)_x$. The restriction $\psi|E'' : E'' \to E$ is a fibrewise isomorphism and hence an isomorphism of vector bundles. The desired morphism is the sum of $\phi = (\psi|E'')^{-1} \oplus j : E \oplus \ker(\psi) = E \oplus E' \to X \times F^N$ in (4).

(5) and (6) are equivalent. For the natural basic sections $\sigma_i(x) = (x, e_i)$ of the trivial bundle $X \times F^N$ have images $s_i = \psi \sigma_i$ given by ψ with (5) for each $i = 1, \ldots, N$ and conversely ψ is defined by a set of N sections s_i in (6) by the condition $\psi(x, a_1, \ldots, a_N) = a_1 s_1(x) + \cdots + a_N s_N(x)$. Finally, the equivalence of (5) and (6) follows by observing that $\psi(x,)$ is surjective over $x \in X$ if and only if $s_1(x), \ldots, s_N(x)$ generates E_x. This proves the theorem.

5.2. Definition A vector bundle satisfying any of the equivalent conditions of (5.1) is called finitely generated.

For $N = n$, we have the following corollary.

5.3. Corollary *The following six properties of a vector bundle E over X of dimension n are equivalent:*

(1) There is a continuous $w : E \rightarrow F^n$ with $w|E_x : E_x \rightarrow F^n$, a linear isomorphism for each $x \in X$.

(2) There is a vector bundle morphism $(u, f) : E \rightarrow E^n(n) = \{*\} \times F^n$.

(3) There is an isomorphism $E \rightarrow f^*(E^n(n))$ over X for the continuous map $f : X \rightarrow Gr_n(F^n) = \{*\}$, a point.

(4) or (5) There is an isomorphism $E \rightarrow X \times F^n$ or $X \times F^n \rightarrow E$ between E and the product n-dimensional bundle over X.

(6) There exist continuous sections $s_1, \ldots, s_n \in \Gamma(X, E)$ such that the vectors $s_1(x), \ldots, s_n(x)$ form a basis of E_x for each $x \in X$.

6 Vector Bundles on a Compact Space

6.1. Theorem *Every vector bundle over a compact space is finitely generated.*

Proof. Let $p : E \rightarrow X$ be a F-vector bundle over X. Let $V_1, \ldots, V_m$ be a finite open covering of X such that $E|V_i$ is trivial for all $i = 1, \ldots, m$. Choose open sets $U_1 \subset V_1, \ldots, U_m \subset V_m$ and continuous functions $\xi_i : X \rightarrow [0, 1] \subset F$ such that

(1) the $U_1, \ldots, U_m$ is an open covering of X and

(2) $\mathrm{supp}(\xi_i) \subset V_i$ and $\xi_i|U_i = 1$ for $i = 1, \ldots, m$. This is possible since X is normal. We define a Gauss map $w : E \rightarrow (F^n)^m$ by choosing trivializing Gauss maps $w_i : E|V_i \rightarrow F^n$ and forming the map $w(v) = (\xi_i(p(v))w_i(v))_{1 \leq i \leq m} \in (F^n)^m$, where this means $\xi_i(p(v))w_i(v) = 0$ if $v \in E - E|V_i$. The function w is continuous by the support condition (2) on the ξ_i, and the restriction $w|E_x$ is a linear monomorphism for each $x \in X$. Thus, there exists a Gauss map for E, and hence E is finitely generated.

6.2. *Remark* If $p : E \rightarrow X$ is a vector bundle over a space and if (U_i, V_i) is a sequence of normal pairs with continuous functions $\xi_i : X \rightarrow [0, 1] \subset F$ such that the $U_1, \ldots, U_m, \ldots$ is an open covering of X, then using the properties $\mathrm{supp}(\xi_i) \subset V_i$ and $\xi_i|U_i = 1$ for $i \geq 1$, we have a Gauss map $w : E \rightarrow (F^n)^\infty = F^\infty$ given by the same formula $w(v) = (\xi_i(p(v))w_i(v))_{1 \leq i} \in (F^n)^\infty = F^\infty$. This Gauss map defines a vector bundle morphism $(u, f) : E \rightarrow E^n(\infty)$ which is an isomorphism on each fibre of E by $f(x) = w(E_x)$ and $u(v) = (E_{p(v)}, w(v))$ for $v \in E$. Then, there is an isomorphism $u' : E \rightarrow f^*(E^n(\infty))$, that is, such a vector bundle is induced from the universal bundle over $Gr_n(F^\infty)$.

7 Collapsing and Clutching Vector Bundles on Subspaces

There are two topological operations on vector bundles which play a basic role in the geometric considerations related to vector bundles.

7.1. Definition Let A be a closed subspace of X and form the quotient $q : X \to X/A$, where A is collapsed to a point. Let $p : E \to X$ be a vector bundle with a trivialization $t : E|A \to A \times F^n$ over the subspace A. The collapsed vector bundle $E/t \to X/A$ is the unique vector bundle defined in terms of q-morphism $u : E \to E/t$ of vector bundles such that $t = t_* u$ for an isomorphism of the fibre $t_* : (E/t)_* \to F^n$.

7.2. Definition Let $X = A' \cup A''$ be the union of two closed subspaces with $A = A' \cap A''$. Let $E' \to A'$ and $E'' \to A''$ be two vector bundles, and $\alpha : E'|A \to E''|A$ be an isomorphism of vector bundles over A. The clutched vector bundle $E' \cup_\alpha E''$ is the unique vector bundle $E \to X$ together with isomorphisms $u' : E' \to E|A'$ and $u'' : E'' \to E|A''$ such that on A we have $u' = u'' \alpha$.

In the case of both of the previous definitions, a direct quotient process gives a bundle of vector spaces, and as for the question of local triviality, we have the following remarks.

7.3. Local Considerations in the Previous Two Definitions For the local triviality of the bundles E/t and $E' \cup_\alpha E''$, we have to be able to extend the trivializing t on A or the clutching isomorphism α on A to an open neighborhood of A. Here, we must assume that either A is a closed subspace in a compact X and then use the Tietze extension theorem or assume that A is a subcomplex of a CW-complex X. For this, see *Fibre bundles* (Husemöller 1994) p.123, 135, for more details.

7.4. Functoriality Both the collapsing and clutching are functorial for maps $w : E' \to E''$ such that $t'' w = t'$ over $A \subset X$ inducing $E'/t' \to E''/t''$ and morphisms $f' : E' \to F'$ over A' and $f'' : E'' \to F''$ over A'' commuting with clutching data $f'' \alpha = \beta f'$ inducing a morphism $f : E' \cup_\alpha E'' \to F' \cup_\beta F''$ of vector bundles.

7.5. Remark Let A be a closed subspace of X, and let E' be a vector bundle on X/A. The induced bundle $q^*(E') = E$ has a natural trivialization t over A, and there is a natural isomorphism $E/t \to E'$ from the collapsed vector bundle to the original vector bundle E'. Let $X = A' \cup A''$ and $A = A' \cap A''$, and let E be a vector bundle over X. For $E' = E|A'$ and $E'' = E|A''$, and α the identity on $E|A$. Then, there is a natural isomorphism

$$E' \cup_\alpha E'' \longrightarrow E$$

of the clutched vector bundle to the original vector bundle.

7.6. Commutation with Whitney Sum Over the quotient X/A, we have a natural isomorphism of the Whitney sum

$$(E'/t') \oplus (E''/t'') \longrightarrow (E' \oplus E'')/(t' \oplus t'').$$

Over $X = A' \cup A''$ and $A = A' \cap A''$, we have a natural isomorphism of the Whitney sum of clutched vector bundles

$$(E' \cup_\alpha E'') \oplus (F' \cup_\beta F'') \longrightarrow (E' \oplus F') \cup_{\alpha \oplus \beta} (E'' \oplus F'').$$

7.7. Commutation with Induced Bundles Let $f : (Y,B) \to (X,A)$ be a map of pairs, and let E be a vector bundle with trivialization $t : E|A \to A \times F^n$. For the natural morphism $w : E \to E/t$ over $q : X \to X/A$, we have a trivialization $tw : f^*(E)|B \to B \times F^n$ and a natural morphism of $f^*(E)/tw \to f^*(E/t)$ induced by w.

Let $f : (Y;B',B'',B) \longrightarrow (X;A',A'',A)$ be a map of coverings as in (7.2), let E' (resp. E'') be a vector bundle over A' (resp. A''), and let $\alpha : E'|A \to E''|A$ be an isomorphism. Then there is an natural isomorphism

$$(f|A')^*(E') \cup_\beta (f|A'')^*(E'') \to f^*(E' \cup_\alpha E''),$$

where $\beta = f^*(\alpha)$.

8 Metrics on Vector Bundles

Let $\bar{z}$ denote the conjugation which on $\mathbb{R}$ is the identity, on $\mathbb{C}$ is complex conjugation, and on $\mathbb{H}$ is the quaternionic conjugation.

8.1. Definition Let V be a left vector space over F. An inner product is a function $\beta : V \times V \to F$ such that

(1) For $a,b \in F$ and $x,x',y,y' \in V$, the sesquilinearity of β is

$$\beta(ax + bx',y) = a\beta(x,y) + b\beta(x',y)$$

$$\beta(x,ay + by') = \bar{a}\beta(x,y) + \bar{b}\beta(x,y').$$

(2) For $x,y \in V$, we have conjugate symmetry $\beta(x,y) = \overline{\beta(y,x)}$.
(3) For $x \in V$, we have $\beta(x,x) \geq 0$ in $\mathbb{R}$ and $\beta(x,x) = 0$ if and only if $x = 0$.

Two vectors $x,y \in V$ are perpendicular provided $\beta(x,y) = 0$. For a subspace W of V, the set $W^\perp$ of all $y \in V$ with y perpendicular to all $x \in W$ is a subspace of V and $V = W \oplus W^\perp$.

8.2. Definition An inner product on a vector bundle E over X is a map $\beta : E \oplus E \to F$ such that the restriction for $x \in X$ to each fibre $\beta_x : E_x \times E_x \to F$ is an inner product on E_x.

8.3. Example On the trivial bundles $X \times F^n$ or even $X \times F^\infty$, there exists a natural inner product with

$$\beta(x,u_1,\ldots,u_n,\ldots,x,v_1,\ldots,v_n,\ldots) = \sum_{1 \leq j} u_j \bar{v}_j$$

(always a finite sum). This formula holds both in the finite and in the infinite case where vectors have only finitely many nonzero components. By restriction, every inner product on a vector bundle gives an inner product on each subbundle. For a map $f : Y \to X$ and a vector bundle E with inner product β, we define a unique inner product on $f^*(E)$ such that the natural f-morphism $w : f^*(E) \to E$ is fibrewise an isomorphism of vector spaces with inner product $f^*(\beta)$ by the formula

$$f^*(\beta)(y,z';y,z'') = \beta(z',z'') \quad \text{for} \quad (y,z'),(y,z'') \in f^*(E).$$

In particular, every bundle which is induced from the universal bundle over $G_n(F^n)$ has a metric.

8.4. Remark If E' is a subbundle of a bundle E with a metric β, then the fibrewise union of $E_x'^\perp$ is a vector bundle, denoted by $E'^\perp$, and the natural $E' \oplus E'^\perp \to E$ is an isomorphism.

Reference

Husemöller, D.: *Fibre Bundles*, 3rd ed. Springer-Verlag, New York (1994)

Chapter 3
Relation Between Vector Bundles, Projective Modules, and Idempotents

The main theorem is that there is an equivalence of the category of complex vector bundles on a space X with the category of finitely generated projective $C(X)$-modules, where $C(X)$ is the $\mathbb{C}$-algebra of continuous complex valued functions on X. For this, we need a suitable hypothesis on X. The module associated to a vector bundle E is the $C(X)$-module of cross sections $\Gamma(X,E)$, and the cross section functor $\Gamma : (\text{vect}/X) \to (C(X))$ is the functor which sets up the equivalence of categories between the category of (complex) vector bundles (vect/X) on X and the full subcategory of finitely generated projective modules in the category $(C(X))$ of all $C(X)$-modules.

The first assertion in this direction is the full embedding property of the cross section functor

$$\Gamma : \text{Hom}_X(E', E'') \longrightarrow \text{Hom}_{C(X)}(\Gamma(X,E'), \Gamma(X,E'')),$$

that is, Γ is a bijection. This is proved in Sect. 2 after some local to global preliminaries in Sect. 1 which are typical of normal spaces. Then, the remainder of the argument in Sect. 3 revolves around seeing that under additional hypotheses, $\Gamma(X,E)$ is indeed a finitely generated projective, and every finitely generated projective module over $C(X)$ is isomorphic to some $\Gamma(X,E)$. This is the case for a compact space X, and for a locally compact space, we can modify the assertion in terms of vanishing and triviality at infinity.

The main result in this chapter is due independently to Jean-Pierre Serre and Richard Swan. It is often referred to as the Serre–Swan correspondence between vector bundles over a space X and finitely generated projective modules over the algebra $C(X)$.

In the process of showing that every projective module comes from a vector bundle, we start with relations between idempotent elements in matrix algebras. The algebraic relation leads to another description of vector bundles and projective modules as classes of idempotents.

D. Husemöller et al.: *Relation Between Vector Bundles, Projective Modules, and Idempotents*, Lect. Notes Phys. **726**, 35–44 (2008)
DOI 10.1007/978-3-540-74956-1_4

1 Local Coordinates of a Vector Bundle Given by Global Functions over a Normal Space

In order to go from fibrewise and locally defined sections of a vector bundle, we use the following open sets which are certain neighborhood pairs.

1.1. Definition A numerable pair (U,V) or (U,V,ξ) consists of two open sets $U \subset V$ and a continuous function $\xi : X \to [0,1]$ with $\xi|U = 1$ and the closure of $\xi^{-1}((0,1]) = \mathrm{supp}(\xi) \subset V$.

1.2. Remark Because of the Urysohn lemma, every pair U and V of open sets satisfying $U \subset \bar{U} \subset V$ in a normal space X is numerable.

1.3. Remark The function $\theta : \Gamma(V,E) \to \Gamma(X,E)$ which assigns to a section $s \in \Gamma(V,E)$ of a vector bundle $E|V$ the section $\theta(s)$, where

$$\theta(s)(x) = \begin{cases} \xi(x)s(x) & \text{for } x \in V \\ 0 & x \in X \backslash \mathrm{supp}(\xi) \end{cases}$$

is a homomorphism. Observe that $\theta(s)|U = s|U$.

1.4. Proposition *Let $p : E \to X$ be a vector bundle, where $E|V$ is trivial for a numerable pair (U,V,ξ). Then, there exist cross sections $s_1,\ldots,s_n \in \Gamma(X,E)$ such that*

(1) for each $x \in U$, the vectors $s_1(x),\ldots,s_n(x)$ form a basis of E_x and
(2) for each $s \in \Gamma(X,E)$, there exist $a_1,\ldots,a_n \in C(X)$ such that s equals $a_1 s_1 + \ldots + a_n s_n$ on U.

Proof. Since $E|V$ is trivial, there exists a basis of n sections $s_i' \in \Gamma(V,E)$ such that $s|V = a_1' s_1' + \ldots + a_n' s_n'$, where $a_i' \in C(V)$. Then on U, the section s equals $\theta(s|V) = a_1 s_1 + \ldots + a_n s_n$, where $a_i = \theta(a_i')$ and $s_i = \theta(s_i')$ giving (2). The vectors $s_i(x) = s_i'(x)$ form a basis of E_x. This proves the proposition.

1.5. Corollary *If $v \in E_x$ or if $s' \in \Gamma(V,E)$, where E is trivial over a numerable pair, then there exists a global section $s \in \Gamma(X,E)$ with $s(x) = v$ or $s|U = s'|U$, respectively.*

1.6. Notation Let E be a vector bundle on X and $x \in X$. Let $\varepsilon_x : C(X) \to \mathbb{C}$ and $\varepsilon_x^E : \Gamma(X,E) \to E_x$ denote the evaluation maps defined by $\varepsilon_x(f) = f(x)$ and $\varepsilon_x^E(s) = s(x)$, respectively. Let I_x denote the kernel of $\varepsilon_x : C(X) \to \mathbb{C}$.

1.7. Proposition *Let E be a vector bundle on a normal space X. For every $x \in X$, it follows that $I_x \Gamma(X,E) = \ker(\varepsilon_x^E : \Gamma(X,E) \to E_x)$, that is, $\ker(\varepsilon_x)\Gamma(X,E) = \ker(\varepsilon_x^E)$.*

Proof. An inclusion of $I_x \Gamma(X,E) \subset \ker(\varepsilon_x^E)$ is immediate since $\varepsilon_x(I_x) = 0$. For the opposite inclusion, we proceed in two steps. If $s|W = 0$ for an open neighborhood of x, then there exists a numerable pair (U,V,ξ) with $x \in U$ and $V \subset W$. Then $\xi s = 0$ because $\xi|(X - V) = 0$. Thus, $s = \xi s + (1 - \xi)s = (1 - \xi)s \in I_x \Gamma(X,E)$.

Next, we use (1.4) to obtain a numerable pair (U, V, ξ), where $E|V$ is trivial and cross sections $s_1, \ldots, s_n \in \Gamma(X, E)$ which are associated with a trivialization of $E|V$. For each $s \in \Gamma(X, E)$, there exists functions $a_1, \ldots, a_n \in C(X)$ such that $s|U = (a_1 s_1 + \ldots + a_n s_n)|U$. From 1(1.4), we see that $0 = \varepsilon_x(s) = s(x)$ if and only if $0 = \varepsilon_x(a_j) = a_j(x)$ for all $j = 1, \ldots, n$, that is, $(a_1 s_1 + \ldots + a_n s_n) \in I_x \Gamma(X, E)$. Now the difference

$$s - (a_1 s_1 + \ldots + a_n s_n)|U = 0$$

so that by the first step $s - (a_1 s_1 + \ldots + a_n s_n) \in I_x \Gamma(X, E)$. Thus, if $s(x) = 0$, then also $s \in I_x \Gamma(X, E)$. This proves the other inclusion and the proposition.

2 The Full Embedding Property of the Cross Section Functor

2.1. Theorem *If X is a normal space, then the cross section functor Γ from vector bundles to $C(X)$-modules is fully faithful, that is, the morphism*

$$\Gamma : \mathrm{Hom}_X(E', E'') \longrightarrow \mathrm{Hom}_{C(X)}(\Gamma(X, E'), \Gamma(X, E''))$$

is an isomorphism.

Proof. For injectivity, we consider $f, g : E' \to E''$ with $\Gamma(f) = \Gamma(g)$. For each $v \in E'_x$, we can choose a global section $s \in \Gamma(X, E')$ with $s(x) = v$ by (1.5). Then we calculate

$$f(v) = fs(x) = (\Gamma(f)s)(x) = (\Gamma(g)s)(x) = gs(x) = g(v).$$

Hence, $\Gamma(f) = \Gamma(g)$ implies $f = g$ and Γ is injective.

For surjectivity, we consider a $C(X)$-linear morphism $\psi : \Gamma(X, E') \to \Gamma(X, E'')$ and observe that $\psi(I_x \Gamma(X, E')) \subset I_x \Gamma(X, E'')$ for each $x \in X$. By (1.7), the $C(X)$-linear ψ defines for each $x \in X$ a quotient linear morphism $\psi_x : E'_x \to E''_x$ giving a commutative diagram

$$
\begin{array}{ccc}
\Gamma(X, E') & \xrightarrow{\;\psi\;} & \Gamma(X, E'') \\
{\scriptstyle \varepsilon'_x} \downarrow & & \downarrow {\scriptstyle \varepsilon''_x} \\
E'_x & \xrightarrow{\;\psi_x\;} & E''_x
\end{array}
$$

We define $f : E' \to E''$ by $f|E'_x = \psi_x$. Then, this fibrewise linear map f will have the property $\Gamma(f) = \psi$, and it will show that Γ is surjective if we can see that f is continuous.

For the continuity of f, we consider numerable pairs (U, V), where $E'|V$ is trivial. Then, there exists by (1.4) global sections, $s_1, \ldots, s_n \in \Gamma(X, E')$, which trivialize E' over U. Then, for any $v \in E'|U$, we can write v uniquely as a linear combination of these sections as follows

$$v = \alpha_1(v)s_1(p'(v)) + \cdots + \alpha_n(v)s_n(p'(v)),$$

where the functions $\alpha_i(v)$ are continuous scalar valued on $E'|U$. This leads to the formula

$$f(v) = \alpha_1(v)(\psi s_1)(p'(v)) + \cdots + \alpha_n(v)(\psi s_n)(p'(v))$$

so that $f : E'|U \to E''|U$ is continuous. Thus, f is continuous, and this proves the full embedding property.

In order to characterize further which $C(X)$-modules are isomorphic to $\Gamma(X,E)$ for some vector bundle E over X, we study the behavior of Γ for subbundles.

2.2. Definition A sequence of morphisms of three vector bundles $E' \overset{u}{\to} E \overset{v}{\to} E''$ is exact provided on each fibre over $x \in X$ the sequence of vector spaces $E'_x \overset{u}{\to} E_x \overset{v}{\to} E''_x$ is, that is, $\mathrm{im}(u) = \ker(v)$. An arbitrary sequence of morphisms of vector bundles is exact provided every subthree term sequence is exact.

2.3. Proposition *If $0 \to E' \overset{u}{\to} E \overset{v}{\to} E''$ is an exact sequence of vector bundles, then the following sequence of $C(X)$-modules $0 \to \Gamma(X,E') \overset{\Gamma(u)}{\longrightarrow} \Gamma(X,E) \overset{\Gamma(v)}{\longrightarrow} \Gamma(X,E'')$ is exact.*

Proof. This exactness assertion is the combination of two assertions. For the first, we note that if $s' \in \Gamma(X,E')$, then $\Gamma(u)(s') = 0$ implies that $u(s'(x)) = 0$ in E_x for each $x \in X$ which in turn just means that $s'(x) = 0$ in E_x for $x \in X$. Thus $s' = 0$ and $\Gamma(u)$ is a monomorphism. With u, we can identify E' as a subbundle of E.

For the second, it is immediate that $\Gamma(v)\Gamma(u) = \Gamma(vu) = 0$. If $\Gamma(v)(s) = 0$ for $s \in \Gamma(X,E)$, then $v(s(x)) = 0$ for each $x \in X$, and this means that $s(x) \in E'_x \subset E_x$ for each $x \in X$ under the identification with u of E' as a subbundle of E. Thus, we have $\mathrm{im}(\Gamma(u)) = \ker(\Gamma(v))$, and this proves the proposition.

Again, there is an important point where one has to be rather careful. As we remarked in 2(3.1), the kernel of a vector bundle morphism is not always a vector bundle. This is the reason for formulating the previous assertion in terms of an exact sequence $0 \to E' \overset{u}{\to} E \overset{v}{\to} E''$, where the kernel of v exists and is exactly the image of u.

3 Finitely Generated Projective Modules

In order to study the functor $\Gamma(X,)$ further, we introduce some notation for the related categories.

3.1. Notation Let (vect/X) denote the category of vector bundles on the space X. Let R be a commutative ring, and let (R) denote the category of modules over R. For a ring R, we denote the categories of left and right modules over R, by respectively, $(R\backslash\mathrm{mod})$ and (mod/R).

In the previous section, we derived the full embedding property of $\Gamma(X,)$: $(\mathrm{vect}/X) \to (C(X))$ under the assumption that X is a normal space. Now, we want to restrict $(C(X))$ to a full subcategory in order to obtain an equivalence of categories under suitable conditions on the space X.

Recall that a module M is finitely generated free if and only if there is a finite set $x_1, \ldots, x_n \in M$ with $u : R^n \to M$ defined by

$$u(a_1, \ldots, a_n) = a_1 x_1 + \cdots + a_n x_n$$

is an isomorphism. The finite set $x_1, \ldots, x_n$ is called a basis.

The next most elementary type of module M is a direct summand of R^n for some n. In this case, we have a pair of morphisms $u : R^n \to M$ and $v : M \to R^n$ such that uv is the identity on M and $p = vu$ is the projection onto $v(M) \subset R^n$ which under the restriction morphism $v : M \to v(M) \subset R^n$ is naturally isomorphic to M.

Again, u is given by $u(a_1, \cdots, a_n) = a_1 x_1 + \cdots + a_n x_n$ in terms of certain elements $x_1, \ldots, x_n \in M$, and v is determined by a formula $v(x) = (v_1(x), \ldots, v_n(x))$, where $v_1, \ldots, v_n \in M^\vee = \mathrm{Hom}(M, R)$ the dual module to M. The fact that uv is the identity on M means that $x = \sum_{i=1}^n v_i(x) x_i$ for each $x \in M$. This leads to the following definition.

3.2. Definition A finite bibasis of a module M is a set of $x_1, \ldots, x_n \in M$ and $v_1, \ldots, v_n \in M^\vee$ such that for each $x \in M$ we have

$$x = \sum_{i=1}^n v_i(x) x_i.$$

To analyze these modules which are direct summands of R^n or equivalently which have bibasis, we make the following definitions.

3.3. Definition A module M is finitely generated provided there is a surjective morphism $u : R^n \to M$. In particular, u is defined by $u(a_1, \ldots, a_n) = a_1 x_1 + \cdots + a_n x_n$ for elements $x_1, \ldots, x_n \in M$. These elements $x_1, \ldots, x_n$ are called generators of M.

3.4. Definition A module P is projective provided for each surjective morphism $w : M \to N$ and each morphism $f : P \to N$ there exists a morphism $g : P \to M$ with $wg = f$. The morphism g is called a lifting of the morphism f.

The situation in (3.4) is often described by the commutative diagram of morphisms with the lifting given by dotted arrow.

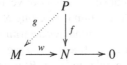

3.5. Example Every free module L is projective. Let E be a basis of L so that each $x \in L$ has the form $x = \sum_{e \in E} a_e(x) e$ with a_e unique and almost all zero. Then, for a surjective morphism $w : M \to N$ and a morphism $f : L \to N$, we have $f(x) = \sum_{e \in E} a_e f(e)$, and in order to lift f to $g : L \to M$ by the surjection $w : M \to N$, we have only to choose $g(e) \in M$ with $g(e) \in w^{-1}(f(e))$ and define $g(x) = \sum_{e \in E} a_e g(e)$. Since $w(g(e)) = f(e)$ for all $e \in E$, it follows that $w(g(x)) = f(x)$ for all $x \in L$.

3.6. Example Let $P = P' \oplus P''$ be a direct sum. Then, P is projective if and only if P' and P'' are projective. Consider a surjective morphism $w : M \to N$ and relate morphisms $f : P \to N$ to their restrictions $f' : P' \to N$ and $f'' : P'' \to N$ by $f = f' \oplus f''$. Hence, if f lifts to g, then f' lifts to $g|P'$ and f'' lifts to $g|P''$, and conversely, if f' lifts $g' : P' \to M$ and f'' lifts $g'' : P'' \to M$, then $f' \oplus f''$ is a lifting of g.

3.7. Proposition *The following are equivalent for M:*

(a) The module M is isomorphic to a direct summand of R^n.
(b) The module M has a bibasis.
(c) The module M is a finitely generated projective module.

Proof. For (a) implies (b) we have seen that the projection $u : R^n \to M$ and the injection $v : M \to R^n$ giving the isomorphism on M onto the direct sum $v(M)$ of R^n in the discussion before (3.2).

For (b) implies (c), we see that the $x_1, \ldots, x_n$ of the bibasis are generators of M. If $f : M \to N''$ is a morphism and $w : N \to N''$ is a surjective morphism, then choose $g(x_i) \in w^{-1}(f(x_i))$, and for $x = \sum_{i=1}^n v_i(x)x_i$ and $f(x) = \sum_{i=1}^n v_i(x)f(x_i)$, we can only define

$$g(x) = \sum_{i=1}^n v_i(x)g(x_i),$$

where g is a lifting of f.

Finally, for (c) implies (a), we have a surjective morphism $u : R^n \to M$ since M is finitely generated. Since M is projective, we can lift the identity $M \to M$ to $v : M \to R^n$, the isomorphism $v : M \to v(M)$, onto a direct sum of R^n. This proves the proposition.

We leave it to the reader to prove the following version of the previous proposition which does not refer to finiteness properties of the modules.

3.8. Remark A module M is projective if and only if M is isomorphic to a direct summand of a free module.

4 The Serre–Swan Theorem

4.1. Notation Let (vect/X) denote the category of vector bundles and vector bundle morphisms over a space X. For a commutative ring R, let (vect/R) denote the full subcategory of the category (R) of R-modules determined by modules which are direct summands of a finitely generated free module (i.e., which are finitely generated projective R-modules). For a general ring (vect/R) is defined the same way with left R-modules.

Now, we combine theorems $(2.1), 2(3.6)$, and $2(5.1)$ together with another argument to arrive at an equivalence of category assertion over a space where every vector bundle is finitely generated, see $2(5.2)$.

4.2. Theorem *Let X be a space over which each vector bundle is finitely generated. Then, the cross section functor*

$$\Gamma = \Gamma(X,) : (\mathrm{vect}/X) \longrightarrow (\mathrm{vect}/C(X))$$

is an equivalence of categories.

Proof. In view of criteria 1(6.7) and (2.1) we only have to show that every finitely generated projective module M over $C(X)$ is isomorphic to $\Gamma(X,E)$ for a vector bundle E. Since M is given by a matrix $e^2 = e \in M_n(C(X))$, we can use $e(x)$ to construct the related vector bundle $E \subset B \times F^n$. This proves the theorem.

As a corollary, we have the following theorem.

Theorem 4.3 (Serre–Swan Theorem) *The cross section functor*

$$\Gamma = \Gamma(X,) : (\mathrm{vect}/X) \longrightarrow (\mathrm{vect}/C(X))$$

is an equivalence of categories for a compact space X.

Proof. Use (4.2) and 2(6.1) for the proof of this assertion.

4.4. Remark For any ring morphism $w : R' \to R''$, we have a function $w_* : (\mathrm{vect}/R') \to (\mathrm{vect}/R'')$ given by

$$w_*(P') = R'' \otimes_{R'} P'.$$

This function defines $(\mathrm{vect}/\ )$ as a functor from the category $(c\backslash rg)$ of rings to the category (set) of sets. There is a functoriality and a naturality related to this equivalence of categories.

4.5. Remark Let $f : X \to Y$ be a map. By right composition of functions on Y with f, we have an induced morphism of rings $C(f) : C(Y) \to C(X)$ and a corresponding morphism $C(f)_* : (\mathrm{vect}/C(Y)) \to (\mathrm{vect}/C(X))$. The function f also induces a functor $f^* : (\mathrm{vect}/Y) \to (\mathrm{vect}/X)$ by sending a vector bundle E to its pullback $f^*(E)$. Then, the cross section functor is contained in the following commutative diagram.

$$
\begin{array}{ccc}
(\mathrm{vect}/Y) & \xrightarrow{\ f^*\ } & (\mathrm{vect}/X) \\
{\scriptstyle \Gamma(Y,)}\downarrow & & \downarrow{\scriptstyle \Gamma(X,)} \\
(\mathrm{vect}/C(Y)) & \xrightarrow{\ C(f)_*\ } & (\mathrm{vect}/C(X)).
\end{array}
$$

Recall that surjective morphisms $R \to R''$ of rings are given by the ideal $I = \ker(R \to R'')$ up to natural factorization by an isomorphism $R/I \to R''$.

4.6. Remark Each inclusion $j : A \to X$ of a closed subspace A into a compact space X defines a surjective $C(j) : C(X) \to C(A)$ by the Tietze extension theorem, and this corresponds to an ideal I, the ideal of functions on X which vanish on A. Using the identification $p : X \to X/A$, we have $I = C(p)(C(X/A, *))$, where $C(X/A, *)$ is the ideal in $C(X/A)$ of all functions equal to zero at $* \in X/A$, where $*$ is the class of A in the quotient space X/A.

5 Idempotent Classes Associated to Finitely Generated Projective Modules

Let R be a ring and form the matrix algebra $M_n(R)$ over R of n by n matrices. We consider direct sums of R-modules $M \oplus N$ isomorphic to R^n as left R-modules, or in other words, we consider finitely generated projective modules M which are quotients of R^n.

5.1. Remark Let $u : R^n \to P$ be an epimorphism onto a projective module P. Then, there exists a morphism $u' : P \to R^n$ with $uu' = \mathrm{id}$, the identity on P and $p = u'u$ is an idempotent, that is, it satisfies $p = p^2$, leading to a direct sum decomposition

$$R^n \cong \mathrm{im}(p) \oplus \ker(p).$$

The restriction $u|\mathrm{im}(p) : \mathrm{im}(p) \to P$ is an isomorphism, and P is finitely generated. We consider an equivalence relation on idempotents $p \in M_n(R)$ such that p and $q \in M_m(R)$ are equivalent if and only if $\mathrm{im}(p)$ and $\mathrm{im}(q)$ are isomorphic.

5.2. Notation Let $(R\backslash\mathrm{mod})$ denote the category of left R-modules. Let $u : R^m \to P$ and $v : R^n \to Q$ be two epimorphisms onto projective objects P and Q in $(R\backslash\mathrm{mod})$, and choose splittings $u' : P \to R^m$ of u and $v' : Q \to R^n$ of v. Then $p = u'u$ is an idempotent in the matrix ring $M_m(R)$ with $\mathrm{im}(p) = P$ and $q = v'v$ is an idempotent in $M_n(R)$ with $\mathrm{im}(q) = Q$. Let $f : Q \to P$ be an isomorphism with inverse $g : P \to Q$. We study related morphisms

$$y = v'gu : R^m \longrightarrow R^n \quad \text{and} \quad x = u'fv : R^n \longrightarrow R^m.$$

We have two properties related to the fact that P and Q are isomorphic:

(A) $xy = p$ and $yx = q$.
  For this, calculate $xy = u'fvv'gu = u'fgu = u'u = p$ and $yx = v'guu'fv = v'gfv = v'v = q$.
(B) $x = px = xq = pxq$ and $y = qy = yp = qyp$.
  For this, calculate $pxq = u'u(u'fv)v'v = u'fv = x$, and a similar calculation gives the other relations.

5.3. Remark Now, we study the relation between two idempotents $p^2 = p \in M_m(R)$ and $q^2 = q \in M_n(R)$ which have $x'y' = p$ and $y'x' = q$, where $x' : R^n \to R^m$ and $y' : R^m \to R^n$, that is, (5.2)(A) holds. We can replace x' by $x = px'q : R^n \to R^m$ and y' by $y = qy'p : R^m \to R^n$. Then, condition (5.2)(A) still holds, that is, $xy = p$ and $yx = q$, and (5.2)(B) now holds. To see this, we calculate $xy = px'qqy'p = px'y'x'y'p = p^5$, and hence $xy = p$. Similarly, $yx = q$. Finally, we calculate $qyp = qqy'pp = qy'p = y$, and the other relations of (5.2)(B) follow by similar calculations.

5.4. Definition Two idempotents $p \in M_m(R)$ and $q \in M_n(R)$ are algebraically equivalent provided there exists $x : R^n \to R^m$ and $y : R^m \to R^n$ satisfying (5.2)(A) and (5.2)(B) or equivalently by (5.3) just (5.2)(A):

$$p = xy \quad \text{and} \quad q = yx.$$

5.5. Proposition *The relation of algebraic equivalence is an equivalence relation between idempotents in matrix algebras.*

Proof. To see that p is equivalent to p, we use $x = y = p$. If p is equivalent to q with $p = xy$ and $q = yx$, then by interchanging x and y, we deduce that q is equivalent to p. For transitivity suppose that p, q, and r are idempotents with $p = xy, q = yx, q = uv$, and $r = vu$ satisfying (5.2)(B). Then, $(xu)(vy) = xqy = xyxy = pp = p$ and $(vy)(xu) = vqu = vuvu = rr = r$. This proves the proposition.

5.6. Proposition *If $p \in M_m(R)$ and $q \in M_n(R)$ are algebraically equivalent idempotents, then $P = \operatorname{im}(p)$ and $Q = \operatorname{im}(q)$ are isomorphic.*

Proof. For this, choose $p = xy$ and $q = yx$ satisfying (5.2)(A) and (5.2)(B). Let $u : R^m \to \operatorname{im}(p)$ be the epimorphism on the direct factor with splitting morphism $u' : \operatorname{im}(p) \to R^m$ so that $p = u'u$, and let $v : R^n \to \operatorname{im}(q)$ be the epimorphism on the direct factor with splitting morphism $v' : \operatorname{im}(q) \to R^n$ so that $q = v'v$. Then form $g = vyu'$ and $f = uxv'$ reversing the definition of x and y from f and g in (5.2).

Clearly, we have

$$g : \operatorname{im}(p) \longrightarrow \operatorname{im}(q) \quad \text{and} \quad f : \operatorname{im}(q) \longrightarrow \operatorname{im}(p),$$

and $gf = vyu'uxv' = vyxv' = vqv' = vv'vv' = \operatorname{im}(q)$, the identity on $\operatorname{im}(q)$. Similarly, $fg = \operatorname{im}(p)$, the identity on $\operatorname{im}(p)$. This proves the proposition.

5.7. Definition Two idempotents $p, p' \in M_n(R)$ are conjugate provided there exists $w \in GL(n, R) = \operatorname{Aut}(R^n)$ with $p' = wpw^{-1}$.

5.8. Proposition *Two idempotents $p, p' \in M_n(R)$ are conjugate if and only if p and p' are algebraically equivalent and $1 - p$ and $1 - p'$ are algebraically equivalent.*

Proof. If $p' = wpw^{-1}$, then choose $x = pw^{-1}$ and $y = wp$. It follows that $xy = pp = p$ and $yx = wppw^{-1} = p'$ so that p and p' are algebraically equivalent. Since $1 - p' = w(1 - p)w^{-1}$, it follows that $1 - p$ and $1 - p'$ are algebraically equivalent.

Conversely, suppose that $p = xy$ and $p' = yx$ and also that $1 - p = uv$ and $1 - p' = vu$ such that (5.2)(B) holds. We calculate

$$
\begin{aligned}
(x + u)(y + v) &= p + uy + xv + (1 - p) \\
&= p + (1 - p)u(1 - p')p'yp + pxp'(1 - p')v(1 - p) + (1 - p) \\
&= p + 1 - p = 1
\end{aligned}
$$

Let $z = y + v$ and so $z^{-1} = x + u$. Using (5.2)(B)

$$v = v - vp \quad \text{so} \quad vp = 0 \quad \text{and} \quad u = u - pu \quad \text{so} \quad pu = 0,$$

we calculate the conjugate idempotent

$$zpz^{-1} = (y+v)p(x+u) = ypx + vpx + ypu + vpu = yxyx = p'p' = p'$$

so p and p' are conjugate idempotents. This proves the proposition.

5.9. Definition Let $\text{Idem}(R)$ denote the set of equivalence classes of idempotents in the various $M_n(R)$ with respect to the algebraic equivalence relation. Then, a ring morphism $w : R' \to R''$ induces a ring morphism $M_n(w) : M_n(R') \to M_n(R'')$ which in turn induces a natural function

$$\text{Idem}(w) : \text{Idem}(R') \longrightarrow \text{Idem}(R'').$$

5.10. Remark The process of assigning to an idempotent $e \in M_n(R)$ its image $\text{im}(e)$ as an object in (vect/R) leads to the following commutative diagram for any ring morphism $w : R' \to R''$

$$
\begin{array}{ccc}
\text{Idem}(R') & \xrightarrow{\;\text{Idem}(w)\;} & \text{Idem}(R'') \\
\text{im}\downarrow & & \downarrow\text{im} \\
\text{Vect}/R' & \xrightarrow{\quad w_*\quad} & \text{Vect}/R''
\end{array}
$$

where Vect/R denotes the isomorphism classes of finitely generated projective R-modules, see 4(3.1). For a continuous map $f : X \to Y$, we can put this commutative diagram together with the one in (4.5) to obtain the functoriality related to the three versions of vector bundle theory

$$
\begin{array}{ccc}
\text{Vect}/Y & \xrightarrow{\quad f^*\quad} & \text{Vect}/X \\
\Gamma(Y,)\downarrow & & \downarrow\Gamma(X,) \\
\text{Vect}/C(Y) & \xrightarrow{\;C(f)_*\;} & \text{Vect}/C(X) \\
\text{im}\uparrow & & \uparrow\text{im} \\
\text{Idem}(C(Y)) & \xrightarrow{\;\text{Idem}(C(f))\;} & \text{Idem}(C(X))
\end{array}
$$

where now Vect/X is the set of isomorphism classes of vector bundles over X, see 4(2.1). See also 3(5.6).

Chapter 4
K-Theory of Vector Bundles, of Modules, and of Idempotents

We wish to consider the isomorphism classes of vector bundles over a space X or in view of the correspondences in the previous chapter, either the isomorphism classes of finitely generated modules over $C(X)$ or algebraic equivalence classes of idempotents in the matrix algebras $M_n(C(X))$ for all n. In each of the three cases, the direct sum leads to a natural addition on this sets of classes, and with the tensor product, there is a second algebraic operation of product. Unfortunately, these sets of isomorphism classes are not a ring under these operations, but only a semiring. The unfulfilled ring axiom is that each element must have a negative.

In the first section, we see how these semirings map universally into rings in the same way as the natural numbers map into the integers $\mathbb{N} \to \mathbb{Z}$, and the resulting class groups or rings are called the K-groups of vector bundles, K-groups of finitely generated projective modules, or the K-groups of matrix ring idempotents. Using the results of the previous chapter, we can show that these semirings and the corresponding K-rings are all isomorphic. Finally, we consider some special features of topological algebras, especially C^*-algebras.

1 Generalities on Adding Negatives

There is a universal mapping of a commutative monoid (or semigroup) into a commutative group and a universal mapping of a commutative semiring in a ring. This is an elementary exercise in the study of functors with universal properties, and it is the basic construction needed to go from isomorphism classes to stable classes of vector bundles, projective modules, and idempotents.

1.1. Notation Let $(s\backslash ab)$ denote the category of abelian semigroups with composition written additively and morphisms $f : A' \to A''$ satisfying $f(0) = 0$ and $f(x+y) = f(x) + f(y)$ for $x, y \in A'$. Let (ab) denote the full subcategory of $(s\backslash ab)$ determined by the abelian groups.

D. Husemöller et al.: *K-Theory of Vector Bundles, of Modules, and of Idempotents*, Lect. Notes Phys. **726**, 45–54 (2008)
DOI 10.1007/978-3-540-74956-1_5

Note for the general semigroup, the axiom on the existence of $-z$ for all z is not necessarily fulfilled, but on the other hand, if $-z$ exists in A' and $f : A' \to A''$, then $f(-z) = -f(z)$ exists in A''.

1.2. Notation Let (s\rg) denote the category of semirings with unit or 1 element and morphisms $f : R' \to R''$ of semirings preserving the 1 element, that is, $f(1) = 1$. Let (rg) denote the full subcategory of (s\rg) determined by the rings with 1.

1.3. Remark Stripping off the multiplication on a semiring (resp. ring) gives an abelian semigroup (resp. abelian group). This leads to a commutative diagram of categories.

$$
\begin{array}{ccc}
(\text{rg}) & \longrightarrow & (\text{s\rg}) \\
{\scriptstyle\text{stripping functor}}\big\downarrow & & \big\downarrow{\scriptstyle\text{stripping functor}} \\
(\text{ab}) & \longrightarrow & (\text{s\ab})
\end{array}
$$

Now, the elementary process of extending to a group or a ring is contained in the next construction, where $[a,x] = a - x$ intuitively.

1.4. Construction The group localization of an abelian semi-group A is the abelian group $T(A)$ whose elements $[a,x]$ are equivalence classes of elements $(a,x) \in A \times A$ by the relation (a',x') is equivalent to (a'',x'') provided there exists $z \in A$ with

$$a' + x'' + z = a'' + x' + z.$$

The group law is given by $[a,x] + [b,y] = [a+b,\ x+y]$, the zero element is $[0,0]$, and the negative $-[a,x] = [x,a]$. In the case that A is a semiring, we define a ring structure on $T(A)$ by the relation

$$[a,x].[b,y] = [ab+xy,\ ay+xb]$$

with 1 element $1 = [1,0]$. We have a natural morphism $\theta : A \to T(A)$ preserving the semigroup or semiring structure. If A is a commutative semiring, then $T(A)$ is a commutative ring.

1.5. Example For the natural numbers $\mathbb{N}$ either as a commutative semigroup or as a semiring, the localization is $T(\mathbb{N}) = \mathbb{Z}$, the abelian group of integers or the ring of integers.

1.6. Proposition *The natural morphism $\theta : A \to T(A)$ is an isomorphism if and only if for every $z \in A$ the negative $-z$ exits in the semigroup or semiring A.*

Proof. The direct implication is immediate, and for the converse, note that for $[a,x] = \theta(a-x)$ with a and x well defined the inverse of θ is well defined by $\theta^{-1}([a,x]) = a - x$. This proves the proposition.

Now we state the universal property of the morphism $\theta : A \to T(A)$.

1.7. Proposition *Let $f : A \to G$ be a morphism of commutative semigroups, where G is a group. Then, there exists a unique morphism of groups $h : T(A) \to G$ with $h\theta = f$. Let $f : A \to R$ be a morphism of semirings, where R is a ring. Then, there exists a unique morphism of rings $h : T(A) \to R$ with $h\theta = f$.*

Proof. In both cases $h([a,x]) = f(a) - f(x)$, and it must satisfy this formula showing the uniqueness.

2 *K*-Groups of Vector Bundles

2.1. Definition Let Vect/X denote the isomorphism classes $[E]$ of vector bundles E over X with the semiring structure

$$[E'] + [E''] = [E' \oplus E''] \quad \text{and} \quad [E'].[E''] = [E' \otimes E''].$$

The zero vector bundle is the zero class, and the trivial line bundle is the unit in the commutative semiring.

2.2. Remark Let $f : X \to Y$ be a map between spaces. The inverse image of vector bundles defines a semiring morphism

$$\mathrm{Vect}/f : \mathrm{Vect}/Y \longrightarrow \mathrm{Vect}/X$$

given by the formula $\mathrm{Vect}/f([E]) = [f^*(E)]$. For a second map $g : Y \to Z$, we have the contravariant functor property

$$\mathrm{Vect}/gf = \mathrm{Vect}/f \circ \mathrm{Vect}/g : \mathrm{Vect}/Z \longrightarrow \mathrm{Vect}/X.$$

2.3. Definition The Grothendieck *K*-functor on the category (top) of spaces and maps is defined by ring localization

$$K(X) = T(\mathrm{Vect}/X).$$

2.4. Remark The universal property makes K into a functor $K : (\mathrm{top}) \to (\mathrm{c\backslash rg})^{\mathrm{op}}$ into the opposite category of the category of commutative rings. The functoriality of K follows from the functoriality of Vect/ and the induced morphisms on the K-rings is defined by applying (1.7) to the following commutative diagram for a map $f : X \to Y$ relating a semiring and its continuation as a ring.

$$
\begin{array}{ccc}
\mathrm{Vect}/Y & \xrightarrow{\ f^*\ } & \mathrm{Vect}/X \\
{\scriptstyle \theta}\big\downarrow & & \big\downarrow{\scriptstyle \theta} \\
K(Y) & \xrightarrow{\ K(f)\ } & K(X).
\end{array}
$$

The functoriality of K follows from that of Vect/and the uniqueness in the universal factorization property.

In order to begin the study of the functor K, we observe that the algebraic morphism given by dimension or rank: $K(*) \to \mathbb{Z}$ is an isomorphism.

2.5. Remark The map $X \to *$ induces a homomorphism $\mathbb{Z} = K(*) \to K(X)$. The image of $1 \in \mathbb{Z}$, which is represented by the trivial line bundle, yields the unit in $K(X)$.

2.6. Remark For each base point $* \to X$, we have a morphism $K(X) \to K(*) = \mathbb{Z}$ which composed with the unit homomorphism $\mathbb{Z} \to K(X)$ of (2.5) is the identity. It gives a splitting of $K(X) = \mathbb{Z} \oplus K(X,*)$, where $K(X,*)$ is an ideal in the supplemented ring called the reduced K-theory of the pointed space $(X,*)$.

Now, we consider an elementary property of the K-functor which plays a basic role later.

2.7. Proposition *Let $j : A \to X$ be the injection of a closed subspace in a compact space X, and let $q : X \to X/A$ denote the quotient map. Then, there is the sequence of abelian groups*

$$K(X/A, *) \xrightarrow{K(q)} K(X) \xrightarrow{K(j)} K(A)$$

and $\mathrm{im}(K(q))$ is an ideal in $K(X)$ which is the kernel of the ring morphism $K(j)$.

Proof. Since the composition $qj : A \to X/A$ is the constant map with value the point $*$, it follows that $K(qj) = K(j)K(q)$ is zero on $K(X/A, *)$. If a class $c \in K(X)$ is mapped to zero by $K(j)$ in $K(A)$, then we represent $c = [E] - [F]$, where F is the trivial vector bundle. After replacing E by $E \oplus G$ with G a trivial vector bundle, we can assume that $j^*(E)$ is trivial, and we choose t a trivialization of $E|A = j^*(E)$. Then $E = q^*(E/t)$, and c is in the image of $K(q)$. This proves the proposition.

3 *K*-Groups of Finitely Generated Projective Modules

3.1. Definition Let Vect$/R$ denote the isomorphism classes $[P]$ of finitely generated projective modules over R with the semigroup structure

$$[P'] + [P''] = [P' \oplus P''].$$

In the case that R is commutative, we have the additional structure

$$[P'].[P''] = [P' \otimes_R P'']$$

making Vect$/R$ into a commutative semiring. The zero module is the zero class, and the R-module R is the unit in the commutative semiring for R commutative.

3.2. Remark Let $f : R \to R'$ be a morphism of rings. The tensor product of modules defines a semigroup morphism, which in the case of commutative rings is a semiring morphism,

$$f_* : \mathrm{Vect}/R \longrightarrow \mathrm{Vect}/R'.$$

It is given by the formula $f_*([P]) = [R' \otimes_R P]$. For a second ring morphism $g : R' \to R''$, we have the covariant functor property

$$(gf)_* = g_* f_* : \mathrm{Vect}/R \longrightarrow \mathrm{Vect}/R''.$$

3.3. Definition The Grothendieck K-functor on the category (rg) of rings is defined by $K(R) = T(\mathrm{Vect}/R)$.

3.4. Remark The universal property makes K into a functor $K : (\mathrm{rg}) \to (\mathrm{ab})$ into the category of abelian groups. When restricted to the category (c\rg) of commutative rings, it yields a functor $K : (\mathrm{c\backslash rg}) \to (\mathrm{c\backslash rg})$. The functoriality of K follows from the functoriality of Vect/ and the universal property (1.7) of the functor T which for a ring morphism $f : R \to R'$ yields a morphism $K(f) : K(R) \to K(R')$ which makes the following diagram commutative

$$
\begin{array}{ccc}
\mathrm{Vect}/R & \xrightarrow{\ f_* \ } & \mathrm{Vect}/R' \\
\theta \downarrow & & \downarrow \theta \\
K(R) & \xrightarrow{\ K(f) \ } & K(R')
\end{array}
$$

Recall from 3(4.2) that in the case of $R = C(X)$ for a compact space X, we have an equivalence of categories given by the cross section functor $\Gamma : (\mathrm{vect}/X) \to (\mathrm{vect}/C(X))$.

3.5. Theorem *For a compact space X and $C(X)$, the ring of complex valued continuous functions on X the equivalence of categories under Γ induces an isomorphism of semirings, denoted by $\Gamma : \mathrm{Vect}/X \to \mathrm{Vect}/C(X)$, and further it induces an isomorphism of the K-rings, again denoted by $\Gamma : K(X) \to K(C(X))$. For a continuous map $f : X \to Y$, we have the commutative diagram*

$$
\begin{array}{ccc}
K(Y) & \xrightarrow{\ K(f) \ } & K(X) \\
\Gamma \downarrow & & \downarrow \Gamma \\
K(C(Y)) & \xrightarrow{\ K(C(f)) \ } & K(C(X)).
\end{array}
$$

In particular, $\Gamma : K \to K \circ C$ is an isomorphism of functors.

In general, we denote $K(f)$ and $K(C(f))$ by simply f^*.

4 K-Groups of Idempotents

4.1. Definition Recall from 3(5.9) the definition of Idem(R), the set of algebraic equivalence classes $[p]$ of idempotents $p = p^2$ in some $M_n(R)$. We define a semigroup structure on Idem(R) by

$$[p'] + [p''] = [p' \oplus p''].$$

In the case that R is commutative we have the additional structure

$$[p'].[p''] = [p' \otimes_R p'']$$

making Idem(R) into a semiring. The zero idempotent $p = 0$ is the zero class, and the identity on the R-module R is the unit in the commutative semiring for R commutative.

4.2. Remark Let $f : R \to R'$ be a morphism of rings. The induced morphism of matrices defines a semigroup morphism, which preserves algebraic equivalence

$$\text{Idem}(f) : \text{Idem}(R) \longrightarrow \text{Idem}(R')$$

given by the formula $\text{Idem}(f)([e]) = [M_n(f)(e)]$ for $e \in M_n(R)$. For a second ring morphism $g : R' \to R''$, we have the covariant functor property

$$\text{Idem}(gf) = \text{Idem}(g)\text{Idem}(f) : \text{Idem}(R) \longrightarrow \text{Idem}(R'').$$

4.3. Definition The Grothendieck idempotent I-functor on the category (rg), resp. (c\rg), to the category (ab) of abelian groups, resp. (c\rg) of commutative rings, is defined by $I(R) = T(\text{Idem}(R))$.

4.4. Remark The universal property makes I into a functor $I : (\text{rg}) \to (\text{ab})$ into the category of abelian groups. The functoriality of I follows from the functoriality of Idem( ) and the universal property (1.7) of the functor T for which the morphism of semigroups $\text{Idem}(f) : \text{Idem}(R) \to \text{Idem}(R')$ induced by a morphism $f : R \to R'$ yields a map $I(f)$ which makes the following diagram commutative

$$
\begin{array}{ccc}
\text{Idem}(R) & \xrightarrow{\ \text{Idem}(f)\ } & \text{Idem}(R') \\
{\scriptstyle \theta}\downarrow & & \downarrow{\scriptstyle \theta} \\
I(R) & \xrightarrow{\ \ I(f)\ \ } & I(R').
\end{array}
$$

4.5. Theorem *For a commutative ring R, the map im, which we have introduced in 3(5.10), induces an isomorphism of semigroups (semirings) denoted by im: Idem(R) $\to$ Vect/R, and further it induces an isomorphism of K-groups (K-rings), again denoted by im:I(R) $\to$ K(R). For a ring morphism $f : R \to R'$, we have a commutative diagram*

$$\begin{CD} I(R) @>I(f)>> I(R') \\ @V\text{im}VV @VV\text{im}V \\ K(R) @>K(f)>> K(R'). \end{CD}$$

In particular, im $: I \to K$ is an isomorphism of functors.

Proof. We have only to check the multiplicative property of the morphism im in case R is commutative. For this, we consider $p \in M_m(R)$ with $p^2 = p$ and $q \in M_n(R)$ with $q^2 = q$ and form $P = p(R^m)$ and $Q = q(R^n)$. The tensor product of the two direct sum decompositions $R^m = P \oplus P'$ and $R^n = Q \oplus Q'$ has the form

$$\begin{aligned} R^{mn} &= (P \otimes Q) \oplus [(P \otimes Q') \oplus (P' \otimes Q) \oplus (Q \otimes Q')] \\ &= \text{im}(p \otimes q) \oplus [(P \otimes Q') \oplus (P' \otimes Q) \oplus (Q \otimes Q')]. \end{aligned}$$

This proves the theorem.

5 K-Theory of Topological Algebras

In Chap. 3, Sect. 4, we related vector bundles over X to finitely generated projective modules over $A = C(X)$, the algebra of continuous complex valued functions on X. We get the same results for the algebra of bounded continuous functions which we also denote by $C(X)$. In the algebraic setting, these modules were related to idempotents in matrix algebras over the matrix algebras $M_n(A)$ for a general algebra in Sect. 5 of Chap. 3. In the first four sections of this chapter, the class group or K-groups of vector bundles, finitely generated projective modules, and idempotents were introduced.

5.1. Remark For the study of class groups, we see that there is a large interface between the topology of bundles and the algebra of certain modules. There is another important connection namely with the area of functional analysis, and it starts in the following two elementary ways:

(1) The algebra $C(X)$ has a natural $*$-algebra structure. This leads to the possibility of using only projections, that is, just selfadjoint idempotents. The $*$-structure is given by defining f^* of an element f by $f^*(x) = \overline{f(x)}$ for $x \in X$.
(2) The algebra $C(X)$ of bounded continuous functions has a norm $\|f\| = \sup_{x \in X} |f(x)|$ making $C(X)$ into a special topological algebra, namely a Banach algebra, that is, a complete normed algebra. At a very early stage in the development of topological K-theory, Banach algebras played an important role giving alternative proofs of Bott periodicity, which is a fundamental subject considered in Chap. 15.

5.2. Remark At the center of this interface between K-theory coming from topology and the use of analysis in K-theory is the Atiyah–Singer index theorem. Indices like

Euler numbers can be interpreted as elements of class groups, and it was natural that the mature form of the theorem would use K-theory in its formulation and proof. In the process, the homotopy description of K-theory given in Chap. 15, Sect. 2, in terms of limits of classifying spaces for fibre bundles, has a version in terms of Fredholm operators. This homotopy classification due independently to Atiyah and Jänich has an important extension to twisted K-theory.

In order to carry this theme a little further, we give two definitions related to the first remark (5.1).

5.3. Definition A $*$-algebra A over the complex numbers $\mathbb{C}$ is an algebra A over the complex numbers together with an additive map $* : A \rightarrow A$, where besides additivity $(x+y)^* = x^* + y^*$, it satisfies $(cx)^* = \bar{c}x^*, (xy)^* = y^*x^*$, and $(x^*)^* = x$ for all $x, y \in A, c \in \mathbb{C}$. An element x in A is selfadjoint provided $x^* = x$. An element $p \in A$ is a projection provided it is a selfadjoint idempotent, that is

$$p^2 = p = p^*.$$

5.4. Example The complex numbers are a $*$-algebra with $*$-operation complex conjugation. The matrix algebra $M_n(A)$ is a $*$-algebra with the $*$-operation given by transposition applied to the $*$ of the matrix elements, that is, for $X = (x_{i,j})$, we define $X^* = Y = (y_{i,j})$, where

$$y_{i,j} = x^*_{j,i}.$$

For K-theory from an algebraic point of view, it is the role of the matrix algebra and idempotents that plays the basic role.

5.5. Remark In the context of topological algebra, we have a new phenomenon that the algebra is so large that there are enough representatives of idempotents or projections in the algebra itself to determine K-theory. This happens for special cases in algebraic K-theory where up to direct summands with finitely generated free modules, every finitely generated projective is a projective submodule of a Dedekind ring R.

For a class of $*$-algebras called C^*-algebras, we return to this question in (5.15).

5.6. Definition A normed algebra is an algebra A with a function $\| \ \| : A \rightarrow \mathbb{R}$ satisfying the following properties:

(1) $\|x\| \geq 0$ for all $x \in A$, and $\|x\| = 0$ if and only if $x = 0$.
(2) $\|cx\| = |c| \cdot \|x\|$ for all $x \in A, c \in \mathbb{C}$.
(3) $\|x+y\| \leq \|x\| + \|y\|$ for all $x, y \in A$.
(4) $\|xy\| \leq \|x\| \cdot \|y\|$ for all $x, y \in A$.

$\| \ \|$ is a norm satisfying (4).

A norm defines a metric $d(x,y) = \|x - y\|$ making an algebra A with a norm a metric space. A norm is complete provided this metric is complete. A Banach algebra A is a normed algebra which is complete, and a $*$-Banach algebra A is a normed ∗-algebra with complete norm such that the relation $\|x^*\| = \|x\|$ for all $x \in A$ is satisfied.

There is a special norm property coming from the norm on $\mathscr{B}(H)$, the ∗-algebra of all bounded operators on a Hilbert space H.

5.7. Definition A norm $\|\ \ \|$ on a ∗-algebra A is a C^*-norm provided $\|xx^*\| = \|x\|^2$ holds for all $x \in A$. A C^*-algebra is a ∗-Banach algebra whose norm is a C^*-norm.

To elements in an algebra A over $\mathbb{C}$, we associate a subset of the complex numbers called the spectrum.

5.8. Definition For an algebra A with unit 1 over $\mathbb{C}$ and $x \in A$, a scalar $\lambda \in \mathbb{C}$ is in the spectrum $sp(x)$ of x provided $x - \lambda \cdot 1$ is not invertible in A. The spectral radius $\rho(x)$ of $x \in A$ is

$$\rho(x) = \sup_{\lambda \in sp(x)} |\lambda|.$$

Note that the function which assigns to $\lambda \in \mathbb{C} - sp(x)$ the element $(x - \lambda \cdot 1)^{-1} \in A$ maps the exterior of the spectrum into the algebra A.

5.9. Remark If A is a Banach algebra, then $sp(x)$ is a closed subset of the closed disc around 0 of radius $\|x\|$ in $\mathbb{C}$. If x is selfadjoint in a ∗-algebra, then $sp(x) \subset \mathbb{R}$, and if $x = yy^*$ in a ∗-algebra, then $sp(x)$ is contained in the positive real axis. It is always the case that $\rho(x) \leq \|x\|$ in a Banach algebra.

5.10. Basic Properties of C^*-Algebras Observe that the spectral radius of an element is defined just from the algebraic data of the algebra. A ∗-algebra A is a C^*-algebra if and only if the spectral radius is a C^*-norm on A. In particular, there is at most one C^*-norm on a ∗-algebra. For a ∗-algebra to be a C^*-algebra is an axiom, not a new structure as it looks like in the initial definition. This remark is proved in the context of the characterization of commutative C^*-algebras, see (5.13).

5.11. Example Every closed ∗-subalgebra of a C^*-algebra is again a C^*-algebra. The space $C(X)$ of bounded continuous complex valued functions on a space X is a commutative C^*-algebra. The algebra $\mathscr{B}(H)$ of bounded linear operators on a Hilbert space H is also a C^*-algebra.

In fact, these examples are related to the following embedding theorem and structure theorem for commutative C^*-algebras due to Gel'fand and his coworkers.

5.12. Theorem *Every commutative C^*-algebra A with a unit is isomorphic to the C^*-algebra of continuous complex valued functions on a compact space X. Moreover, the space X can be taken to be the subspace of continuous ∗-algebra morphisms $\phi : A \to \mathbb{C}$ with the subspace topology of the weak topology on the unit ball in the Banach space dual to A.*

5.13. Remark This bijective relation between isomorphism classes of commutative C^*-algebras and of compact spaces is a consequence of an anti-equivalence of categories which has been an important motivation for many developments relating geometry to algebras. A similar relation exists between commutative rings and affine schemes and also between algebras of holomorphic functions and Stein varieties. One current idea, resulting from this, is that a general C^*-algebra should be regarded as a "noncommutative space."

In Chap. 23, we consider the K-theory of general topological algebras in a bivariant setting. We finish this section with some K-theory considerations.

5.14. Special Features of the K-Theory of C^*-Algebras For a unital C^*-algebra A we can describe the K-theory in terms of classes of projections, that is, elements $p \in A$ satisfying $p^* = p = p^2$. The relation of algebraic equivalence is replaced by p is similar to q provided there exists an element $u \in A$ with $p = uu^*$ and $q = u^*u$.

Secondly, if A is a C^*-algebra, then all matrix algebras $M_n(A)$ are C^*-algebras, and the direct sum with the zero matrix gives an embedding $M_n(A) \to M_{n+q}(A)$ preserving all the C^*-structure except any unit. The union

$$M_\infty(A) = \bigcup_{0 \le n} M_n(A)$$

is a normed $*$-algebra whose norm satisfies the C^*-condition. The completion $\mathscr{K}(A)$ of $M_\infty(A)$ is a C^*-algebra which is naturally isomorphic to the C^*-algebra tensor product $A \otimes \mathscr{K}$, where $\mathscr{K} = \mathscr{K}(\mathbb{C})$ is the $\mathbb{C}$-algebra of compact operators on a separable Hilbert space.

Finally, through the work of Cuntz and others, there was a development of $K(A)$ for a C^*-algebra A including Cuntz's proof of Bott periodicity in the context of C^*-algebras. This theory follows in many ways the lines of the topological theory in the next chapter even though $K(A)$ is the algebraic K-theory of A.

We recommend the following two references. The first, Murphy (1990) carries out the K-theory of C^*-algebras up to Cuntz's proof of Bott periodicity in the last chapter. The second is Blackadar (1998). In Sects. 4.6.2 and 4.6.4 of this book, the relation between algebraic equivalence and similarity is worked out. This book also takes the reader to many considerations taken up in Chap. 23.

References

Blackadar, B.: K-Theory for Operator Algebras, 2nd ed, MSRI Publications 5, Cambridge University Press, Cambridge (1998)
Murphy, G.J.: C-algebras and Operator Theory. Academic Press, San Diego (1990)

Chapter 5
Principal Bundles and Sections of Fibre Bundles: Reduction of the Structure and the Gauge Group I

In this chapter, we consider bundles $p : E \to B$ where a topological group G acts on the fibres through an action of G on the total space E. This is just an action $E \times G \to E$ of G on E such that $p(xs) = p(x)$ for all $x \in E$ and $s \in G$. In particular, there is a restriction $E_b \times G \to E_b$ of the globally defined action to each fibre E_b, $b \in B$, of the bundle. A principal G-bundle is a bundle $p : P \to B$ with an additional algebraic and continuity action property implying, for example, that all fibres are isomorphic to G by any map $G \to P_b$ of the form $s \mapsto us$ for any $u \in P_b$ and all $s \in G$. For a principal G-bundle $p : P \to B$ and a left G-space Y, we have the fibre bundle construction $q : P[Y] = P \times^G Y \to B$. Vector bundles are examples of fibre bundles where $G = GL(n)$, the general linear group, and Y is an n-dimensional vector space. The characterization of sections in $\Gamma(B, P[Y])$ by certain maps $P \to Y$ plays a fundamental role in applications of principal bundle theory. We outline two important aspects within principal bundle theory in this chapter. At first, the reduction of the structure group G of a principal bundle P. The second aspect is the study of $\mathrm{Aut}_G(P)$, the automorphism group of the principal G-bundle P.

Chapter 4 of *Fibre Bundles* (Husemöller 1994) is a reference for this chapter.

1 Bundles Defined by Transformation Groups

1.1. Definition A topological group G is a set G together with a group structure and topology on G such that the function $(s, t) \mapsto st^{-1}$ is a continuous map $G \times G \to G$. A morphism $f : G' \to G''$ of topological groups is a function which is both a group morphism and a continuous map.

This continuity condition on $(s, t) \mapsto st^{-1}$ is equivalent to $G \times G \to G$, $(s, t) \mapsto st$ and $G \to G$, $s \mapsto s^{-1}$ being continuous maps.

D. Husemöller et al.: *Principal Bundles and Sections of Fibre Bundles: Reduction of the Structure and the Gauge Group I*, Lect. Notes Phys. **726**, 55–62 (2008)
DOI 10.1007/978-3-540-74956-1_6

1.2. Example The vector spaces $\mathbb{R}^n$ and $\mathbb{C}^n$ with the addition of vectors as well as the matrix groups $GL(n, \mathbb{R})$ and $GL(n, \mathbb{C})$ with the usual composition are topological groups. A subgroup H of a topological group G with the subspace topology is again a topological group. The determinant det : $GL(n) \to GL(1)$ is an example of a morphism of topological groups, and its kernel $SL(n)$ with the subspace topology is a closed subgroup.

1.3. Definition Let G be a topological group. A right G-space X is a space X together with a map $X \times G \to X$ denoted by $(x, s) \mapsto xs$ satisfying the algebraic axioms:

(1) $x(st) = (xs)t$ for $x \in X$ and $s, t \in G$.
(2) $x1 = x$ for $x \in X$ and 1 the identity of the group.

A morphism $f : X \to Y$ of right G-spaces is a map $f : X \to Y$ of spaces satisfying $f(xs) = f(x)s$ for $x \in X$ and $s \in G$. Such morphisms are also called G-equivariant maps or G-maps. The category of right G-spaces we denote $(G\backslash \text{top})$.

A left G-space X is one with a map $G \times X \to X$ having the algebraic properties $(st)x = s(tx)$ as in (1) and $1x = x$. A left G-space has a natural right G-space structure given by $xs = s^{-1}x$, and a right G-space has a natural left G-space structure given by $sx = xs^{-1}$. Of course, these two concepts are equivalent under this involutionary correspondence $G \to G, s \mapsto s^{-1}$.

1.4. Example The group $GL(n, F)$ acts on the vectors F^n by matrix multiplication leaving 0 fixed, and hence, it also acts on $F^n - \{0\}$. We will return to other actions later. The two-element group $\{\pm 1\}$ acts on each sphere S^n, and the circle $T = \{e^{i\theta}\}$ acts on all odd dimensional spheres $S^{2n-1} \subset \mathbb{C}^n$ via complex multiplication.

1.5. Definition Let X be a right (left) G-space. A point $x \in X$ is called a fixed point if $xs = x$ (or $sx = x$ resp.) for all $s \in G$. We denote the subspace of fixed points by either X^G or $\text{Fix}(X)$

For example, $\text{Fix}(F^n) = \{0\}$ is an example for the matrix groups $GL(n, F)$ action on F^n.

1.6. Definition For a G-space, the orbit of $x \in X$, denoted by xG, is equal to the set of all xs for a right action and Gx equal the set of all sx for a left action. Let X/G denote the set of orbits for a right action with projection $q : X \to X/G$ given by $q(x) = xG$, and let $G\backslash X$ denote the set of orbits for a left action with projection $q : X \to G\backslash X$. The orbit space has the quotient topology under the projections.

1.7. Remark For a G-space X, the map $x \mapsto xs$ is an isomorphism, with inverse $x \mapsto x(s^{-1})$. Also the projection $\pi : X \to X/G$ is an open map since $\pi(W)$ is open in X/G for each open set W of X from the formula $\pi^{-1}\pi(W) = \bigcup_{s \in G} Ws$.

1.8. Definition A G-bundle $p : E \to B$ is a bundle with a right G-space structure $E \times G \to E$ on E such that $p(xs) = p(x)$ for all $x \in E$ and $s \in G$. A morphism $(u, f) : E' \to E''$ of G-bundles $p' : E' \to B'$ and $p'' : E'' \to B''$ is a morphism of bundles, so $p''u = fp'$, and u is a G-map, so $u(xs) = u(x)s$ for all $s \in G$ and $x \in E'$. The category of G-bundles is denoted by $(G-\text{bun})$. Note that in later chapters we consider more general bundles and call them still G-bundles. These are bundles where G acts on the base B as well and p is equivariant, see 13(2.1).

1.9. Remark There is the functor $(G\backslash\text{top}) \to (G-\text{bun})$ which assigns to a right G-space X the G-bundle $q : X \to X/G = B$. Most G-bundles that we will consider come by this functor from G-spaces.

2 Definition and Examples of Principal Bundles

2.1. Definition A free G-space X is a right G-space X such that the subspace X^* of $X \times X$ consisting of all $(x, xs) \in X \times X$ has a unique continuous function $\tau : X^* \to G$ with the property that $x'\tau(x', x'') = x''$.

If $x's = x''$, then $\tau(x', x'') = s$, and in particular $xs' = xs''$ implies that $s' = s''$ in G.

2.2. Definition A principal bundle is a G-bundle $p : P \to B$ in the sense of (1.8) such that P is a free G-space and the natural morphism $P/G \to B$ is an isomorphism. A principal bundle morphism $(u, f) : P' \to P''$ is a morphism of G-bundles which are principal bundles $p' : P' \to B'$ and $p'' : P'' \to B''$.

2.3. Example A basic example comes from a closed subgroup $G \subset H$ and $p : H \to H/G$ the quotient map is an example of a principal G-bundle. If $p' : P' \to B'$ is a principal G'-bundle and if $p'' : P'' \to B''$ is a principal G''-bundle, then the product bundle $p' \times p'' : P' \times P'' \to B' \times B''$ is a principal $G' \times G''$-bundle. In the case where $B = B' = B''$, the fibre product $q : P' \times_B P'' \to B$ is a principal $G' \times G''$-bundle.

2.4. Theorem *Every morphism of principal bundles over B is an isomorphism.*

2.5. Remark It is easy to see that a morphism $u : P' \to P''$ is a bijection on each fibre, and hence, it has an inverse function. In *Fibre Bundles* (Husemöller 1994), 4(3.2) on p. 43, it is proved to be continuous using the translation maps τ of P' and P'' (Husemöller 1994, p. 43).

2.6. Example Let $p : P \to B$ be a principal G-bundle, and let $f : B' \to B$ be a continuous map. The induced bundle $f^{-1}P \to B'$ has a G-bundle structure with $(b', x)s = (b', xs)$, and it is a principal bundle with $\tau((b', x_1), (b', x_2)) = \tau(x_1, x_2)$ for P. The morphism $(w, f) : f^{-1}P \to P$ is a morphism of principal bundles.

2.7. Proposition *If $p : P \to B$ is a principal G-bundle and if $f : B' \to B$ is a continuous map, then $f^{-1}P \to B'$ is a principal bundle. If $(u, f) : P' \to P$ is a morphism of principal G-bundles, then u factors as a principal G-bundle isomorphism $v : P' \to f^{-1}P$ and the canonical f-morphism $w : f^{-1}P \to P$.*

Again, observe that $v(x') = (p'(x'), u(x'))$ is a formula for the desired factorization as in the case of vector bundles.

3 Fibre Bundles

3.1. Definition Let P be a principal G-bundle P over B and let Y be a left G-space Y. Form the quotient $P \times Y \to P[Y] = P \times^G Y$, where the right action of G on $P \times Y$ is given by $(x,y)s = (xs, s^{-1}y)$. The associated fibre bundle to P with fibre Y is the space $P[Y]$ together with the projection $p_Y : P[Y] \to B$ given by $p_Y((x,y)G) = p(x)$.

In the literature, the notation $P \times_G Y$ is also used to denote $P[Y]$, but this notation conflicts with the usual notation of fibre product, so that we will not use it here.

3.2. Remark There is a map $Y \to p_Y^{-1}(b)$ for each choice of x with $p(x) = b$ given by $y \mapsto (x,y)G$. Since τ is continuous, it follows that this map is an isomorphism.

3.3. Example Each n-dimensional vector bundle E over B is isomorphic to a fibre bundle $V_n(E)[F^n]$ for the group $GL(n,F)$. The principal $GL(n,F)$-bundle $V_n(E)$ is called the frame bundle associated with E, and it is the subbundle of $E \times_B \overset{(n)}{\ldots} \times_B E$ whose fibre over a point b consists of n-tuples of vectors $(v_1, \ldots, v_n)$ in E_b which are linearly independent.

3.4. Example Each n by n matrix algebra bundle A over B is isomorphic to a fibre bundle $W_n(A)[M_n(F)]$ for the group $PGL(n,F) = GL(n,F)/GL(1,F)$. The principal $PGL(n,F)$-bundle $W_n(A)[M_n(F)]$ is called the matrix frame bundle associated with A, and it is the subbundle of $\text{Hom}_{\text{alg}}(M_n(F),A)$ of algebra morphisms $M_n(K)$ into the fibres of A and is a subspace of n^2 fibre product of A with itself over B.

The description of cross sections of a fibre bundle is very basic because the result is in terms of equivariant maps from the principal bundle space to the fibre of the fibre bundle.

3.5. Theorem *Let $p : P \to B$ be a principal G-bundle, and let Y be a left G-space. The set of cross sections $\Gamma(B, P[Y])$ of the fibre bundle $P[Y] \to B$ are in bijective correspondence with maps $\phi : P \to Y$ satisfying $\phi(xs) = s^{-1}\phi(x)$.*

3.6. Remark Let $\phi : P \to Y$ be a map satisfying $\phi(xs) = s^{-1}\phi(x)$, and form the map $\sigma_\phi(xG) = (x, \phi(x))G$ which is a well-defined continuous section.

If σ is a section of $P[Y]$, then it has the form $\sigma(xG) = (x, \phi_\sigma(x))G = (xs, s^{-1}\phi_\sigma(x))G$ which implies that ϕ_σ satisfies the relation $\phi_\sigma(xs) = s^{-1}\phi_\sigma(x)$. To deduce the continuity of ϕ_σ from the continuity of σ, we use the continuity of the translation function, see *Fibre Bundles* (Husemöller 1994), 4(8.1) on page 48, or leave it as an exercise for the reader.

4 Local Coordinates for Fibre Bundles

4.1. Trivial Bundles For the product G-principal bundle $B \times G$ and for any product fibre bundle $B \times Y$ over B, the automorphisms are each given by maps $g : B \to G$

and have the form $\alpha_g(b,s) = (b,g(b)s)$ or $\alpha_g(b,y) = (b,g(b)y)$, where Y is a left G-space.

4.2. Remark We consider open coverings U_i of a space B indexed by $i \in I$, and we form the coproduct or disjoint union space $U = \sqcup_{i \in I} U_i$. The open cover arises from a fibre bundle $p : E \to B$ such that $E|U_i$ is trivial for each $i \in I$. For the natural projection $q : U \to B$, we see that each $E|U_i$ is trivial if and only if $q^*(E)$ is trivial on U. The projection $q : U \to B$ is an example of an étale map, and we can make the following definition independent of indexed coverings.

4.3. Definition A bundle $p : E \to B$ whether a vector bundle, a principal G-bundle, or a fibre bundle $E = P[Y]$, where P is a principal G-bundle, is locally trivial provided there exists an étale map $q : U \to B$ with $q^*(E)$ trivial, that is, $q^*(E)$ is isomorphic to $pr : U \times Y \to U$.

When we start with a trivial bundle over U, we can ask what extra data do we need to recover the bundle E on B. In the case of $B = \bigcup_{i \in I} U_i$, we consider the two trivializations $E|U_i \to U_i \times F$ and $E|U_j \to U_j \times F$ and compare them on the intersection $U_i \cap U_j$ with a map $g_{i,j} : U_i \cap U_j \to G$ as in (4.1). There is a compatibility relation on the triple intersection $U_i \cap U_j \cap U_k$ called a cocycle condition or descent condition of the form

$$g_{i,k}(b)g_{k,j}(b) = g_{i,j}(b) \quad \text{for} \quad b \in U_i \cap U_j \cap U_k. \tag{4.1}$$

4.4. Remark For a double indexed family $g_{i,j} : U_i \cap U_j \to G$ of maps satisfying the cocycle condition $(*)$, there exists a principal bundle over $B = \bigcup_{i \in I} U_i$, denoted by $p : P \to B$, with $P|U_i$ isomorphic to the product bundle $U_i \times G \to U_i$ such that the change of coordinates are given by the maps $g_{i,j}$.

We can give a version of this construction without reference to the indexing of the covering, only starting with the map $q : U = \sqcup_{i \in I} U_i \to B$.

4.5. Definition Let $q : U \to B$ be an étale map. Let $Z_0 = U$, let $Z_1 = U \times_B U$, and let $Z_2 = U \times_B U \times_B U$, and this defines our related two-stage pseudosimplicial space Z with d_i, the projection deleting the ith coordinate in the fibre product

$$Z_2 \xrightarrow{\ d_0, d_1, d_2\ } Z_1 \xrightarrow{\ d_0, d_1\ } Z_0 \quad \text{and with} \quad d_i d_j = d_{j-1} d_i \quad \text{for} \quad i < j.$$

Descent data for principal G-bundles of Z is the following:

(a) a principal G-bundle Q on Z_0,
(b) an isomorphism $\alpha : d_1^*(Q) \to d_0^*(Q)$ of principal G-bundles over Z_1, and
(c) a compatibility of three induced versions of α on Z_2 which is just the commutativity of the following diagram,

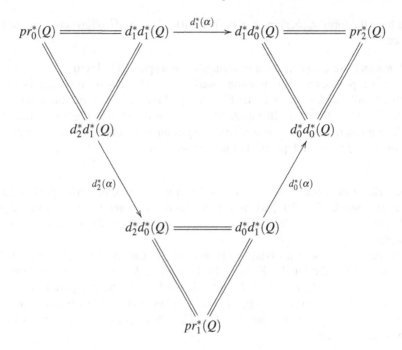

that is, we have the following formula

$$d_0^*(\alpha)d_2^*(\alpha) = d_1^*(\alpha).$$

4.6. Theorem *For an étale map* $q : U \to B$, *principal G-bundles P on B up to iso-morphism are in bijective correspondence with principal G-bundles Q together with descent data on Z under the map which assigns to P on B the bundle* $Q = q^*(P)$ *on* $Z_0 = U$.

4.7. Remark If $U = \sqcup_{i \in I} U_i = Z_0$, then we have

$$Z_1 = \sqcup_{(i,j) \in I^2} (U_i \cap U_j) \quad \text{and} \quad Z_2 = \sqcup_{(i,j,k) \in I^3} (U_i \cap U_j \cap U_k)$$

with d_j, the inclusion of an intersection into the intersection with the index j deleted. The isomorphism α is the collection of isomorphisms given by $g_{i,j} : U_i \cap U_j \to G$ and the descent commutative diagram is the cocycle condition (1) in (4.3) when $Q = U \times G$ is the product bundle. For general $Q \to U$, this descent formulation is a gluing statement.

5 Extension and Restriction of Structure Group

5.1. Definition Let $G' \subset G''$ be a closed subgroup G' in a topological G''. Let P' be a G'-principal bundle over B, and P'' be a G''-principal bundle with an inclusion $j : P' \to P''$ map. Then P'' is an extension of P', and P' is a restriction of P'' provided $j(xs) = j(x)s$ for all $x \in P', s \in G'$.

It is always possible to extend the structure group.

5.2. Proposition *Let P' be a G'-principal bundle, where G' is a closed subgroup of G''. Then, the fibre bundle $P'[G'']$ has a principal G''-bundle structure, and the natural morphism given by $P' = P'[G'] \to P'[G''] = P''$ is an extension of P'.*

The restriction of the structure group is not always possible, and it depends on the existence of a cross section for a related fibre bundle.

5.3. Proposition *Let P'' be a G''-principal bundle, where G' is a closed subgroup of G''. Then, there exists a restriction of P'' to a G'-principal bundle $P' \subset P''$ if and only if $P''[G''/G']$ has a cross section.*

Proof. If $P' \to P''$ exists, then $P'[G'/G'] = B \to P''[G''/G']$ is the desired section, and if $\sigma : B \to P''[G''/G']$ is a cross section, then it corresponds to a map $\phi : P'' \to G''/G'$ with $\phi(xs) = s^{-1}\phi(x)$. The subbundle P' has total space $\phi^{-1}(eG') \subset P''$.

5.4. Remark Using the notation of (5.1), we consider local coordinates $g_{i,j} : U_i \cap U_j \to G'$ of P', and these can be taken to be the local coordinates with values in $G'' \supset G'$. Starting with local coordinates $g_{i,j} : U_i \cap U_j \to G''$ of P'', we can restrict the bundle to G' if and only if there exists $h_i : U_i \to G''$ with $h_i g_{i,j} h_j^{-1} : U_i \cap U_j \to G' \subset G''$.

These concepts extend from a general inclusion $G' \subset G''$ to the case of a general continuous homomorphism $w : G' \to G''$.

5.5. Definition Let $w : G' \to G''$ be a morphism of topological groups, and let B be a base space.

(a) The extension of a principal G'-bundle P' over B is $P'' = P'[G''] = P' \times^{G'} G''$ with the G''-principal bundle structure given by

$$[x', s'']t'' = [x', s''t''],$$

where $[x't', s''] = [x', w(t')s'']$ for $x' \in P'$, $t' \in G'$, and $s'', t'' \in G''$.
(b) A restriction of a principal G''-bundle P'' over B is a principal G'-bundle P' over B such that the extension of P' to a principal G''-bundle $P'[G'']$ is isomorphic to P'' as a principal G''-bundle.
(c) A G'-structure on a principal G''-bundle P'' over B is an equivalence class of pairs (P', v), where P' is a G'-principal bundle restriction of P'', and $v : P'[G''] \to P''$ is an isomorphism of principal G''-bundles. Two pairs (P'_1, v_1) and (P'_2, v_2) defining a G'-structure on P'' are equivalent provided there exists an isomorphism $u : P'_1 \to P'_2$ of principal G'-bundles with $v_2 \circ u[G''] = v_1$.

5.6. Terminology When $G' = SO(n)$ in (5.5), a G'-structure is called an orientation, and when $G' = Spin(n)$ (cf. 12 (1.5) and 12 (3)), it is called a spin structure.

6 Automorphisms of Principal Bundles and Gauge Groups

6.1. Definition Let $p : P \to B$ be a principal G-bundle. The gauge group $\mathrm{Aut}_B(P)$ of P is the space of all gauge transformations of P, that is, all maps $u \in \mathrm{Map}(P, P)$ with

$$pu = p \text{ and } u(xs) = u(x)s \text{ for all } s \in G, x \in P.$$

In particular, we see that a gauge transformation is just an automorphism of the principal bundle since $u^{-1} : P \to P$ is defined and continuous. There are two other ways of looking at gauge transformations.

6.2. Proposition *For a principal bundle P, there is a bijective correspondence between the following three sets:*

(1) gauge transformations $u : P \to P$,
(2) continuous maps $\phi : P \to G$ with $\phi(xs) = s^{-1}\phi(x)s$ for $x \in P, s \in G$, and
(3) cross sections of the fibre bundle $P[\mathrm{Ad}(G)]$.

Proof. The bijection between the sets defined by the condition (2) and condition (3) is just the description of the set of cross sections of a fibre bundle (5.1). If ϕ is given by (2), then we define $u_\phi(x) = x\phi(x)$, and we calculate $u_\phi(xs) = xss^{-1}\phi(x)s = u_\phi(x)s$. Conversely, if $u : P \to P$ is a gauge transformation, then $u(x) = x\phi_u(x)$, and the automorphism property $u(xs) = u(x)s$ implies that $u(xs) = xs\phi_u(xs) = x\phi_u(x)s$, and hence it follows that ϕ_u satisfies $\phi_u(xs) = s^{-1}\phi_u(x)s$. This sets up the correspondence and proves the proposition.

6.3. Summary Relations In (a) we summarize the bijection between (1) and (2), and in (b) we summarize the bijection between (2) and (3):

(a) From $\phi : P \to G$ we form $u_\phi(x) = x\phi(x)$, and from $u : P \to P$ we form $\phi_u : P \to G$ by $u(x) = x\phi_u(x)$.
(b) From $\phi : P \to G$ we form $\sigma_\phi(xG) = (x, \phi(x)) \mod G$, and from $\sigma : B \to P[\mathrm{Ad}(G)]$ we have $\sigma(xG) = (x, \phi_\sigma(x)) \mod G$.

Reference

Husemöller, D.: *Fibre Bundles*, 3rd ed. Springer-Verlag, New York (1994)

Part II
Homotopy Classification of Bundles and Cohomology: Classifying Spaces

In the first part, Chaps. 1–5, principal bundles, fibre bundles, and vector bundles over a space were introduced and studied. Of special importance was the fact that a vector bundle is a particular example of a fibre bundle which comes with a principal bundle in the background. In this part, Chaps. 6–12, the homotopy theory and classification of bundles is considered. For this, we begin with the basic definition of homotopy between two maps $f', f'' : X \to Y$ in terms of a map $h : X \times [0, 1] \to Y$.

If $p : E \to B$ is a bundle over B, then the product with the unit interval $p \times [0, 1] :$ $E \times [0, 1] \to B \times [0, 1]$ is a bundle over $B \times [0, 1]$ with the same fibres in the sense that $(E \times [0, 1])_{(b,t)} = E_b$. The vector bundles, principal bundles, and fibre bundles have a converse property, that is, under suitable local triviality in these three cases a bundle $q : E' \to B \times [0, 1]$ is isomorphic to

$$p \times [0, 1] : E \times [0, 1] \longrightarrow B \times [0, 1],$$

where $E = E'|(B \times \{0\})$. This is the homotopy invariance of bundles.

Two maps $f', f'' : X \to Y$ are homotopic provided there is a map $h : X \times [0, 1] \to Y$, called a homotopy from f' to f'', and if $q : E \to Y$ is a type of bundle with homotopy invariance property, then $f'^{-1}(E)$ and $f''^{-1}(E)$ are isomorphic over X

There is a converse for this theorem for the Milnor construction contained in Chap. 7, that is, if two bundles are isomorphic, then the maps are homotopic. These homotopy questions lead to fibre space questions. In Chaps. 9–12, the cohomology properties of bundles are considered, especially the theory of characteristic classes.

Chapter 6
Homotopy Classes of Maps
and the Homotopy Groups

In this chapter, we prepare the basic definitions on the homotopy relation between maps. These ideas apply everywhere in geometry, and it is usually the case that invariants of maps which are interesting are those which are the same for two homotopic maps.

A homotopy from X to Y is a map $h : X \times [0,1] \to Y$, and in terms of the variables $x \in X$ and $t \in [0,1]$, it can be written either as $h(x,t)$, as $h(x)(t)$, or as $h_t(x)$. This gives three interpretations of homotopy which are discussed in the first sections. For example, $h_t(x)$ suggests a deformation from $h_0 : X \to Y$ to $h_1 : X \to Y$ which is another name for a homotopy from h_0 to h_1. The form $h(x)(t)$ suggests that for each $x \in X$ the function $t \mapsto h(x)(t)$ is a path in the space Y and a homotopy from this perspective is just a map from X to the space of paths on Y. The homotopy groups $\pi_n(X, x_0)$ of a space X with base point x_0 are defined either as certain (equivalence classes of) maps of the n-sphere S^n or of the n-cube I^n into the space X. We show that they are in general commutative groups except for the fundamental group $\pi_1(X)$.

The cylinder space $X \times [0,1]$ over X plays clearly a basic role in the study of homotopy properties. For bundles, we will consider those classes of vector bundles, principal bundles, and fibre bundles which have the property that a bundle $p : E \to B \times [0,1]$ is isomorphic to $(E|(B \times \{0\})) \times [0,1]$. This is the homotopy invariance property of bundles. This implies that two homotopic maps induce isomorphic bundles.

1 The Space Map(X,Y)

As a set, Map(X,Y) is the set of all continuous functions $f : X \to Y$. For $L \subset X$ and $M \subset Y$, we use the notation $\langle L, M \rangle$ for the subset of all $f \in \text{Map}(X,Y)$ such that $f(L) \subset M$. A space or set is compact provided it satisfies the Heine–Borel covering property and the Hausdorff separation condition.

D. Husemöller et al.: *Homotopy Classes of Maps and the Homotopy Groups*, Lect. Notes Phys. **726**, 65–74 (2008)
DOI 10.1007/978-3-540-74956-1_7 © Springer-Verlag Berlin Heidelberg 2008

1.1. Definition The compact open topology on $\mathrm{Map}(X,Y)$ is the topology generated by all sets $\langle K,V \rangle$, where K is a compact subset of X and V is an open subset of Y.

Recall this means that $T \subset \mathrm{Map}(X,Y)$ is open in the compact open topology if and only if for all $f \in T$ there exists compact subsets $K_i \subset X$ and open subsets $V_i \subset Y$ for $i = 1,\ldots,m$ such that $f \in \langle K_1,V_1 \rangle \cap \ldots \cap \langle K_m,V_m \rangle \subset T$. The compact open topology works best for separated (Hausdorff) spaces, and we will assume that our spaces are separated. It is needed in the following remark.

1.2. Remark If Φ is a family of open sets generating the topology of Y, then the set of all $\langle K,W \rangle$, where K is compact in X and $W \in \Phi$ generates the topology of $\mathrm{Map}(X,Y)$.

1.3. Functoriality Let $u : X' \to X$ and $v : Y \to Y'$ be two continuous maps. Then, $u^* : \mathrm{Map}(X,Y) \to \mathrm{Map}(X',Y)$ defined by $u^*(f) = fu$ and $v_* : \mathrm{Map}(X,Y) \to \mathrm{Map}(X,Y')$ defined by $v_*(f) = vf$ are continuous. In addition, let $u' : X'' \to X'$ and $v' : Y' \to Y''$ be two continuous maps. Then, we have the following relations for $f \in \mathrm{Map}(X,Y)$: $u^*(v_*(f)) = vfu = v_*(u^*(f)), (uu')^* = (u')^*u^*$, and $(v'v)_* = v'_*v_*$.

2 Continuity of Substitution and $\mathrm{Map}(X \times T, Y)$

2.1. Definition For three spaces X, T, and Y we define a map

$$c : \mathrm{Map}(X \times T, Y) \longrightarrow \mathrm{Map}(X, \mathrm{Map}(T,Y))$$

by the requirement that $c(f)(x)(t) = f(x,t)$ for $f \in \mathrm{Map}(X \times T, Y)$. This is a well-defined continuous map by (1.2), and it is also clearly injective. Moreover, it restricts to a topological isomorphism $c : \mathrm{Map}(X \times T, Y) \to \mathrm{im}(c) \subset \mathrm{Map}(X, \mathrm{Map}(T,Y))$, where the image of c, denoted by $\mathrm{im}(c)$, has the subspace topology.

2.2. Definition For two spaces T and Y, the evaluation function

$$e : \mathrm{Map}(T,Y) \times T \to Y$$

is given by the formula $e(f,t) = f(t)$.

2.3. Proposition *Let T be an arbitrary space. The evaluation function $e : \mathrm{Map}(T,Y)$ $\times T \to Y$ is continuous for every Y if and only if the map $c : \mathrm{Map}(X \times T, Y) \to \mathrm{Map}(X, \mathrm{Map}(T,Y))$ is surjective for all spaces X.*

Proof. If e for T is continuous, then for $g \in \mathrm{Map}(X, \mathrm{Map}(T,Y))$ we have $g = c(f)$, where $f(x,t) = e(g(x),t)$. Conversely if c is surjective, then $e = c^{-1}(\mathrm{id}_{\mathrm{Map}(T,Y)})$ is continuous for the case $X = \mathrm{Map}(T,Y)$. This proves the proposition.

2.4. Example If T is a locally compact space, then the evaluation function $e : \mathrm{Map}(T,Y) \times T \to Y$ is continuous for all spaces Y, and hence by (2.3), the map $c : \mathrm{Map}(X \times T, Y) \to \mathrm{Map}(X, \mathrm{Map}(T,Y))$ is an isomorphism of spaces (also called a homeomorphism).

3 Free and Based Homotopy Classes of Maps

For many considerations, especially those related to algebraic structures, it is important to consider spaces X together with a given point in X.

3.1. Definition A pointed space X is a pair consisting of a space X together with a point $x_0 \in X$, called the base point. Let X and Y be two pointed spaces. A pointed map $f : X \to Y$ is a map f of the underlying spaces such that $f(x_0) = y_0$.

Another name for a pointed map is a base-point-preserving map. In many cases we use the symbol $*$ for the base point, and then base-point-preserving is the relation $f(*) = *$. Clearly, the composition of two pointed maps is again a pointed map.

3.2. Definition As before, (top) denotes the category of spaces and continuous functions. The category of pointed spaces and pointed maps is denoted by $(\text{top})_*$. The natural stripping (or forgetful) functor $(\text{top})_* \to (\text{top})$ assigns to a pointed space its underlying space and to a pointed map the same map.

There are two notions of homotopy corresponding to the two categories: (top), where (free) homotopies are defined, and $(\text{top})_*$, where pointed homotopies or base-point-preserving homotopies are defined.

3.3. Definition Two maps $f', f'' : X \to Y$ in (top) are homotopic provided there exists a map $h : X \times [0,1] \to Y$ with $h(x,0) = f'(x)$ and $h(x,1) = f''(x)$. Two maps $f', f'' : X \to Y$ in $(\text{top})_*$ are homotopic provided there exists a map $h : X \times [0,1] \to Y$ with $h(*,t) = *$, $h(x,0) = f'(x)$ and $h(x,1) = f''(x)$.

3.4. Example If $f', f'' : X \to \mathbb{R}^m$ are two maps, then

$$h_t(x) = (1-t)f'(x) + tf''(x)$$

is a homotopy between f' and f''. If f' and f'' preserve a base point, then $h_t(*) = *$.

If in the previous definition we denote $h(x,t)$ by $f_t(x)$, then $f_t(x)$ is either a one-parameter family of maps $f_t : X \to Y$ which preserve base point in the case $(\text{top})_*$ or it is map

$$X \longrightarrow \text{Map}([0,1], Y)$$

equal to f' at $t = 0$ and f'' at $t = 1$. In other words, $f_0 = f'$ and $f_1 = f''$.

3.5. Remark The homotopy relation is an equivalence relation, that is,

(1) it is reflexive meaning that $f : X \to Y$ is homotopic to itself by the homotopy $h(x,t) = f(x)$ for all $x \in X, t \in [0,1]$,
(2) it is symmetric meaning that if f is homotopic to f' say with a homotopy $h(x,t)$, then f' is homotopic to f with, for example, $h'(x,t) = h(x, 1-t)$, and
(3) it is transitive meaning that if f is homotopic to f' and f' is homotopic to f'', then f is homotopic to f''. If $h(x,t)$ is a homotopy from f to f' and if $h'(x,t)$ is a homotopy from f' to f'', then we define a homotopy $h''(x,t)$ by the relation

$$h''(x,t) = \begin{cases} h(x,2t) & \text{for} \quad t \in [1,\frac{1}{2}] \\ h'(x,2t-1) & \text{for} \quad t \in [\frac{1}{2},1] \end{cases}$$

The above formula for h'' in terms of h and h' comes up often in many constructions, but there is a small point to check, namely continuity of h''. This follows from the following general continuity result which is also used in bundle theory for many purposes.

3.6. Proposition *Let X be a space with a covering Φ of subsets, and let $f : X \to Y$ be a function with values in the space Y. Assume that all the restrictions $f|M : M \to Y$ for $M \in \Phi$ are continuous.*

(1) If all $M \in \Phi$ are open sets, then f is continuous or
(2) If all $M \in \Phi$ are closed sets and Φ is a finite set, then f is continuous.

Proof. If W is a subset of Y, then $f^{-1}(W) = \bigcup_{M \in \Phi}(f|M)^{-1}(W)$, and for (1) take W, any open set, then $f^{-1}(W)$ is an open set and for (2) take W, any closed set, then $f^{-1}(W)$ is a closed set. Thus, f is continuous in both cases.

4 Homotopy Categories

Category theory is useful as an organization of some of the basic composition properties of homotopic maps. For this, we use the following proposition.

4.1. Proposition *Let $f',f'' : X \to Y$ and $g',g'' : Y \to Z$ be two pairs of maps (resp. pointed maps) which are homotopic (resp. pointed homotopic). Then, $g'f',g''f'' : X \to Z$ are homotopic (resp. pointed homotopic).*

Proof. Let $h : X \times [0,1] \to Y$ be a homotopy (resp. pointed homotopy) from f' to f'', and let $k : Y \times [0,1] \to Z$ be a homotopy (resp. pointed homotopy) from g' to g''. Then, the function $q : X \times [0,1] \to Z$ given by the formula $q(x,t) = k(h(x,t),t)$ is continuous, and q is a homotopy (resp. pointed homotopy) from $g'f'$ to $g''f''$. This proves the proposition.

This proposition means that we can form homotopy quotient categories of the category (top) of spaces and (top)$_*$ of pointed spaces.

4.2. Definition The quotient categories and related quotient functors, (top) $\to$ (htp) and (top)$_* \to$ (htp)$_*$, are defined such that the quotient functors are the identity on the objects, and a map (resp. pointed map) f is carried to its homotopy class (resp. pointed homotopy class) $[f]$. The set of $[f]$ where $f : X \to Y$ is denoted by $[X,Y]$.

The standard category language has a special terminology when dealing with homotopy categories.

4.3. Definition A map or pointed map $f : X \to Y$ is a homotopy equivalence or pointed homotopy equivalence provided $[f]$ is an isomorphism in (htp) or (htp)$_*$, respectively. Two spaces or pointed spaces are homotopically equivalent provided they are isomorphic in (htp) or (htp)$_*$, respectively.

In terms of the quotient homotopy categories, we can speak of homotopy invariant functors. This is a property of a functor which is common in geometry.

4.4. Definition A functor $F : (\text{top}) \to \mathscr{C}$ or $F : (\text{top})_* \to \mathscr{C}$ is homotopy invariant provided $F(f') = F(f'')$ for any pair of homotopic morphisms f' and f''. This is equivalent to F factoring by the quotient functor $(\text{top}) \to (\text{htp})$ as a functor $F : (\text{htp}) \to \mathscr{C}$ or in the base-point-preserving case by the quotient functor $(\text{top})_* \to (\text{htp})_*$ as a functor $F : (\text{htp})_* \to \mathscr{C}$.

Now, we return briefly to Chap. 1, Sect. 4, and general category theory.

4.5. Definition The opposite category $\mathscr{C}^{\text{op}}$ of a category $\mathscr{C}$ has the same objects as $\mathscr{C}$, and the morphisms $X \to Y$ in $\mathscr{C}^{\text{op}}$ are the morphisms $f \in \text{Hom}_{\mathscr{C}}(Y,X)$ which are denoted by $f^{\text{op}} : X \to Y$ in $\mathscr{C}^{\text{op}}$. With this, notation composition of $f^{\text{op}} : X \to Y$ and $g^{\text{op}} : Y \to Z$ is given by the contravariant law $g^{\text{op}} f^{\text{op}} = (fg)^{\text{op}}$ for $fg : Z \to X$ in $\mathscr{C}$.

4.6. Examples of Functors For each category $\mathscr{C}$ and object T, we have two functors $\text{Hom}(T,) : \mathscr{C} \to (\text{set})$ and $\text{Hom}(,T) : \mathscr{C}^{\text{op}} \to (\text{set})$. For $v : Y' \to Y''$, the set $\text{Hom}(T,)(Y') = \text{Hom}(T,Y')$, the set of morphisms in $\mathscr{C}$, and $\text{Hom}(T,v) : \text{Hom}(T,Y') \to \text{Hom}(T,Y'')$ are left composition by v. For $u : X'' \to X'$ the set $\text{Hom}(,T)(X') = \text{Hom}(X',T)$ and $\text{Hom}(u,T) : \text{Hom}(X',T) \to \text{Hom}(X'',T)$ are right composition by u. Viewing the object T in $\mathscr{C}^{\text{op}}$ just changes the roles of these two functors.

4.7. Remark The functor $[,T]$ arises in bundle theory and in cohomology, where T is a classifying space. It is a functor $(\text{top})^{\text{op}} \to (\text{set})$, or since it is homotopy invariant by definition $(\text{htp})^{\text{op}} \to (\text{set})$, it is an example of (4.6). For $n > 0$, the homotopy group functor $[S^n,] : (\text{htp})_* \to (\text{gr})$ starts as $[S^n,] : (\text{top})_* \to (\text{set})$, where it is homotopy invariant and has a natural group structure from the geometry of S^n. This is considered in the next section.

5 Homotopy Groups of a Pointed Space

For $n > 0$, the homotopy groups $\pi_n(X,x_0)$ of a pointed space $X = (X,x_0)$ have a very direction definition as the group of homotopy classes of base-point-preserving maps from S^n to X. For the multiplication or group structure, we use a little geometry, that is, the sphere S^n, which is the space of points $t = (t_0,\dots,t_n) \in \mathbb{R}^{n+1}$ with $||t|| = 1$, is mapped onto the join of two spheres $S' \vee S''$.

5.1. One-Point Join of Spheres Let S' and S'' denote the following two spheres in $\mathbb{R}^{n+1}$.

(1) S' has only points $t_n \geq 0$ with equation

$$||t - (0,\dots,0, 1/2)|| = 1/2,$$

and

(2) S'' has only points $t_n \leq 0$ with equation

$$||t - (0,\ldots,0,-1/2)|| = 1/2.$$

Observe that the two spheres S' and S'' have just the origin $0 = (0,\ldots,0) \in \mathbb{R}^{n+1}$ in common, and the union of the two spheres, denoted by $S' \vee S''$, is called the one point union or one point join of the two spheres. Of course, all three spheres S^n, S', and S'' are isomorphic.

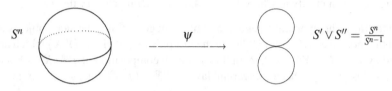

$S^n \qquad\qquad\qquad\qquad\qquad \psi \qquad\qquad\qquad\qquad\qquad S' \vee S'' = \frac{S^n}{S^{n-1}}$

5.2. Remark There is a natural map $\psi : S^n \to S' \vee S''$ which preserves the last coordinate t_n for $n > 0$. This means that S^{n-1}, equal to all points on S^n with $t_n = 0$, is mapped to $0 = (0,\ldots,0)$ by ψ, and further ψ has the additional property that the restriction $\psi|(S^n - S^{n-1})$ is a homeomorphism $S^n - S^{n-1} \to S' \vee S'' - \{0\}$.

If X/A denotes the space X with the closed subspace $A \subset X$ pinched or collapsed to a point, then ψ induces an isomorphism $S^n/S^{n-1} \to S' \vee S''$.

5.3. Definition Let $n > 0$ and denote $\pi_n(X) = [S^n, X]$ for pointed X. Let $a, b \in \pi_n(X)$ be represented by $a = [\alpha]$ and $b = [\beta]$, where $\alpha, \beta : S^n \to X$ are base-point-preserving maps. We define the operation $a * b = [\gamma]$, where $\gamma = (\alpha \perp \beta)\psi$, and where $\alpha \perp \beta : S' \vee S'' \to X$ is α on the first sphere S' and β on the bottom sphere S''.

This simple geometric construction has many algebraic properties due to the flexibility resulting from deforming maps $S^n \to X$ by homotopies.

5.4. Assertion The map $* : \pi_n(X) \times \pi_n(X) \to \pi_n(X)$ is a group composition law on the set $\pi_n(X)$ with neutral element $e = [0]$, where $0 : S^n \to X$ is the map carrying the sphere to the base point of X. If $a = [\alpha]$, then the inverse a' of a is given by $a' = [\alpha']$, where $\alpha'(t_0,\ldots,t_n) = \alpha(t_0,\ldots,t_{n-1},-t_n)$. If $f : X \to Y$ is a base-point-preserving map, then $\pi_n(f)([\alpha]) = [f\alpha]$ is a well-defined map $\pi_n(f) : \pi_n(X) \to \pi_n(Y)$ which is a group morphism, and further, if $f', f'' : X \to Y$ are base-point-preserving homotopic maps, then $\pi_n(f') = \pi_n(f'') : \pi_n(X) \to \pi_n(Y)$ is equal as group morphisms.

The associative law is a construction of an interesting homotopy. If we parametrize $S^n = I^n/\partial I^n$ as the unit cube I^n of $(s_1,\ldots,s_n)$ with $0 \leq s_i \leq 1$ with the boundary points ∂I^n collapsed to a point, then the group law can be described by the following formulas. We have $[u] * [v] = [w]$, where $w : I^n \to X$ is defined by

$$w(s_1,\ldots,s_n) = \begin{cases} u(s_1,\ldots,s_{n-1},2s_n) & \text{for} \quad s_n \leq 1/2 \\ v(s_1,\ldots,s_{n-1},2s_n - 1) & \text{for} \quad s_n \geq 1/2 \end{cases}$$

Recall that ∂I^n consists of all (s_i) with some s_i equal to either 0 or 1. Now, the group axioms are easily checked.

5.5. *Remark* We can summarize the properties of π_n for $n > 0$ as follows, namely that $\pi_n : (\text{htp})_* \to (\text{gr})$ is a functor with values in the category of groups (gr). In the special case $n = 0$, we have $\pi_0 : (\text{htp})_* \to (\text{set})_*$ as just a functor with values in the category of pointed sets. The group $\pi_1(X)$ is also called the fundamental group or Poincaré group while in general $\pi_n(X)$ is called the nth homotopy group of the pointed space X.

5.6. *Example* The group $\pi_n(S^n) = \mathbb{Z}$, where the isomorphism is given by the degree of the map $\deg : \pi_n(S^n) \to \mathbb{Z}$. For $n = 1$ and $S^1 \subset \mathbb{C}$, the unit circle in the complex plane

$$\deg[\alpha] = \frac{1}{2\pi i} \int_\alpha \frac{dz}{z}$$

is the winding number. For the wedge of two circles, the group $\pi_1(S^1 \vee S^1) = \mathbb{Z} * \mathbb{Z}$, the free group on two generators.

This brings up the question of commutativity of the homotopy groups, and for this we go back to the second definition of the homotopy groups introduced in the context of the associative law.

5.7. Second definition of $\pi_n(X)$ In (5.4), we parametrized the n-sphere S^n as $I^n/\partial I^n$ as the unit cube I^n of $(s_1, \ldots, s_n)$ with $0 \leq s_i \leq 1$ with the boundary points ∂I^n collapsed to a point. For the ith variable, we have a group law $*_{(i)}$ described by the following formulas. We have $[u] *_{(i)} [v] = [w_i]$, where $w_i : I^n \to X$ is given by the formulas

$$u *_{(i)} v(s_1, \ldots, s_n) = w_i(s_1, \ldots, s_n) = \begin{cases} u(s_1, \ldots, 2s_i, \ldots, s_n) & \text{if} \quad s_n \leq 1/2 \\ v(s_1, \ldots, 2s_i - 1, \ldots, s_n) & \text{if } s_n \geq 1/2 \end{cases}$$

Recall that ∂I^n consists of all (s_i) with some s_j equal to either 0 or 1. Observe that for $i < j$, we have the following distributive formula $(u *_{(i)} v) *_{(j)} (u' *_{(i)} v') = (u *_{(j)} u') *_{(i)} (v *_{(j)} v')$ and for the related homotopy classes

$$([u] *_{(i)} [v]) *_{(j)} ([u'] *_{(i)} [v']) = ([u] *_{(j)} [u']) *_{(i)} ([v] *_{(j)} [v']).$$

This leads to the following algebraic lemma which applies to only homotopy classes, and it is not true on the level of functions since there is no unit property.

5.8. Lemma *Let (E, e) be a pointed set with two laws of composition*

$$*' : E \times E \to E \quad \text{and} \quad *'' : E \times E \to E$$

with e as unit and satisfying the following distributive law

$$(a *' b) *'' (a' *' b') = (a *'' a') *' (b *'' b') \quad \text{for all} \quad a, b, a', b' \in E.$$

Then $a *' b = a *'' b$, and the law of composition is commutative, that is, $a * b = b * a$ for all $a, b \in E$.

Proof. If we set $b = e$ and $a' = e$, then we have

$$a *'' b' = (a *' e) *'' (e *' b') = (a *'' e) *' (e *'' b') = a *' b'$$

so that $*'' = *'$. If we set $a = e$ and $b' = e$, then we have

$$b *'' a' = (e *' b) *'' (a' *' e) = (e *'' a') *' (b *'' e) = a' *' b$$

so that $b * a = a * b$ for the law of composition. This proves the lemma.

5.9. Proposition *For $n > 1$, the homotopy groups are abelian and define functors $\pi_n : (\text{htp})_* \to (\text{ab})$ with values in the category of abelian groups (ab). Moreover, the group law on $\pi_n(X)$ can be calculated using any of the n-coordinates of representative maps $u : I^n \to X$ of elements of $\pi_n(X)$.*

5.10. Proposition *The projection maps from a product of pointed spaces induce an isomorphism*

$$\pi_n(X_1 \times \ldots \times X_r) \longrightarrow \pi_n(X_1) \oplus \ldots \oplus \pi_n(X_r).$$

Proof. A function $f : S^n \to X_1 \times \ldots \times X_r$ into a product decomposes as an r-tuple $f = (f_1, \ldots, f_r)$ of functions $f_i : S^n \to X_i$, and this is true of homotopy classes also. This proves the proposition.

5.11. Corollary *For the r-dimensional torus $T^r = S^1 \times \overset{(r)}{\ldots} \times S^1$, we have $\pi_1(T^r) = \mathbb{Z}^r$.*

5.12. Proposition *If (G, e) is a pointed space with a continuous multiplication $a.b$ having e as unit, then $\pi_1(G)$ is commutative, and addition in $\pi_n(G)$ can be calculated as $[w'] + [w''] = [w]$, where $w(t) = w'(t).w''(t)$ for $w', w'' : S^n \to G$.*

Proof. This is another application of the Lemma (5.8). Of course this proposition applies to a topological group. Of special interest are Lie groups, and here, we state the following basic results for a compact Lie group.

5.13. Example Let G be a compact, connected Lie group. If G is simple, then $\pi_1(G)$ is finite, and if G is simply connected, that is, $\pi_1(G) = 0$, then $\pi_2(G) = 0$. We have also the special cases which we return to in Chap. 12

(1) $\pi_1(SO(n)) = \mathbb{Z}/2$ for $n > 1$, but $\pi_1(S^1) = \pi_1(U(1)) = \mathbb{Z}$.
(2) $\pi_1(SU(n)) = \pi_2(SU(n)) = 0$ and $\pi_3(SU(n)) = \mathbb{Z}$ for $n > 1$.

6 Bundles on a Cylinder $B \times [0, 1]$

In the introduction to this chapter, we mentioned that the homotopy properties of bundles start with the assertion that a bundle on $B \times [0, 1]$ is determined by its restriction to $B \times \{0\}$. The homotopy property of principal G-bundles, and hence of fibre bundles which includes vector bundles, is established only for locally trivial bundles where the open sets in question are of the form $V = \eta^{-1}((0, 1])$, where $\eta : B \to [0, 1]$ is a continuous numerical function. The open sets V form a covering of B in the sense of the next definition.

6.1. Definition A family Φ is a locally finite open covering of a space B provided each $V \in \Phi$ is open, $B = \bigcup_{V \in \Phi} V$, and for each $b \in B$, there exists a neighborhood $N(b)$ of b with $N(b) \cap V$ empty except for finitely many $V \in \Phi$.

Now, we bring in the numerical functions.

6.2. Definition A family of continuous functions $\{\eta_i : B \to [0,1]\}_{i \in I}$ is a partition (resp. envelope) of unity provided there is a locally finite open covering Φ of B with the closure of $\eta_i^{-1}((0,1])$ contained in some $V \in \Phi$ for each i and $\sum_{i \in I} \eta_i(b) = 1$ (resp. $\max_{i \in I} \eta_i(b) = 1$).

For establishing the homotopy property of principal G-bundles, we use envelopes of unity, and for the comparison of a bundle with the Milnor construction in the next chapter, we use partitions of unity. A partition of unity defines an envelope of unity and vice versa by just rescaling the functions.

6.3. Definition A principal G-bundle $P \to B$ is a numerable bundle provided it is trivial over the closures of $\eta_i^{-1}((0,1])$ for each $i \in I$ for some family of functions $\{\eta_i\}_{i \in I}$. A corresponding covering Φ then is called a numerable covering of the base space.

Observe that an induced numerable bundle is numerable.

6.4. Theorem *Let $P' \to B \times [0,1]$ be a numerable principal G-bundle, and form the restriction $P = (P'|(B \times \{t\}))$. Then, the extended principal G-bundle $(P'|(B \times \{t\})) \times [0,1] \to B \times [0,1]$ is isomorphic to the given $P' \to B \times [0,1]$ for any $t \in [0,1]$.*

Before we sketch the proof of this theorem, we state the main corollary of this theorem which is the homotopy invariance property for principal G-bundles.

6.5. Corollary *Let $P \to B$ be a numerable principal G-bundle, and let $f', f'' : B' \to B$ be two homotopic maps. Then, the induced G-bundles $(f')^*(P)$ and $(f'')^*(P)$ over B' are isomorphic.*

Sketch of the proof of the theorem The proof is achieved by constructing a morphism over the projection $r : B \times [0,1] \to B \times [0,1]$ given by $r(b,t) = (b,1)$. The proof divides into three steps:

Step 1 This is the special case where we show that a principal G-bundle P over $B \times [a,b]$ is trivial if for $a < c < b$ the two restrictions $P|(B \times [a,c])$ and $P|(B \times [c,b])$ are trivial. For this, we just use the trivializing condition for P of an equivariant map $P \to G$.

Step 2 The bundle P over $B \times [0,1]$ is numerable, and we show that there is a numerable covering Φ of B with $P|(V \times [0,1])$ trivial for $V \in \Phi$. This is done by using step 1 and a finite covering of each $\{b\} \times [0,1]$ and for the envelope of unity $\eta_i(b)$ equal to a maximum of the $\eta_j(b,t)$ for $t \in [0,1]$ for the open sets in the finite covering associated with $\eta_i : B \to [0,1]$.

Step 3 The morphism $(u, r) : P \to P$ over $r : B \times [0, 1] \to B \times [0, 1]$ is a composition of locally defined morphisms $(u_i, r_i) : P \to P$, where

$$r_i(b, t) = (b, \max(\eta_i(b), t)), \ u_i(h_i(b, t, s)) = h_i(b, \max(\eta_i(b), t), s)$$

and $h_i : U_i \times [0, 1] \times G \to P|U_i \times [0, 1]$ is an isomorphism of principal G-bundles related to the local triviality.

For the composition, we consider a well ordering of the indexing set I. In a suitable neighborhood of $b \in B$, there is only a finite number $n(b)$ of indices i with $\eta_i(b) \neq 0$. Let $I(b) = \{i(1) < \ldots < i(n(b))\}$ be the corresponding finite ordered set. On the neighborhood, we form the compositions $r = r_{i(n(b))} \cdots r_{i(1)}$ and $u = u_{i(n(b))} \cdots u_{i(1)}$. The other terms (u_i, r_i) for $i \in I - I(b)$ are all the identity, and this leads to a global definition of the desired morphism (u, r).

For more details, see *Fibre Bundles*, 3(4.1)–3(4.3) and 4(9.4)–4(9.6).

6.6. Remark A convenient class of spaces is the class of paracompact spaces. A Hausdorff space is paracompact if each open covering is numerable. As a consequence, a locally trivial principal G-bundle over a paracompact space is always numerable.

Chapter 7
The Milnor Construction: Homotopy Classification of Principal Bundles

For a given topological group G, we consider all the principal bundles. At first, it looks like finding all principal G-bundles over a space might be a great task, but there is a special construction of a principal G-bundle due to Milnor. It has the property that all other numerable principal G-bundles over all possible spaces are induced from this particular bundle. Thus, it is called the universal principal G-bundle, and its base space is called the classifying space $B(G) = B_G = BG$ of the group G.

Not only do we prove that all principal G-bundles P over a space are induced from the Milnor universal bundle G-bundle $E(G) \to B(G)$ over $B(G)$, see (2.9), but there is a homotopy uniqueness theorem which says that if two maps induce isomorphic bundles, then the maps are homotopic to each other. This should be put side by side with the result of the previous chapter which says that homotopic maps induce isomorphic principal bundles. This follows from the analysis of bundles over a space of the form $B \times [0,1]$.

Main assertion The function which assigns to a homotopy class $[f]$ in $[B, B(G)]$ the isomorphism class of the principal bundle $f^{-1}(E(G))$ over B is a bijection of the set $[B, B(G)]$ onto the set of isomorphism classes of numerable principal G-bundles over B, see (3.3).

We conclude the chapter with some specific examples of the Milnor construction where the involved spaces $E_n(G)$ and $E(G)$ (to be defined in Sect. 2) are just spheres.

Chapter 4, Sect. 11–13 of *Fibre Bundles* (Husemöller 1994) is a reference for this chapter.

1 Basic Data from a Numerable Principal Bundle

1.1. Remark If a principal G-bundle $P \to B$ is trivial over an open set of the form $\eta^{-1}((0,1]) \subset B$, where $\eta : B \to [0,1]$ is a map, then we have an isomorphism of bundles over $\eta^{-1}((0,1])$ of the form

D. Husemöller et al.: *The Milnor Construction: Homotopy Classification of Principal Bundles*, Lect. Notes Phys. **726**, 75–81 (2008)
DOI 10.1007/978-3-540-74956-1_8

$$P|\eta^{-1}((0,1]) \longrightarrow \eta^{-1}((0,1]) \times G,$$

which we can compose with $\eta^{-1}((0,1]) \times G \to (0,1] \times G$ defined simply by (b,s) which is mapped to $(\eta(b),s)$.

One way to think about the Milnor total space construction is that we wish to piece together these image spaces consisting of elements (t,s) with $t > 0$ and $s \in G$ into a global object. The first step is to see how to extend $(0,1] \times G$ at $t = 0$ to $(0,1] \times G \subset [0,1] \times G/(\{0\} \times G)$ by adding one point.

Now, the functions η will come from a partition of unity which is indexed by an arbitrary set. It would be useful to have control over this arbitrary set, and with the next proposition, it is possible to always use a countable set.

1.2. Proposition *Let P be a numerable principal G-bundle over a space B. Then, there exists a countable partition of unity $\{\eta_n\}$ with $P|\eta_n^{-1}((0,1])$, a trivial G-bundle.*

Proof. We start with a partition of unity ξ_i indexed by $i \in I$, I is an arbitrary set. For each $b \in B$, we have the finite set $I(b)$ of $i \in I$ with $\xi_i(b) > 0$, and for each finite subset $J \subset I$, we have the open subset $V(J) \subset B$ of $b \in B$ with $\xi_j(b) > \xi_i(b)$ for all $j \in J$ and $i \in I - J$. Let $\xi_J(b) = \max\{0, \min_{j \in J, i \in I-J}(\xi_j(b) - \xi_i(b))\}$, and then observe that $V(J) = \xi_J^{-1}((0,1])$.

Now, if the number of elements #
$J' = \#J''$ for two distinct finite sets $J', J'' \subset I$, then the intersection $V(J') \cap V(J'')$ is empty since $\xi_{j'}(b) > \xi_{j''}(b)$ and $\xi_{j''}(b) > \xi_{j'}(b)$ cannot hold simultaneously.

Let $V_m = \bigcup_{\#J=m} V(J)$ and let $\xi_m = \sum_{\#J=m} \xi_J$. Again, we have $\xi_m^{-1}((0,1]) = V_m$, and $P|V_m$ is trivial because it is trivial over each open set $V(J)$ in the disjoint union giving V_m. Finally, the desired partition of unity is

$$\eta_m(b) = \frac{\xi_m(b)}{\sum_{n \geq 0} \xi_n(b)}$$

with $\eta_m^{-1}((0,1]) = V_m$.

2 Total Space of the Milnor Construction

In order to see the natural character of the Milnor construction, we return to the local charts of a principal bundle coming from a countable open covering as in (1.2).

2.1. Remark Let $\pi : P \to B$ be a principal G-bundle trivial over the open sets of a covering of the form $\eta_n^{-1}((0,1])$ together with isomorphisms

$$w_n : P|\eta_n^{-1}((0,1]) \longrightarrow \eta_n^{-1}((0,1]) \times G,$$

which we can compose with $\eta_n^{-1}((0,1]) \times G \to (0,1] \times G$. The projection onto the second factor gives maps

$$u_n : P|\eta_n^{-1}((0,1]) \longrightarrow G$$

with $u_n(xs) = u_n(x)s$ for $x \in P$, $s \in G$. Then, we have the relation $w_n(x) = (\pi(x), u_n(x))$. It is now this data (η_n, u_n) which we wish to assemble into the Milnor construction and a G-equivariant map.

Partitions of unity map the space into simplexes.

2.2. Definition The affine n-simplex A^n is the compact subset of $\mathbb{R}^{n+1}$ consisting of all $t = (t_i) \in \mathbb{R}^{n+1}$ satisfying $t_0 + \ldots + t_n = 1$ and $t_i \geq 0$.

For example, A^0 is a point, A^1 is a closed segment, A^2 is a triangle, and A^3 is a tetrahedron. Singular homology is constructed by mappings of affine simplexes into a space, while a map $\eta : B \to A^n$ corresponds to a partition of unity on B of $n+1$ open subsets.

2.3. Definition (Finite Case) We start with the data (η_i, u_i) and over the open set $W_n = \eta_0^{-1}((0,1]) \cup \ldots \cup \eta_n^{-1}((0,1])$. We form the product $A^n \times G^{n+1}$ and take a quotient $E_n(G)$ where elements are written as double $n+1$ tuples $(t_0 : s_0, \ldots, t_n : s_n)$ with $t \in A^n$ and each $s_i \in G$. The equivalence relation is the following as an equality in $E_n(G)$, where

$$(t_0' : s_0', \ldots, t_n' : s_n') = (t_0'' : s_0'', \ldots, t_n'' : s_n'')$$

if and only if $t_i' = t_i''$, and when $t_i' = t_i'' > 0$, we require $s_i' = s_i''$.

In particular, when $t_i' = t_i'' = 0$, there is no relation between s_i' and s_i'' reducing to a single point.

2.4. Principal Action We have an action $E_n(G) \times G \to E_n(G)$ given by the formula $(t_0 : s_0, \ldots, t_n : s_n)s = (t_0 : s_0 s, \ldots, t_n : s_n s)$, and this action is principal with continuous

$$\tau_i((t_0' : s_0', \ldots, t_n' : s_n'), (t_0'' : s_0'', \ldots, t_n'' : s_n'')) = (s_i')^{-1} s_i''$$

on the open subset where $t_i' = t_i'' > 0$.

2.5. Basic Assertion The data (η_i, u_i), where $i \leq n$, is equivalent to a map $v : P|W_n \to E_n(G)$ satisfying $v(xs) = v(x)s$. The relation comes by looking at the coordinates of which gives a formula for $v(x)$, that is,

$$v(x) = (\eta_0(\pi(x)) : u_0(x), \ldots, \eta_n(\pi(x)) : u_n(x)).$$

Now, we extend the basic assertion for a finite part of the local trivializing data to the case of a countable family.

2.6. Definition (Countable Case) We start with the data (η_i, u_i) and the open covering $\{\eta_i^{-1}((0,1])\}$ of B. We see that the natural inclusion of the products $A^n \times G^{n+1} \subset A^{n+1} \times G^{n+2}$ passes to the quotient $E_n(G) \subset E_{n+1}(G)$ as an injection. The elements written as double $n+1$ tuples $(t_0 : s_0, \ldots, t_n : s_n)$ are carried to the double $n+2$ tuples $(t_0 : s_0, \ldots, t_n : s_n, 0 : 1)$ for $1 \in G$, the identity, $s_i \in G$, and $t \in A^n$. The equivalence relation in $E_{n+1}(G)$ induces the equivalence relation in $E_n(G)$, and the inclusion is G-equivariant.

2.7. Definition (Milnor Construction) The total space of the Milnor universal principal G-bundle is $E(G) = \bigcup_{0 \le n} E_n(G)$ with the weak topology, that is, a subset $M \subset E(G)$ is closed if and only if $M \cap E_n(G)$ is closed in $E_n(G)$ for each n. The Milnor classifying space is the quotient $E(G)/G = B(G)$.

2.8. Basic Assertion The data (η_i, u_i) is equivalent to a map $v : P \to E(G)$ satisfying $v(xs) = v(x)s$. The relation comes as in the finite case by looking at the coordinates of which gives a formula for $v(x)$, that is,

$$v(x) = (\eta_0(\pi(x)) : u_0(x), \ldots, \eta_n(\pi(x)) : u_n(x), \ldots).$$

2.9. Basic Universal Property of the Milnor Construction The Milnor principal G-bundle is $\pi : E(G) \to E(G)/G$. The bundle is numerable with partition of unity given by

$$\omega_i(t_0 : s_0, \ldots, t_n : s_n, \ldots) = t_i$$

and a principal action with translation functions

$$\tau_i((t_0' : s_0', \ldots, t_n' : s_n', \ldots), (t_0'' : s_0'', \ldots, t_n'' : s_n'', \ldots)) = (s_i')^{-1} s_i''.$$

For each numerable principal G-bundle $\pi : P \to B$ with localizing data, (η_i, u_i) gives arise to a G-equivariant map $v : P \to E(G)$ with formula $v(x) = (\eta_0(\pi(x)) : u_0(x), \ldots, \eta_n(\pi(x)) : u_n(x), \ldots)$. The map $v : P \to E(G)$ defines a quotient map $f : B \to B(G)$, and the pair $(v, f) : P \to E(G)$ is a morphism of principal G-bundles.

Finally, we have an isomorphism $P \to f^{-1}(E(G))$ over B induced by v using the general result 5(2.4). In particular, every numerable principal G-bundle P over B is of the form $f^{-1}(E(G))$ for some map $f : B \to B(G)$. From 6(6.5), two homotopic maps $f : B \to B(G)$ give isomorphic-induced bundles. This property holds for $B(G)$ replaced by any space B'. In the case of maps $B \to B(G)$ the converse holds, that is, if two maps $f', f'' : B \to B(G)$ have the property that the two induced bundles $f'^{-1}(E(G))$ and $f''^{-1}(E(G))$ are isomorphic, then f' and f'' are homotopic. We sketch this in the next section.

3 Uniqueness up to Homotopy of the Classifying Map

We consider various maps $B(G) \to B(G)$ induced by G-equivalent maps $E(G) \to E(G)$.

3.1. Notation Let $E(G, \text{ev})$ (resp. $E(G, \text{odd})$) be the subspace of $E(G)$ consisting of $(t_0 : s_0, \ldots, t_n : s_n, \ldots)$ with $t_{2i} = 0$ (resp. $t_{2i+1} = 0$). Let $B(G, \text{ev}) = E(G, \text{ev})/G \subset B(G)$ and $B(G, \text{odd}) = E(G, \text{odd})/G \subset B(G)$. Now, we define $h_s^{\text{ev}} : E(G) \to E(G)$ with image $E(G, \text{ev})$ for $s = 0$ and $h_s^{\text{odd}} : E(G) \to E(G)$ with image $E(G, \text{odd})$ for $s = 0$ such that these in turn induce the homotopies

$$g_s^{\text{ev}} : B(G) \longrightarrow B(G) \quad \text{and} \quad g_s^{\text{odd}} : B(G) \longrightarrow B(G)$$

which for $s = 0$ have disjoint images, namely $B(G,\mathrm{ev})$ and $B(G,\mathrm{odd})$, and which for $s = 1$ are both the identity.

The details we leave to the reader who can find them in the paragraph just before $4(12.3)$ of *Fibre Bundles*. These deformations are used in the following theorem which completes the last step in the homotopy classification theorem announced in the introduction.

3.2. Theorem *Let* $f', f'' : B \to B(G)$ *be maps such that the induced bundles* $(f')^{-1}(E(G))$ *and* $(f'')^{-1}(E(G))$ *are isomorphic. Then, the maps* f' *and* f'' *are homotopic.*

Proof. The first step is to use the homotopies described in (3.1) to modify f' and f'' so that $f'(B) \subset B(G,\mathrm{odd})$ and $f''(B) \subset B(G,\mathrm{ev})$. Assuming that f' and f'' have this property, we have a principal bundle P over B and principal bundle morphisms $(u', f') : P \to E(G,\mathrm{odd}) \subset E(G)$ and $(u'', f'') : P \to E(G,\mathrm{ev}) \subset E(G)$.

Now, we define a principal G-bundle morphism $(u, f) : P \times I \to E(G)$ with $f|X \times \{0\} = f'$ and $f|X \times \{1\} = f''$ in order to prove the theorem. This is done by a formula for u starting with desired properties for $t = 0$ and $t = 1$ where for $x \in P$, we have from u' and u'' the values

$$u(x,0) = (t_0(x) : s_0(x), 0, t_2(x) : s_2(x), 0, \ldots)$$

and

$$u(x,1) = (0, t_1(x) : s_1(x), 0, t_3(x) : s_3(x), 0, \ldots).$$

Now, prolong to $u(x,t)$ by the formula

$$u(x,t) = ((1-t)t_0(x) : s_0(x), tt_1(x) : s_1, (1-t)t_2(x) : s_2(x), tt_3(x) : s_3(x), \ldots).$$

Clearly, we have $u(xs,t) = u(x,t)s$ for $(x,t) \in P \times [0,1]$ and $s \in G$. Hence, the G-equivariant map $u : P \times [0,1] \to E(G)$ defines the homotopy $f : B \times [0,1] \to B(G)$ with $f(b,0) = f'(b)$ and $f(b,1) = f''(b)$. This proves the theorem.

Now, we return to the introduction of this chapter.

3.3. Main Assertion The function which assigns to a homotopy class $[f] \in [B, B(G)]$ the isomorphism class of the principal bundle $f^{-1}(E(G))$ over B is a bijection of the set $[B, B(G)]$ onto the set of isomorphism classes of numerable principal G-bundles over B.

Proof. It is a well-defined function by $6(6.5)$. It is injective by (3.2), and it is surjective by (2.9). This proves the homotopy classification of numerable principal G-bundles in terms of the Milnor construction.

3.4. Remark Let $w : G' \to G''$ be a morphism of topological groups inducing a map $Bw : BG' \to BG''$ on the classifying spaces, and let B be a base space.

(a) Let P' be the principal G'-bundle $(f')^*(EG')$ for a map $f' : B \to BG'$ into the classifying space. The extension of P' to the principal G''-bundle is given by $(f'')^*(EG'')$, where $f'' = (Bw) \circ f'$.

(b) A restriction of a principal G''-bundle $P'' = (f'')^*(EG'')$ for $f'' : B \to BG''$ is any $P' = (f')^*(EG'')$, where $(Bw) \circ f'$ and f'' are homotopic. In particular, extension and restriction of structure groups of principal bundles on B is determined by the function

$$[B, Bw] : [B, BG'] \to [B, BG'']$$

on homotopy classes of maps into classifying spaces.

(c) Let $w : G' \to G''$ be a morphism of topological groups inducing a map $Bw : BG' \to BG''$ on the classifying spaces. Let $P'' = (f'')^*(EG')$ be a principal G''-bundle on B, where $f'' : B \to BG''$. A G'-structure on P'' is an equivalence class of maps $f' : B \to BG'$ such that $(Bw) \circ f'$ and f'' are homotopic. Two maps $f_0', f_1' : B \to BG'$ are equivalent provided there is a homotopy $h_t : B \to BG'$ with $h_0 = f_0'$, $h_1 = f_1'$, and $(Bw) \circ h_t = (Bw) \circ f_0'$ for all $t \in [0, 1]$.

4 The Infinite Sphere as the Total Space of the Milnor Construction

4.1. Unit Sphere in the Real Numbers The zero dimensional sphere S^0 is the space of $t \in \mathbb{R} = \mathbb{R}^1$ with $|t| = 1$, that is, $S^0 = \{+1, -1\}$, and under multiplication of these two real numbers, it is the group of two elements. Every $x \in \mathbb{R}$ is of the form $x = r(\pm 1)$ for some $r \geq 0$, and for $x \neq 0$, the strictly positive r is unique and is the absolute value $|x| = r$.

4.2. Unit Sphere in the Complex Numbers The one-dimensional sphere S^1 is the space of $z \in \mathbb{C} = \mathbb{R}^2$ with $|z| = 1$, that is, $S^1 = \{e^{2\pi i t} : t \in [0, 1]\}$ the circle of angles t. Under multiplication of complex numbers, it is a group called the circle group.

There is one more sphere which is the topological group of numbers in a number system, and this is S^3, the group of unit quaternions.

4.3. Quaternions The skew field $\mathbb{H}$ of quaternions is given by $\mathbb{H} = \mathbb{R}1 \oplus \mathbb{R}i \oplus \mathbb{R}j \oplus \mathbb{R}k$, where the basis elements i, j, k of the quaternions satisfy the relations $i^2 = j^2 = k^2 = -1$ and

$$ij = k = -ji, \quad jk = i = -kj, \quad \text{and } ki = j = -ik.$$

We can write $\mathbb{H} = \mathbb{C} \oplus \mathbb{C}j$ with $k = ij$ and multiplication given by $zj = j\bar{z}$ for $z \in \mathbb{C}$ and $\bar{z} = x - iy$ the complex conjugate of $z = x + iy$. For a quaternion $q = a + bi + cj + dj$, the quaternion conjugate is given by $\bar{q} = a - bi - cj - dk$. It has properties similar to the complex conjugate, namely $q\bar{q} = a^2 + b^2 + c^2 + d^2$ is the norm squared $|q|^2 = q\bar{q}$ and $\bar{\bar{q}} = q$. Moreover, quaternionic conjugation is antimultiplicative, that is, one has $\overline{q'q''} = \overline{q''}\,\overline{q'}$.

4.4. Unit Sphere in the Quaternions The three-dimensional sphere can be parametrized by the unit quaternions $q \in \mathbb{H}$ so with $|q| = 1$. Moreover, under quaternionic multiplication, it is a topological group isomorphic to $SU(2)$. Every $q \in \mathbb{H}$ is of the form $q = |q|u$, where $u \in S^3$ is a unit quaternion.

Now, we show how to view certain spheres as the total space in the Milnor construction in a very concrete way.

4.5. Spheres as $E_n(G)$ and Infinite Spheres as $E(G)$

(a) (Real case) For a point $(x_0, \ldots, x_{m-1}) \in S^{m-1}$, we introduce $t_i = x_i^2$ and $s_i = +1$ if $x_i \geq 0$ and $s_i = -1$ if $x_i < 0$. Then, the map from $(x_0, \ldots, x_{m-1}) \in S^{m-1}$ to $(t_0 : s_0, \ldots, t_{m-1} : s_{m-1}) \in E_{m-1}(\{\pm 1\})$ is a topological isomorphism, and with the same formulas, we have a topological isomorphism $S^\infty \to E(\{\pm 1\})$, where S^∞ is the increasing union $\bigcup_n S^n$ with the inductive limit topology. These mappings are S^0 or $\{\pm 1\}$-equivariant.

(b) (Complex case) For a point $(z_0, \ldots, z_{m-1}) \in S^{2m-1}$, we introduce $t_i = |z_i|^2$ and $s_i = z_i/|z_i|$. Then, the map from $(z_0, \ldots, z_{m-1}) \in S^{2m-1}$ to $(t_0 : s_0, \ldots, t_{m-1} : s_{m-1}) \in E_{m-1}(S^1)$ is a topological isomorphism, and with the same formulas, we have a topological isomorphism $S^\infty \to E(S^1)$. These mappings are S^1-equivariant.

(c) (Quaternionic case) For a point $(q_0, \ldots, q_{m-1}) \in S^{4m-1}$, we introduce $t_i = |q_i|^2$ and $s_i = q_i/|q_i|$. Then, the map from $(q_0, \ldots, q_{m-1}) \in S^{4m-1}$ to $(t_0 : s_0, \ldots, t_{m-1} : s_{m-1}) \in E_{m-1}(S^3)$ is a topological isomorphism, and with the same formulas, we have a topological isomorphism $S^\infty \to E(S^3)$. These mappings are S^3-equivariant.

4.6. Other Constructions of $B(G)$

There are other constructions of a numerable principal G-bundle $E(G) \to B(G)$ with the key property that $E(G)$ is contractible, and hence, by a theorem of Steenrod (1951, Sect. 19), it can be used to induce and classify up to isomorphism numerable principal G-bundles. In particular, there also is the construction of a classifying space BG which is functorial for group morphisms $G' \to G''$ in all cases and which in contrast to the Milnor construction has the property that the projections $G' \times G'' \to G'$ and $G' \times G'' \to G''$ induce a homeomorphism $B(G' \times G'') = B(G') \times B(G'')$. If G is an abelian group, then the group law $G \times G \to G$ is a morphism of groups and induces a map $B(G \times G) = B(G)$ which when composed with the inverse of the product isomorphism gives an abelian group structure $B(G) \times B(G) = B(G)$, and then the classifying space construction can be iterated to obtain $B^n(G) = B(B^{n-1})$ inductively.

References

Husemöller, D.: *Fibre Bundles*, 3rd ed. Springer-Verlag, New York (1994)
Steenrod, N.E.: Topology of Fibre Bundles. Princeton University Press, Princeton, NJ (1951)

Chapter 8
Fibrations and Bundles: Gauge Group II

The fundamental theorem in fibre bundle theory is the main assertion 7(3.3) which says that numerable principal G-bundles are not only induced from the Milnor construction but also classified up to homotopy by maps into the classifying base space of the universal bundle. In fact, for a given topological group G, this universal property is true for other principal G-bundles than the Milnor construction, and in this chapter, we investigate which bundles have this property. The base space of each universal bundle is a new model for the classifying space BG of the group G, and it is homotopy equivalent to the base space of the Milnor construction. For this analysis, we introduce the notion of fibre map and fibre mapping sequence.

In this chapter, we consider loop spaces and the related path space. These are *not* principal bundles, but they have important bundle properties relative to homotopy. We relate and compare these loop space bundles to universal principal G-bundles. The key concept of fibre map is common to both path space bundles and the universal bundles.

For a fibre space fibration $p : E \to B$ with fibre $F = p^{-1}(b_0)$ and choice of base point $x_0 \in F \subset E$, we have a homotopy exact triangle

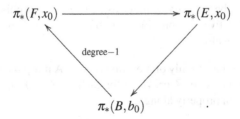

For a given topological group G, we consider all the principal bundles over a space B which is reduced to homotopy theory by the universal bundle. We will see that a principal G-bundle $P \to B$ is a universal bundle when P is contractible, and this leads to other versions of the $E(G) \to B(G)$. These considerations are applied to $B(\mathrm{Aut}(P))$ for the gauge group of $\mathrm{Aut}(P)$ of $P \to B$.

Chapters 6 and 7 of *Fibre Bundles* (Husemöller 1994) is a reference for this chapter.

D. Husemöller et al.: *Fibrations and Bundles: Gauge Group II*, Lect. Notes Phys. **726**, 83–96 (2008)
DOI 10.1007/978-3-540-74956-1_9   © Springer-Verlag Berlin Heidelberg 2008

1 Factorization, Lifting, and Extension in Square Diagrams

1.1. Remark In a category $\mathscr{C}$, it is frequently useful to consider commutative square diagrams of morphisms (i/p)

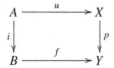

so that $pu = fi$. In fact, we can speak of the category of squares $Sq(\mathscr{C})$ over a category $\mathscr{C}$, where a morphism is a morphism on each corner giving a commutative cube.

1.2. Definition A factorization k of the square (i/p) is a morphism $k : B \rightarrow X$ leading to a commutative diagram (i/p)

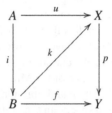

that is, $ki = u$ and $pk = f$. We can also speak of k as a lifting of f along p or as an extension of u along i.

Relative to classes of morphisms in $\mathscr{C}$, we can define properties on morphisms with respect to the existence of factorization. These are lifting and extension properties.

1.3. Definition Let $\mathscr{E}$ be a family of morphisms in $\mathscr{C}$. A morphism $p : X \rightarrow Y$ has the lifting property relative to $\mathscr{E}$ provided for all $i : A \rightarrow B$ in $\mathscr{E}$ every diagram (i/p) has the lifting property along p.

1.4. Definition Let $\mathscr{F}$ be a family of morphisms in $\mathscr{C}$. A morphism $i : A \rightarrow B$ has the extension property relative to $\mathscr{F}$ provided for each $p : X \rightarrow Y$ in $\mathscr{F}$ every diagram (i/p) has the extension property along i.

These properties are dual to each other in $\mathscr{C}$ and its opposite category $\mathscr{C}^{\mathrm{op}}$. Examples are generated by the following two pairs of dual statements.

1.5. Example Let $\mathscr{E}$ be a family of morphisms in $\mathscr{C}$, and let $p : X \rightarrow Y$ and $q : Y \rightarrow Z$ be two morphisms in $\mathscr{C}$. If p and q have the lifting property relative to $\mathscr{E}$, then the composite $qp : X \rightarrow Z$ has the lifting property relative to $\mathscr{E}$.

1.6. Example Let $\mathcal{F}$ be a family of morphisms in $\mathcal{C}$, and let $i : A \to B$ and $j : B \to C$ be two morphisms in $\mathcal{C}$. If i and j have the extension property relative to $\mathcal{F}$, then the composite $ji : A \to C$ has the extension property relative to $\mathcal{F}$.

1.7. Example Let $\mathcal{E}$ be a family of morphisms in $\mathcal{C}$, and let $p : X \to Y$ be a morphism with the lifting property relative to $\mathcal{E}$. If $Y' \to Y$ is any morphism in $\mathcal{C}$, then the induced morphism

$$p' : X' = Y' \times_Y X \longrightarrow Y'$$

has the lifting property relative to $\mathcal{E}$. Recall that the induced morphism is the projection from the fibre product to the first factor.

1.8. Example Let $\mathcal{F}$ be a family of morphisms in $\mathcal{C}$, and let $i : A \to B$ be a morphism with the extension property relative to $\mathcal{F}$. If $A \to A'$ is any morphism in $\mathcal{C}$, then the coinduced morphism

$$i' : A' \longrightarrow B' = A' \sqcup_A B$$

has the extension property relative to $\mathcal{F}$. Recall that the coinduced morphism is the injection into the cofibre coproduct of the first factor.

1.9. Remark We will apply these general lifting properties to homotopy theory. This was first done by Quillen in *LN 43* where he presented an axiomatic version of homotopy theory which seems to be the most promising approach to an axiomatic version of homotopy theory. This has become especially clear in recent years.

2 Fibrations and Cofibrations

Now, we use the elementary factorization formalism of the previous section to describe fibrations, also called fibre maps, and cofibrations, also called cofibre maps. For this, we need the classes $\mathcal{E}$ of elementary cofibrations and $\mathcal{F}$ of elementary fibrations.

2.1. Definition The elementary fibration associated with any space Y is the map $\varepsilon : \mathrm{Map}([0,1],Y) \to Y$ given by evaluation $\varepsilon(\gamma) = \gamma(0)$ for a path $\gamma \in \mathrm{Map}([0,1],Y)$. The elementary cofibration associated with any space A is the map $\beta : A \to A \times [0,1]$ given by inclusion $\beta(a) = (a,0)$ on the bottom of the cylinder over A.

2.2. Definition A map $p : X \to Y$ is a fibration provided any square (β/p) with any elementary cofibration $\beta : A \to A \times [0,1]$ has the lifting property along p. A map $i : A \to B$ is a cofibration provided any square (i/ε) with any elementary fibration $\varepsilon : \mathrm{Map}([0,1],Y) \to Y$ has the extension property along i.

2.3. Remark The fibration or fibre map part of the previous definition has a square
of the form

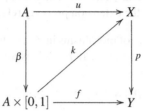

The fibration axiom is also called the homotopy-lifting property since the homotopy
f is required to lift to a homotopy k.

The cofibration or cofibre map part of the previous definition has a square of the
form

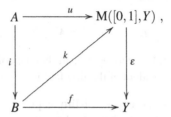

but in this case, we do not see the homotopy extension property for u so clearly
in this concept. In fact, we can rewrite the above diagram in a context of a map
$v : A \times [0,1] \to Y$, which must extend to $B \times [0,1]$.

2.4. Remark In the previous diagram defining a cofibration $i : A \to B$, we consider $v : A \times [0,1] \to Y$ the adjoint map to u. The condition for the existence of
the lifting k along ε is equivalent to the existence of a map $k' : B \times [0,1] \to Y$
satisfying

$$k'(i \times [0,1]) = v : A \times [0,1] \longrightarrow Y \quad \text{and} \quad k'\beta = f : B \longrightarrow Y.$$

The related diagram is a colimit diagram of the form

$$
\begin{array}{ccc}
A & \longrightarrow & A \times [0,1] \\
\downarrow & & \downarrow{\scriptstyle i \times [0,1]} \quad \searrow{\scriptstyle v = \text{adjoint of } u} \\
B & \underset{\text{id} \times [0,1]}{\longrightarrow} & B \times [0,1] \xrightarrow{k'} Y \\
\end{array}
$$

Conversely, given k' the homotopy extension of $f : B \to Y$ and $v : A \times [0,1] \to Y$,
then the adjoint $k : B \to \text{Map}([0,1],Y)$ gives an extension along i.

2.5. Remark A commutative square of the form (β, ε) for two spaces A and Y has
the form

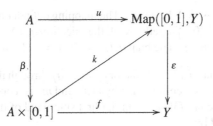

where u and f are related maps with $f(a,0) = u(a)(0)$, all $a \in A$. The natural lifting k is given as follows. We introduce the map $k'' : A \times [0,1] \times [0,1]$ by

$$k''(a,t,s) = \begin{cases} (u(a))(s(t+1)) & \text{for} \quad 0 \le s \le \frac{1}{t+1} \\ f(a,s(t+1)-1) & \text{for} \quad \frac{1}{t+1} \le s \le 1, \end{cases}$$

k'' defines the natural lifting $k : A \times [0,1] \to \text{Map}([0,1],Y)$.

If $s = 0$, then $k(a,t)(0) = u(a)(0) = f(a,0)$. If $t = 0$, then $k(a,0)(s) = u(a)(s)$. This leads to the following assertion.

2.6. Proposition The elementary fibrations ε are fibrations which are homotopy equivalences. The elementary cofibrations δ are cofibration which are homotopy equivalences

Proof. A right inverse c of ε is given by $c : Y \to \text{Map}([0,1],Y)$, where $c(y)(t) = y$ such that εc is homotopic to the identity by

$$h'_s : \text{Map}([0,1],Y) \longrightarrow \text{Map}([0,1],Y)$$

given by $h'_s(\gamma)(t) = \gamma(st)$ with $h'_1(\gamma) = \gamma$ and $h'_0 = \varepsilon c$. A left inverse q of β is given by $q : A \times [0,1] \to A$ such that βq is homotopic to the identity by $h''_s : A \times [0,1] \to A \times [0,1]$ given by $h''_s(a,t) = (a,st)$ with $h''_1(a,t) = (a,t)$ and $h''_0 = \beta q$. This proves the proposition.

Let $\varepsilon' : \text{Map}([0,1],Y) \to Y$ be defined by $\varepsilon'(\gamma) = \gamma(1)$. Using the elementary fibration ε' which is also a homotopy equivalence, we can replace every map by a fibration up to a homotopy equivalence.

2.7. Definition Let $f : X \to Y$ be a map. The mapping track of f is the fibre product $T_f = X \times_Y \text{Map}([0,1],Y)$ together with the injection $t_f : X \to T_f$ and projection $p_f : T_f \to Y$ to the base. The map t_f satisfies $t_f(x) = (x,c(f(x)))$, where $c(y)(t) = y$.

2.8. Remark Every map $f : X \to Y$ factors $f = p_f t_f$ through the mapping track $X \to T_f \to Y$, where p_f is a fibre map by (1.7) and t_f is a homotopy equivalence with homotopy inverse $t'(f)$ given by projection on the first factor and homotopy from the identity to $t_f t'(f)$ given by $h_s(x,\gamma) = (x,\gamma((1-s)t+s))$.

Let $\beta' : X \to X \times [0,1]$ be defined by $\beta'(x) = (x,1)$. Using the elementary cofibration β' which is also a homotopy equivalence, we can replace every map by a cofibration up to a homotopy equivalence.

2.9. Definition Let $f : X \to Y$ be a map. The mapping cylinder of f is the cofibre co-product $M_f = Y \sqcup_X X \times [0,1]$ together with the injection $q_f : X \to M_f$ and projection $m_f : M_f \to Y$ to the factor Y. The map $q_f(x) = f(x) = (x,1)$ in M_f.

2.10. Remark Every map $f : X \to Y$ factors $f = m_f q_f$ through the mapping cylinder $X \to T_f \to Y$, where q_f is a cofibre map by (1.8) and m_f is a homotopy equivalence with homotopy inverse $m'(f)(y) = y$ in the first summand of the cofibre coproduct and homotopy defined by

$$h_s(y) = y \quad \text{and} \quad h_s(x,t) = (x,(1-s)t+s).$$

At $s = 0$, it is the identity and $h_1 = m'(f)m_f$.

2.11. Remark There are essentially two versions of the definition of fibre maps or fibration. The first is due to Hurewicz, where the lifting property is required for all elementary $\beta : A \to A \times [0,1]$. The second is due to Serre, where the lifting property is required only for all finite complexes A or equivalently all n-cubes $A = [0,1]^n$.

3 Fibres and Cofibres: Loop Space and Suspension

The definition of fibration and cofibration do not refer to base points, and the same is true of the replacement of $f : X \to Y$ by a fibre map $p_f : T_f \to Y$ or by a cofibre map $q_f : X \to M_f$. Now base points arise naturally when we consider fibres and cofibres of a map.

3.1. Definition Let $f : X \to Y$ be a map. For each $y \in Y$, the fibre of f over $y \in Y$ is the subspace $X_y = f^{-1}(y) \to X$ with inclusion in X. The cofibre of f is the quotient space $Y/f(X)$, where the image $f(X)$ in Y is reduced to a point. When $f : X \to Y$ is base point preserving, the fibre F of f is defined to be the fibre over the base point of Y with base point from X. When f is base point preserving, the base point of the cofibre $C = Y/f(X)$ is the class of $f(X)$ in the cofibre C. There are two three term sequences which we wish to study

$$F \to X \xrightarrow{f} Y \quad \text{and} \quad X \xrightarrow{f} Y \to C.$$

Fibres are related to fibrations and cofibres to cofibrations by the following exact-ness properties.

3.2. Theorem *Let $F \to X$ be the fibre of a pointed fibre map $f : X \to Y$. For each pointed space T, we have an exact sequence of pointed sets*

$$[T,F]_* \longrightarrow [T,X]_* \longrightarrow [T,Y]_*.$$

When the space T is a finite complex like a sphere, then the assertion holds also for a Serre fibration.

Let $Y \to C$ *be the cofibre of a pointed cofibre map* $f : X \to Y$. *For each pointed space* Z, *we have an exact sequence of pointed sets*

$$[C,Z]_* \longrightarrow [Y,Z]_* \longrightarrow [X,Z]_*.$$

A sequence $M' \xrightarrow{u} M \xrightarrow{v} M''$ of pointed sets is a exact provided $\operatorname{im}(u) = u(M') = \ker(v) = v^{-1}(*)$. In the previous theorem, the assertion that $\operatorname{im}(u) \subset \ker(v)$ is just the immediate assertion that $vu = *$, and it holds both for maps and homotopy classes of maps. The reverse inclusion uses the fibration and cofibration conditions.

Taking the fibre or the cofibre of a map in special situations leads to new constructions to which theorem (3.2) applies.

3.3. Definition Let Y be a pointed space. The path space $P(Y)$ of Y is the fibre $P(Y)$ of $\varepsilon : \operatorname{Map}([0,1],Y) \to Y$ together with the restriction of ε' to $\pi : P(Y) \to Y$. The fibre of π is the loop space $\Omega(Y)$ of the pointed space Y.

3.4. Remark The loop space $\Omega(Y)$ is the subspace of $\gamma \in \operatorname{Map}([0,1],Y)$ with $\gamma(0) = \gamma(1) = *$, and path space $P(Y)$ is the subspace of $\gamma \in \operatorname{Map}([0,1],Y)$ with $\gamma(0) = *$ and $\pi(\gamma) = \gamma(1)$. The space $P(Y)$ is contractible by $h_s : P(Y) \to P(Y)$ with $h_s(\gamma)(t) = \gamma(st)$.

3.5. Definition Let X be a pointed space. The cone $C(X)$ on X is the cofibre of $\beta : X \to X \times [0,1]$ with $* \times [0,1]$ also collapsed to the base point together with composite of β' with the quotient morphism $q : X \to C(X)$. The cofibre of q is the suspension $S(X)$ of the pointed space X.

3.6. Remark The suspension $S(X)$ is the quotient of $X \times [0,1]$ with the subspace $(X \times \{0,1\}) \cup (\{*\} \times [0,1])$ reduced to a point, and the cone $C(X)$ is the quotient of $X \times [0,1]$ with the subspace $(X \times \{0\}) \cup (\{*\} \times [0,1])$ reduced to a point, and $q : X \to C(X)$ and $q(x)$ is the map which maps a point x in X to the class of $(x,1)$ in $C(X)$.

3.7. Corollary *For pointed spaces* T, X, Y, *and* Z, *we have the following sequences of pointed sets*

$$[T, \Omega(Y)]_* \longrightarrow [T, P(Y)]_* \to [T, Y]_*$$

and

$$[S(X), Z]_* \longrightarrow [C(X), Z]_* \to [X, Z]_*.$$

The sets $[T, P(Y)]_*$ *and* $[C(X), Z]_*$ *reduce to a single point.*

Now, we have a mapping space and a mapping cone associated with a map $f : X \to Y$ generalizing $P(Y)$ and $C(X)$ and built from the mapping track and mapping cylinder, respectively. We sketch this topic in the next section where it allows one to extend the sequences of the theorem (3.2) infinity far to the left for fibrations and infinity far to the right for cofibrations.

4 Relation Between Loop Space and Suspension Group Structures on Homotopy Classes of Maps [X,Y]$_*$

We use the notation $\mathrm{Hom}(X,A;Y,*)$ for the subspace of $\mathrm{Map}(X,Y)$ of all maps $f : X \to Y$ with $f(A) = *$. The quotient map $X \to X/A$ induces a homeomorphism $\mathrm{Map}_*(X/A,Y) \to \mathrm{Map}(X,A;Y,*)$ under suitable compactness assumptions on (X,A).

4.1. Remark The adjunction morphism 6(2.1) for pointed spaces X and Y and a compact pointed space T becomes the following by restriction

$$
\begin{array}{ccc}
\mathrm{Map}(X \times T,Y) & \xrightarrow{\ c\ } & \mathrm{Map}(X,\mathrm{Map}(T,Y)) \\
\cup\uparrow & & \cup\uparrow \\
\mathrm{Map}_*(X \times T, X \vee T; Y,*) & \longrightarrow & \mathrm{Map}_*(X,\mathrm{Map}_*(T,Y)) \\
=\uparrow & & \\
\mathrm{Map}_*(X \wedge T,Y) & &
\end{array}
$$

where $X \wedge T = (X \times T)/(X \vee T)$ is the reduced or smash product where $X \vee T$ is the coproduct of X and T.

4.2. Remark The adjunction morphism 6(2.1) for $T = [0,1]$ is the isomorphism $c : \mathrm{Map}(X \times [0,1],Y) \to \mathrm{Map}(X,\mathrm{Map}([0,1],Y))$. Using the modifications in the previous Sect. 4.1, we obtain two basic adjunctions

$$
\begin{array}{ccc}
\mathrm{Map}_*(C(X),Y) & \longrightarrow & \mathrm{Map}_*(X,\mathrm{Map}_*(T,Y)) \\
\cup\uparrow & & \cup\uparrow \\
\mathrm{Map}_*(S(X),Y) & \longrightarrow & \mathrm{Map}_*(X,\Omega(Y))
\end{array}
$$

From another point of view, we can describe the spaces in these two adjunctions as $S(X) = X \wedge S^1$ and $\Omega(Y) = \mathrm{Map}_*(S^1,Y)$ and as $C(X) = X \wedge [0,1]$ and $P(Y) = \mathrm{Map}_*([0,1],Y)$. Here, the circle S^1 is $[0,1]/\{0,1\}$, the interval with its end points reduced to one point.

4.3. Unit Circle Parametrizing the circle with $[0,1] \to S^1 \subset \mathbb{C}$ in the complex plane with the exponential function $e(t) = \exp(2\pi i t)$, we can return to the map used to define $\pi_1(Y)$ in 6(5.1)–6(5.4). The map $\psi : S^1 \to S^1 \vee S^1$ can be used to defined the following.

4.4. Definition The coH-space structure on $S(X)$ is defined via the following diagram

$$
\begin{array}{ccc}
S(X) & \xrightarrow{\;\;\;\psi\;\;\;} & S(X) \vee S(X) \\
\downarrow{\scriptstyle\cong} & & \downarrow{\scriptstyle\cong} \\
X \wedge S^1 \xrightarrow{id \wedge \psi} X \wedge (S^1 \vee S^1) & \xrightarrow{\;\cong\;} & (X \wedge S^1) \vee (X \wedge S^1)
\end{array}
$$

The H-space structure on $\Omega(Y)$ is the transpose $\phi : \Omega(Y) \times \Omega(Y) \to \Omega(Y)$ of ψ using the topological isomorphism $\mathrm{Map}_*(S^1 \vee S^1, Y) \cong \Omega(Y) \times \Omega(Y)$.

In concrete terms, we have for $t \in [0,1]$ the comultiplication on $S(X)$

$$
\psi(x,t) = \begin{cases} ((x,2t),*) & \text{for} \quad t \leq 1/2 \\ (*,(x,2t-1)) & \text{for} \quad 1/2 \leq t \end{cases}
$$

and the multiplication on $\Omega(Y)$ is

$$
\phi(u,v)(t) = \begin{cases} u(2t) & \text{for} \quad t \leq 1/2 \\ v(2t-1) & \text{for} \quad 1/2 \leq t. \end{cases}
$$

The questions of homotopy associativity, homotopy unit, and homotopy inverse carry over from the same considerations which lead to a group structure on $\pi_n(Y)$ for $n > 0$. As a result, we have the following homotopy level extension of the adjunction of (4.2).

4.5. Theorem *The natural adjunction bijection*

$$
[S(X),Y]_* \longrightarrow [X,\Omega(Y)]_*
$$

is an isomorphism of groups with group structure on the left defined by

$$
\psi : S(X) \longrightarrow S(X) \vee S(X)
$$

and on the right by

$$
\phi : \Omega(Y) \times \Omega(Y) \to \Omega(Y).
$$

The two structures on $[S(X),\Omega(Y)]_*$ are equal and abelian.

For the last statement we use 6(5.8).

5 Outline of the Fibre Mapping Sequence and Cofibre Mapping Sequence

5.1. Definition Let $f : X \to Y$ be a map. The homotopy fibre of f is the fibre product $E_f = X \times_Y P(Y)$ together with the projection $a(f) : E_f \to X$ to X.

5.2. Remark In concrete terms, the elements of E_f are pairs

$$
(x,\gamma) \in X \times \mathrm{Map}([0,1],Y)
$$

such that $f(x) = \gamma(1)$ and $\gamma(0) = y_0$. The fibre of the projection $a(f)$ is just the elements (x_0, γ) with $\gamma(0) = \gamma(1) = y_0$, that is, the loop space $\Omega(Y)$. Moreover, $a(f)$ is a fibre map.

5.3. Assertion There is a natural map $j : f^{-1}(y_0) \rightarrow E_f$ defined by $j(x) = (x, c(f(x)))$, where $c(y)(t) = y$ for all $t \in [0,1]$. A fundamental property of j is that it is a homotopy equivalence if f is a fibre map.

This is used in comparing two possible sequences in the following diagram which also relates E_f to the mapping track T_f.

5.4. Definition The fibre mapping sequence is the following diagram which starts with the homotopy fibre

$$E_f \xrightarrow{a(f)} X \xrightarrow{f} Y$$

of a map $f : X \rightarrow Y$ and continues both by iteration and with the role of the loop space

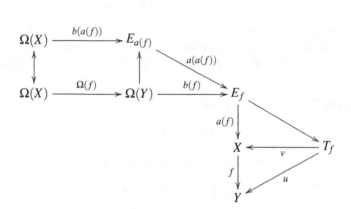

with $u = fv$.

5.5. Theorem *Let T be a pointed space, and let $f : X \rightarrow Y$ be a pointed fibre map. The following sequence is exact*

$$\cdots \rightarrow [T, \Omega^2 Y]_* \rightarrow [T, E_{\Omega f}]_* \rightarrow [T, \Omega X]_* \rightarrow [T, \Omega Y]_* \rightarrow [T, E_f]_* \rightarrow [T, X]_* \rightarrow [T, Y]_* \rightarrow 0.$$

Proof. Use (5.4) and (3.2).

5.6. Definition Let $f : X \rightarrow Y$ be a map. The homotopy cofibre of f is the cofibre coproduct $C_f = Y \sqcup_X C(X)$ together with the inclusion $a(f) : Y \rightarrow C_f$ from Y.

5.7. Remark In concrete terms, the elements of C_f is the union of $y \in Y$ with (x, t) with (x_0, t) and $(x, 0)$ all identified to the base point and $(x, 1)$ identified to $f(x) \in Y$. The cofibre of the injection $a(f)$ is just the further identification of all of Y to the base point, that is, the suspension $S(X)$. Moreover, $a(f)$ is a cofibre map.

5.8. Assertion There is a natural map $q : C_f \to Y/f(X)$ defined by collapsing all points (x,t) to the base point. A fundamental property of q is that it is a homotopy equivalence if f is a cofibre map.

This is used in comparing two possible sequences in the following diagram which also relates C_f to the mapping cylinder Z_f.

5.9. Definition The cofibre mapping sequence starts with the homotopy cofibre $X \xrightarrow{f} Y \xrightarrow{a(f)} C_f$ of a map $f : X \to Y$

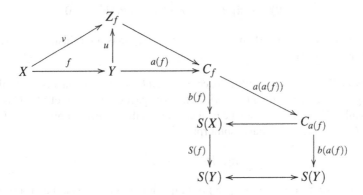

with $v = uf$.

5.10. Theorem *Let Z be a pointed space, and let $f : X \to Y$ be a pointed cofibre map. Then, the following sequence is exact*

$$0 \leftarrow [X, Z]_* \leftarrow [Y, Z]_* \leftarrow [C_f, Z]_* \leftarrow [S(X), Z]_* \leftarrow [S(Y), Z]_* \leftarrow [C_S(f), Z]_* \leftarrow [S^2(X), Z]_* \leftarrow \ldots$$

Proof. Use (5.9) and (3.2).

6 From Base to Fibre and From Fibre to Base

6.1. Exact Homotopy Sequence of a Fibration Let $p : E \to B$ be a fibration with base points and fibre $F \to E$. The exact homotopy sequence of the fibration is the following triangle

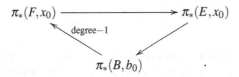

In the homotopy exact triangle, two morphisms are induced by $F \to E$ and $p : E \to B$, respectively, while the third, called the boundary morphism and sometimes denoted by ∂,

$$\pi_i(B, b_0) \longrightarrow \pi_{i-1}(F, x_0)$$

has degree -1 and is defined from the fibration property.

6.2. Elementary Corollaries We use the notation of (6.1)

(1) If F and B are path connected, then E is path connected, and if E is path connected, then B is path connected for surjective p, that is, $\pi_0(B) = 0$.
(2) Assume that F, E, and B are path connected. The last four possibly nonzero terms of the above sequence are

$$\pi_2(B) \longrightarrow \pi_1(F) \longrightarrow \pi_1(E) \longrightarrow \pi_1(B) \longrightarrow 0.$$

In particular, if $\pi_1(F) = 0$, then $\pi_1(E) \to \pi_1(B)$ is an isomorphism, and $\pi_2(E) \to \pi_2(B)$ is surjective.

6.3. Remark If E is a contractible space, then $\pi_*(E) = 0$, and conversely, if all homotopy groups are zero, then for a wide class of spaces E is contractible. This class includes all CW-complexes, see the next chapter. In this case, a fibration $F \to E \to B$ with $\pi_*(E) = 0$ has a boundary isomorphism

$$\pi_{i+1}(B) \longrightarrow \pi_i(F).$$

In particular, we have $\pi_1(B) \to \pi_0(F)$ is a bijection. Even if F is not a group, its set of connected components has a natural group structure coming from the fundamental group of the base.

This leads to a general idea which ties the homotopy theory considerations to fibre bundle theory.

6.4. Acyclic Fibrations An acyclic fibration $F \to E \to B$ is defined by the property that E is contractible. There are two constructions:

(1) *Base to fibre*: Given a path connected, pointed space B, the acyclic fibration associated to B is the loop space fibration

$$\Omega(B) \longrightarrow P(B) \to B,$$

and the boundary isomorphism $\pi_{i+1}(B) \to \pi_i(\Omega(B))$ shows that the homotopy of B in degree $i+1$ is the homotopy of loop space $\Omega(B)$ in degree i.
(2) *Fibre to base*: Given a topological group G, the acyclic fibration associated to G is the principal bundle fibration

$$G \to E(G) \longrightarrow B(G)$$

coming from the Milnor construction, and the corresponding boundary isomorphism $\pi_{i+1}(B(G)) \to \pi_i(G)$ shows that the homotopy of a topological group G in degree i is the homotopy of the classifying space $B(G)$ in degree $i+1$.

7 Homotopy Characterization of the Universal Bundle

7.1. Remark We classify an arbitrary principal G-bundle $P \to B$ with a map $f : B \to B(G)$ into the classifying space of the Milnor universal bundle. If all $\pi_i(P) = 0$, then we have again an isomorphism $\pi_{i+1}(B) \to \pi_i(G)$. Since f is part of an morphism of principal G-bundles $(u,f) : P \to E(G)$, it commutes with the boundary map in the homotopy exact triangle and hence

$$\pi_{i+1}(f) : \pi_{i+1}(B) \longrightarrow \pi_{i+1}(B(G))$$

is an isomorphism. Now, in the next chapter, we study for which spaces X and Y a map $f : X \to Y$ with the property that $\pi_*(f)$ is an isomorphism has a homotopy inverse $g : Y \to X$. Any space which is homotopy equivalent to a space which admits the structure of a CW-complex (to be defined in 9(1.2)) has this property, which is the assertion of the Whitehead mapping theorem 9(2.2). Moreover, the Milnor construction $B(G)$ can be equipped with the structure of a CW-complex when G can be given the structure of a CW-complex, which includes most groups of interest in geometry.

7.2. Theorem *Let G be a group which is a CW-complex, and let $E \to B$ be a numerable principal G-bundle such that B is a CW-complex and $\pi_*(E) = 0$. Then, the function which assigns to $[f] \in [X,B]$ the principal bundle $f^*(E)$ over X is a bijection from $[X,B]$ to the isomorphism classes of numerable principal G-bundles over X.*

Proof. The classification map $w : B \to B(G)$ is a homotopy equivalence with inverse v, and $w^{-1}(E(G))$ is isomorphic to E and $v^{-1}(E)$ is isomorphic to $E(G)$. Now, apply the same theorem that holds for the Milnor construction 7(3.3).

7.3. Remark Now, we can speak of the universal bundle and the classifying space in the more general setting as a principal G-bundle with contractible total space and base space as classifying space.

8 Application to the Classifying Space of the Gauge Group

8.1. Notation Let $p : P \to B$ be a principal G-bundle, and let $E(G) \to B(G)$ denote a universal principal G-bundle. Let $\mathrm{Map}_G(P,E(G))$ denote the subspace of $w \in \mathrm{Map}(P,E(G))$ such that $w(xs) = w(x)s$ for all $s \in G$. Let $\mathrm{Map}_P(B,B(G))$ denote the subspace of $f \in \mathrm{Map}(B,B(G))$ such that $f^{-1}(E(G))$ and P are isomorphic principal G-bundles over B. Each $w \in \mathrm{Map}_G(P,E(G))$ determines modulo G, a map $f \in \mathrm{Map}_P(B,B(G))$, and this is a continuous map $p_P : \mathrm{Map}_G(P,E(G)) \to \mathrm{Map}_P(B,B(G))$. The maps w and f are related by the commutative diagram

$$P \xrightarrow{\ w\ } E(G)$$

$$p \downarrow \qquad\qquad \downarrow$$

$$B \xrightarrow{\ f\ } B(G)$$

8.2. Remark The gauge group $\mathrm{Aut}_B(P)$ of automorphisms of P acts on the right of $\mathrm{Map}_G(P, E(G))$ with $u \in \mathrm{Aut}_B(P)$ acting on w equal to wu. Moreover, if $w', w'' \in \mathrm{Map}_G(P, E(G))$ with $p_P(w') = p_P(w'')$, then we have $\tau(w', w'') = (w')^{-1}w'' \in \mathrm{Aut}_B(P)$ with $w'\tau(w', w'') = w''$ which is the translation function mapping $\mathrm{Map}_G(P, E(G))$ into a principal $\mathrm{Aut}_B(P)$ bundle over $\mathrm{Map}_P(B, B(G))$.

8.3. Theorem *With the above notations, the space $\mathrm{Map}_G(P, E(G))$ is contractible, and the classifying space for the gauge group $\mathrm{Aut}_B(P)$ is just $\mathrm{Map}_P(B, B(G))$.*

Since we will not need this theorem, we refer to *Fibre Bundles*, (Husemöller 1994) 7(3.4), for the proof which uses (7.2). The proof is based on the isomorphism

$$\mathrm{Map}_G(P \times Y, E(G)) \longrightarrow \mathrm{Map}(Y, \mathrm{Map}_G(P, E(G)), .)$$

These spaces with Y having a trivial G-action are path connected by the homotopy classification theorem. For $Y = \mathrm{Map}_G(P, E(G))$, the identity is path connected to the trivial map giving the contracting homotopy on $\mathrm{Map}_G(P, E(G))$.

8.4. Corollary *The gauge group of the product (trivial) principal bundle is*

$$\mathrm{Aut}_B(B \times G) = \mathrm{Map}(B, G).$$

9 The Infinite Sphere as the Total Space of a Universal Bundle

We return to the ideas of the previous chapter and the identification of the Milnor construction as S^∞. The space S^∞ is seen to be contractible for it is $E(S^i)$ for $i = 0, 1$, or 3.

9.1. Remark The construction of 7(4.4) carries over to $E(S^i)$ for any i, and the result is S^∞. If a finite group G acts freely on S^i, then G acts diagonally on $E(S^i)$ freely, and in this way, we have a universal principal G-bundle $E(S^i) \to E(S^i)/G$ by (7.2).

9.2. Example For a finite cyclic group G, we can embed $G \subset S^1$ so that it acts freely on S^1 and with complex coordinates on the infinite sphere S^∞. The quaternion group of order 8 consisting of $\{\pm 1, \pm i, \pm j, \pm k\}$ is a subgroup of S^3 so that is acts freely on S^3 and with quaternionic coordinates on the infinite sphere S^∞.

Reference

Husemöller, D.: *Fibre Bundles*, 3rd ed. Springer-Verlag, New York (1994)

Chapter 9
Cohomology Classes as Homotopy Classes: CW-Complexes

We consider filtered spaces and especially CW-complexes. Using the cofibre constructions, we discuss the Whitehead mapping theorem. This characterization of homotopy equivalence was already used in the study of the uniqueness properties of classifying spaces. This completes a question left open in the previous chapter.

The usual definition of cohomology arises in terms of the dual of homology either directly for coefficients in a field or in terms of chains and cochains as linear forms on chain groups. These chains can arise very geometrically as cell chains or more generally as simplicial chains. The cochains can be algebraic linear functionals or as in the case of manifolds they can be differential forms which become linear functionals on chains of simplexes upon integration over the simplexes and summing over the chain.

There is another perspective on the cohomology of X discovered by Eilenberg and MacLane in terms of homotopy classes of maps of X into an Eilenberg–MacLane space $K(G,n)$. This is a topological space which has the homotopy type of a CW-complex and is characterized by its homotopy groups

$$\pi_i(K(G),n)) = \begin{cases} G & \text{if} \quad n = i \\ 0 & \text{if} \quad n \neq i \end{cases}$$

There is a canonical cohomology class $\iota_n \in H^n(K(G,n),G)$ corresponding to the identity morphism in $\mathrm{Hom}(G,G)$, where

$$\mathrm{Hom}(G,G) = \mathrm{Hom}(H_n(G,n),G) = H^n(K(G,n),G)$$

is part of the universal coefficient theorem.

In this chapter, we use the notion of cofibre maps and the cofibre sequence for a map. This is needed to establish cohomology properties of the K-theory functor and in the discussion of the above result that cohomology is defined also by maps into a classifying space which in this case it is a $K(\pi,n)$.

Spanier (1963) is a reference for this chapter.

D. Husemöller et al.: *Cohomology Classes as Homotopy Classes: CW-Complexes*, Lect. Notes Phys. **726**, 97–109 (2008)
DOI 10.1007/978-3-540-74956-1_10

1 Filtered Spaces and Cell Complexes

In analyzing a space X, it is often useful to have in addition some kind of decomposition of the space. Frequently, this comes as an increasing sequence of subspaces with union the entire space.

1.1. Definition A filtered space X is a space X with an increasing union of closed subspaces X_n such that a set M in X is closed if and only if $M \cap X_n$ is closed in X_n. This topology on X from the X_n is called the weak topology, and we have already used the topology on, for example, $S^\infty = \bigcup_n S^n$, the infinite sphere which was our first example of a filtered space. If $f_n|X_n : X_n \to Y$ is a sequence of continuous functions satisfying $f_n|X_{n-1} = f_{n-1}$ for each n, then there is a function $f : X \to Y$ with $f|X_n = f_n$, and the weak topology condition is exactly what is needed to show that f is continuous. In order to extend f_{n-1} to f_n, we need further properties on a filtered space.

1.2. Definition A CW-complex is a filtered space such that each $X_{n-1} \to X_n$ is a cofibration of the form

$$\vee_{I(n)} S^{n-1} \xrightarrow{\alpha} X_{n-1} \longrightarrow C_\alpha = X_n.$$

For each $i \in I(n)$, we have a restriction $\alpha_i : S^{n-1} \to X_{n-1}$ which is used to attach a closed n-cell $e^n(i)$ to X_{n-1}. The set X_n is a disjoint union of X_{n-1} and the interiors of each of the cells $e^n(i)$. Then the cofibre X_n/X_{n-1} is a wedge of n-spheres

$$\vee_{i \in I(n)} e^n(i) / \partial e^n(i) = \vee_{i \in I(n)} S^n$$

consisting of the closed n-cells with all boundary points reduced to a single point which is topologically isomorphic to the one-point union of a family of n-spheres.

1.3. Remark Coming back to the question of extending f_{n-1} to f_n for a CW-complex X, we see that it is possible when $f_{n-1}\alpha$ is null homotopic on each of the disjoint spheres. Up to the question of base points, this corresponds to a family of elements $[\alpha_i] \in \pi_{n-1}(X_{n-1})$ indexed by $i \in I(n)$ which must be zero.

The above considerations are part of general obstruction theory where the elements $[\alpha_i]$ are the obstructions to extending the continuous map. Here is a basic application.

1.4. Theorem *Let X be a CW-complex with all homotopy groups $\pi_i(X) = 0$ for $i \geq 0$. Then, X is contractible.*

Idea of the Proof. We must construct a map $h : X \times [0,1] \to X$ step by step with the property that $h(x,0) = x$ and $h(x,1) = *$, a fixed point in X_0. For $x \in X_0$, we define $h_0(x,t)$ as a path from x to $* \in X_0$ which is possible since $\pi_0(X) = 0$ means any two points can be joined by a path, that is, X is path connected. We define a CW-complex structure on $X \times [0,1]$ by the requirement that

$$(X \times [0,1])_n = (X_n \times \{0,1\}) \cup (X_{n-1} \times [0,1]).$$

For the inductive step assume that we have a deformation $h_{n-1} : (X \times [0,1])_{n-1} \to X$ with $h_{n-1}(x,0) = x$ and $h_{n-1}(x,1) = *$. We define $h_n : (X \times [0,1])_n \to X$ using the cofibre sequence going from $(X \times [0,1])_{n-1}$ to $(X \times [0,1])_n$

$$\coprod_{I(n)} S^{n-1} \xrightarrow{\alpha} (X \times [0,1])_{n-1} \longrightarrow C_\alpha = (X \times [0,1])_n.$$

The obstruction is given by maps α_i which are zero as in the discussion of (1.3). This proves the theorem.

1.5. Remark The construction of the space $E_n(G)$ of 7(2.3) does not require G to be a group and can be defined as before for a general topological space T. The resulting space $E_n(T)$ is also called the join of T with itself $n+1$ times, and if T is a CW-complex, then $E_n(T)$ and $E(T)$ are CW-complexes. In particular, for a topological group G which is a CW-complex, $E(G)$ and $B(G)$ are also CW-complexes. The CW-filtration $E(G)_n$ on $E(G)$ in general differs from the Milnor construction filtration $E_n(G)$ on $E(G)$.

1.6. Example The filtration of subspheres $S^{i-1} \subset S^i \subset \ldots$ on either S^m or S^∞ is a CW-filtration with $S^n/S^{n-1} = S^n \vee S^n$. This last quotient of the n-sphere (for $n > 0$) was used to define the multiplication in the nth homotopy groups of a pointed space. Since $E(\{\pm 1\}) = S^\infty$, the group S^0 appears in the fact that $E_n(S^0)/E_{n-1}(S^0)$ has two components S^n one for each element of $S^0 = \{\pm 1\}$.

2 Whitehead's Characterization of Homotopy Equivalences

The previous homotopy extension theorem has an analog with the Whitehead homotopy equivalence characterization for mappings between spaces.

2.1. Remark Since taking the homotopy groups is a functor on the homotopy category with base points and since a homotopy equivalence is an isomorphism in the homotopy category, a homotopy equivalence $f : X \to Y$ has the property that $f_* : \pi_*(X,x) \to \pi_*(Y,f(x))$ is an isomorphism for each $x \in X$.

2.2. Theorem *(Whitehead) Let $f : X \to Y$ be a map between path connected CW-complexes. If*

$$f_* : \pi_*(X,x) \longrightarrow \pi_*(Y,f(x))$$

is an isomorphism for some $x \in X$, then f is a homotopy equivalence.

Idea of the proof. By using the mapping cylinder $Z(f)$ of f, we can replace Y by $Z(f)$ and f by the inclusion $X \subset Z(f)$. Thus, we can assume that f is an inclusion $X \subset Y$ and that Y is X with cells adjoined as in a CW-complex. We wish to extend the identity on $A = X \times [0,1] \cup Y \times \{0\}$ to a map $Y \times [0,1] \to A$. The the restriction to $Y \times \{1\}$ defines a homotopy inverse of f.

For further background see the book by Whitehead (1978), especially Chaps. 2 and 5.

3 Axiomatic Properties of Cohomology and Homology

A complete development of homology and cohomology would take us out of the scope of this work. So we will list properties of homology and cohomology which can be used in an axiomatic definition as given in the book of Eilenberg and Steenrod. Let k be a field, or more generally a commutative ring, and let (k) (resp. (vect/k)) denote the category of k-vector spaces or k-modules (resp. finitely generated k-vector spaces or finitely generated projective k-modules). The ring k plays the role of coefficients for the theory.

3.1. Functorial Properties Homology and cohomology are sequences of functors on the category (pairs) of pairs of spaces with values in (k) or in (vect/k). More precisely, homology is given through functors $H_q : (\text{pairs}) \to (k)$, while cohomology is given through functors $H^q : (\text{pairs})^{\text{op}} \to (k)$.

If $f : (X,A) \to (Y,B)$ is a map of pairs, that is, $f(A) \subset B$ as a map $f : X \to Y$, then on homology $H_q(f) = f_* : H_q(X,A) \to H_q(Y,B)$ and on cohomology $H^q(f) = f^* : H^q(Y,B) \to H^q(X,A)$ are k-linear maps. For a second map of pairs $g : (Y,B) \to (Z,C)$, the composition relation takes the form $(gf)_* = g_* f_*$ and $(gf)^* = f^* g^*$. Also $(\text{identity})_* = \text{identity}$ and $(\text{identity})^* = \text{identity}$.

3.2. Absolute Versus Reduced Homology and Cohomology We write $H_q(X) = H_q(X,\emptyset)$ and $H^q(X) = H^q(X,\emptyset)$ for the so-called absolute homology and cohomology modules. These are functors on (top), the category of topology spaces. Reduced homology and cohomology is defined for pointed spaces X relative to the base point, and we write $\widetilde{H}_q(X) = H_q(X,*)$ and $\widetilde{H}^q(X) = H^q(X,*)$ for the corresponding functors on $(\text{top})_*$, the category of pointed spaces.

3.3. Homotopy Properties If $f', f'' : (X,A) \to (Y,B)$ are homotopic maps of pairs, that is, there is a homotopy $h_t : X \to Y$ with $h_0 = f'$, $h_1 = f''$, and $h_t(A) \subset B$ for all t, then $H_q(f') = H_q(f'')$ and $H^q(f') = H^q(f'')$.

In particular, homology and cohomology are defined on a quotient category of pairs. Also $H_q(f)$ and $H^q(f)$ are isomorphisms for a homotopy equivalence, that is, a map which is an isomorphism in the homotopy category. Furthermore, the homology and cohomology of a contractible space is isomorphic to the homology and cohomology of a point.

The next property of excision takes various forms, but since we consider usually pairs (X,A) of the form, where X is obtained from A by attaching cells, we formulate excision in terms of the collapsing map

$$(X,A) \longrightarrow (X/A, *).$$

Further, using the mapping sequences, we can formulate the exact sequence properties.

3.4. Excision and Exactness A cofibrant pair (X,A) is a pair where the inclusion $A \to X$ is a cofibration. If (X,A) is a cofibrant pair, then the excision morphisms

$$H_q(X,A) \longrightarrow H_q(X/A,*) \quad \text{and} \quad H^q(X/A,*) \longrightarrow H^q(X,A)$$

are isomorphisms for all q. Moreover, for the mapping sequence

$$A \xrightarrow{j} X \longrightarrow C_j \longrightarrow S(A) \xrightarrow{S(j)} S(X),$$

we have the following exact sequences of linear mappings for the reduced homology modules

$$\tilde{H}_q(A) \longrightarrow \tilde{H}_q(X) \longrightarrow \tilde{H}_q(C_j) \longrightarrow \tilde{H}_q(S(A)) \longrightarrow \tilde{H}_q(S(X))$$

and for the reduced cohomology modules

$$\tilde{H}^q(S(X)) \longrightarrow \tilde{H}^q(S(A)) \longrightarrow \tilde{H}^q(C_j) \longrightarrow \tilde{H}^q(X) \longrightarrow \tilde{H}^q(A).$$

In particular, we have formulated these two basic properties in terms of the reduced homology and cohomology. At this point, there is no relation between homology and cohomology for different values of q, an integer. In the next axiom, we have this relation.

3.5. Suspension Axiom We have morphisms of functors

$$\tilde{H}_q(X) \longrightarrow \tilde{H}_{q+1}(S(X)) \quad \text{and} \quad \tilde{H}^{q+1}(S(X)) \longrightarrow \tilde{H}^q(X).$$

Again this is for reduced homology and cohomology.

3.6. Classical Form of Exactness For the mapping sequence of a cofibrant pair (X,A)

$$A \xrightarrow{j} X \longrightarrow C_j \longrightarrow S(A) \xrightarrow{S(j)} S(X),$$

we have the following exact sequences of linear mappings for the reduced homology modules

$$\tilde{H}_q(A) \longrightarrow \tilde{H}_q(X) \longrightarrow \tilde{H}_q(X/A) \longrightarrow \tilde{H}_{q-1}(A) \longrightarrow \tilde{H}_{q-1}(X)$$

and for the reduced cohomology modules

$$\tilde{H}^{q-1}(X) \longrightarrow \tilde{H}^{q-1}(A) \longrightarrow \tilde{H}^q(X/A) \longrightarrow \tilde{H}^q(X) \longrightarrow \tilde{H}^q(A).$$

Further, if excision is incorporated into the middle terms, we have the following exact sequences of linear mappings for the reduced homology modules

$$\tilde{H}_q(A) \longrightarrow \tilde{H}_q(X) \longrightarrow H_q(X,A) \longrightarrow \tilde{H}_{q-1}(A) \longrightarrow \tilde{H}_{q-1}(X)$$

and for the reduced cohomology modules

$$\tilde{H}^{q-1}(X) \longrightarrow \tilde{H}^{q-1}(A) \longrightarrow H^q(X,A) \longrightarrow \tilde{H}^q(X) \longrightarrow \tilde{H}^q(A).$$

Also these hold for either the reduced or the absolute homology and cohomology.

3.7. Remark From the homotopy property, exactness, and excision, we can go back to the suspension property using $S(A) = C(A)/A$, for $\widetilde{H}(CA) = 0$ in the reduced theory so we have that

$$0 \longrightarrow \widetilde{H}_q(S(A)) \longrightarrow \widetilde{H}_{q-1}(A) \longrightarrow 0 \quad \text{and} \quad 0 \longrightarrow \widetilde{H}^{q-1}(A) \longrightarrow \widetilde{H}^q(S(A)) \longrightarrow 0$$

are exact, which implies that the suspension property holds.

3.8. Dimension Axiom For ordinary homology and cohomology, we require that the absolute $H_i(*) = H^i(*) = 0$ for $i \neq 0$ and in degree $i = 0$ it is one dimensional, or the reduced homology and cohomology it is

$$\widetilde{H}_i(S^0) = \widetilde{H}^i(S^0) = \begin{cases} 0 & \text{for} \quad i \neq 0 \\ k & \text{for} \quad i = 0. \end{cases}$$

Here, k is the one-dimensional k-module.

Of special interest is $k = \mathbb{Z}$, where $\mathbb{Z}$-modules are just abelian groups and $(\mathbb{Z}) = $ (ab) the category of abelian groups.

3.9. Remark With the dimension axiom, we have the following calculation for the ordinary reduced homology and cohomology of spheres

$$\widetilde{H}_i(S^q) = \widetilde{H}^i(S^q) = \begin{cases} 0 & \text{for} \quad i \neq q \\ k & \text{for} \quad i = q. \end{cases}$$

This is an immediate application of the suspension property and the relation $S(S^q) = S^{q+1}$.

One frequently used tool is the following sequence which is a consequence of the above properties.

3.10. Mayer–Vietoris Sequence Let $A, B \subset X$ such that the interiors of A and B form a covering of X. Then, the natural inclusions yields a long exact sequence in homology

$$H_{q+1}(X) \to H_q(A \cap B) \longrightarrow H_q(A) \oplus H_q(B) \longrightarrow H_q(X) \longrightarrow H_{q-1}(A \cap B)$$

and one in cohomology

$$H^{q-1}(A \cap B) \longrightarrow H^q(X) \longrightarrow H^q(A) \oplus H^q(B) \longrightarrow H^q(A \cap B) \longrightarrow H^{q+1}(X).$$

Here, the homomorphism out of $H_q(A \cap B)$ is the sum of the homomorphisms induced by the inclusions $A \cap B \subset A, B$ while the homomorphism into $H_q(X)$ is given by the difference of the two homomorphisms induced by the inclusions $A, B \subset X$. In the cohomological case, the homomorphism is given in a corresponding manner.

3.11. Remark There is a further extension of all the above considerations to homology $H_q(X,A;M)$ and cohomology $H^q(X,A;M)$ with coefficients in an abelian group M, or more generally, a k-module M which is useful for certain considerations. When we write, for example, $H^q(X;\mathbb{Z}/n)$, this can mean either cohomology as a $\mathbb{Z}/n$-module or cohomology group with coefficients in the abelian group $\mathbb{Z}/n$.

4 Construction and Calculation of Homology and Cohomology

4.1. Singular Theory The most important construction of homology and cohomology starts with the so-called singular simplices into a space as basis elements of chain groups. The linear forms on chain groups are the cochain groups, and the boundary map d on chains under transpose becomes the coboundary map δ on the cochains. Then, homology is the quotient $H_*(X) = \ker(d)/\mathrm{im}(d)$ and cohomology is the quotient $H^*(X) = \ker(\delta)/\mathrm{im}(\delta)$. In the process of defining these concepts precisely, one shows that $dd = 0$ and hence also $\delta\delta = 0$ which yields the relevant inclusions $\mathrm{im}(d) \subset \ker(d)$ and $\mathrm{im}(\delta) \subset \ker(\delta)$.

4.2. deRham Theory Let M be a smooth manifold, and let $A^q(M)$ denote the real or complex vector space of differential q-forms. In local coordinates $x_1,\ldots,x_n$, these are expressions of the form

$$\theta = \sum_{i(1)<\ldots<i(q)} a_{i(1),\ldots,i(q)} dx_{i(1)} \wedge \ldots \wedge dx_{i(q)}.$$

The algebra of differential forms centers around the basic relation $dx_i \wedge dx_j = -dx_j \wedge dx_i$ which is equivalent to $dx_i \wedge dx_i = 0$. The differential $d\theta$ is defined using $df = \sum_i \frac{\partial f}{\partial x_i} dx_i$, and it is given by the formula

$$d\theta = \sum_{i(1)<\ldots<i(q)} (da_{i(1),\ldots,i(q)}) \wedge dx_{i(1)} \wedge \ldots \wedge dx_{i(q)}.$$

Then, the differential on forms $d : A^q(M) \to A^{q+1}(M)$ is a coboundary map with $dd = 0$. From $A^q(M)$ we form the subquotient

$$H^q_{DR}(M) = \frac{\ker(d)}{\mathrm{im}(d)}$$

which is called qth deRham cohomology group. It is a real or complex vector space depending on whether the forms are real or complex.

deRham cohomology is an analytical definition of cohomology while singular theory is topological. Both play a fundamental role in topology and geometry. For further reading see Spanier (1963) or Warner (1971).

As for calculating homology and cohomology of a space, we use the fact that we know the homology and cohomology of the spheres and hence, also the following sum formulas.

4.3. Wedge Product Formulas For homology, we have

$$\tilde{H}_q(X_1 \vee \ldots \vee X_m) = \tilde{H}_q(X_1) \oplus \ldots \oplus \tilde{H}_q(X_m),$$

and for cohomology, we have

$$\tilde{H}^q(X_1 \vee \ldots \vee X_m) = \tilde{H}^q(X_1) \oplus \ldots \oplus \tilde{H}^q(X_m).$$

These formulas follow from the cofibre character of the wedge product.

4.4. Cellular Chains for a Finite Complex Let X be a finite CW-complex so that $X_n = X$ for some n, and the minimal one is the dimension zero and for each m, the number of spheres in the wedge decomposition of X_m/X_{m-1} is finite denoted by $c(m)$. In particular, by (4.3), we have

$$\tilde{H}_q(X_m/X_{m-1}) = \begin{cases} k^{c(m)} & \text{if} \quad q = m \\ 0 & \text{if} \quad q \neq m \end{cases}$$

and

$$\tilde{H}^q(X_m/X_{m-1}) = \begin{cases} k^{c(m)} & \text{if} \quad q = m \\ 0 & \text{if} \quad q \neq m \end{cases}$$

Consider several stages in the filtration of X at once from X_{m-2} to X_{m+1}, and use the exact sequence in homology with the following k-modules to define what is called the cell complex of X

$$H_{m+1}(X_{m+1},X_m) \xrightarrow{d} H_m(X_m,X_{m-1}) \xrightarrow{d} H_{m-1}(X_{m-1},X_{m-2})$$
$$\searrow \quad \nearrow \qquad\qquad \searrow \quad \nearrow ,$$
$$\tilde{H}_m(X_m) \qquad\qquad \tilde{H}_{m-1}(X_{m-1})$$

with $C_m(X) = H_m(X_m,X_{m-1})$ and differential $d : C_m(X) \to C_{m-1}(X)$. We leave it to the reader to check $dd = 0$, prove $H_m(C_*(X)) \cong H_m(X)$, and to set up $C^*(X)$, the corresponding cell complex for cohomology.

4.5. Example Let X be a finite cell complex with the dimension of the cells spaced in the sense that if $X_m - X_{m-1}$ is nonempty, then $X_{m-2} = X_{m-1}$ and $X_m = X_{m+1}$ or in terms of the number of m cells $c(m)$, that if $c(m) \neq 0$, then $c(m-1) = c(m+1) = 0$. Then, the homology and cohomology can be calculated as $H_m(X) = H^m(X) = k^{c(m)}$ since the differentials in the cell complex are zero.

4.6. Relation Between Homology and Cohomology Let M be a k-module, and let X be a space. If either k is a field or if X is a space with $H_{n-1}(X,k) = 0$, then $H^n(X,M) = \text{Hom}_k(H_n(X,k),M)$. The corresponding statement is true on the chain level, and under either of these hypothesis, it holds on the homology level.

5 Hurewicz Theorem

In this section, all homology is over the integers.

5.1. Hurewicz Map Let $\tilde{H}_n(S^n) = \mathbb{Z}i_n$ be a choice of a generator of the infinite cyclic reduced homology group of S^n. For a pointed space X, the Hurewicz map $\phi : \pi_q(X) \to \tilde{H}_q(X)$ is defined by

$$\phi([w]) = H_q(w)(i_q).$$

In other words, a map defining the element in the homotopy group is used homologically to carry the fundamental class i_q to a homology class in $\tilde{H}_q(X)$.

5.2. Remark For $q > 0$, the Hurewicz map is a group morphism.

5.3. Theorem *(Poincaré) Let X be a pointed path connected space. The Hurewicz morphism $\phi : \pi_1(X) \to H_1(X)$ is a surjective group homomorphism whose kernel is the commutator subgroup of $\pi_1(X)$.*

For the proof of (5.3), we use a covering space corresponding to the commutator subgroup of $\pi_1(X)$.

This result was extended by Hurewicz to the following theorem.

5.4. Theorem *For a path connected space with abelian fundamental group $\pi_1(X)$, the following are equivalent: $\pi_i(X) = 0$ for $i < m$ if and only if reduced homology $H_i(X) = 0$ for $i < m$, and moreover, the Hurewicz map*

(1) $\phi : \pi_m(X) \to H_m(X)$ is an isomorphism and
(2) $\phi : \pi_{m+1}(X) \to H_{m+1}(X)$ is an epimorphism.

For the proof of (5.4), we use induction on m starting with the Poincaré theorem together with the loop space fibration $\Omega X \to PX \to X$, where the $\pi_{i-1}(\Omega X) = \pi_i(X)$ and the same holds for homology in a range near m.

6 Representability of Cohomology by Homotopy Classes

6.1. Definition An Eilenberg–MacLane space $K(G,n)$ is a space which can be given the structure of a CW-complex and whose homotopy groups are given by

$$\pi_i(K(G),n)) = \begin{cases} G & \text{if} \quad n = i \\ 0 & \text{if} \quad n \neq i, \end{cases}$$

The above data determine the homotopy type of $K(G,n)$.

By the Hurewicz theorem, $\phi : G \to H_n(K(G,n))$ is an isomorphism of abelian groups for $n > 1$, and for $n = 1$ it is the abelianization of the group G. There is a canonical cohomology class $\iota_n \in H^n(K(G,n),G)$ corresponding to the identity morphism in $\mathrm{Hom}(G,G)$, where

$$\mathrm{Hom}(G,G) = \mathrm{Hom}(H_n(G,n),G) = H^n(K(G,n),G)$$

by the universal coefficient theorem (4.6) for abelian groups.

Now as a sort of dual to the Hurewicz map, we have the following.

6.2. Definition The Eilenberg–MacLane map

$$\psi : [X, K(G,n)] \longrightarrow H^n(X,G)$$

is defined by $\psi([w]) = H^n(w)(\iota_n)$.

In other words, a map defining the element in the cohomology group is used cohomologically to carry the fundamental class ι_n to the corresponding cohomology class in $H^n(X,G)$.

6.3. Theorem *The Eilenberg–MacLane map is an isomorphism for a CW-complex X.*

For the proof of (6.3), we use induction on the skeletons X_m which are CW-complexes of dimension $\leq m$. Then, it is all zero up to $m < n$, and at $m = n$, $n + 1$, the result follows by consideration of the three cofibre maps $X_{n-1} \to X_n$, $X_n \to X_{n+1}$, and $X_{n+1} \to X$ and their commutation properties with the Eilenberg–MacLane morphism.

7 Products of Cohomology and Homology

Let k denote a commutative ring, for example, $\mathbb{Z}$, $\mathbb{Z}/n$, $\mathbb{Q}$, $\mathbb{R}$, or $\mathbb{C}$. The cohomology groups $H^i(X)$ or $H^i(X,A)$ shall refer to cohomology with coefficients in k (cf. 3.11).

7.1. Cup Product There is a natural pairing $H^p(X) \times H^q(X) \to H^{p+q}(X)$ on the cohomology over k. The image of a pair (a,b) is called the cup product of a and b and is denoted ab or $a \smile b$.

The naturality means that for $a \in H^p(X)$, $b \in H^q(X)$, and $f : Y \to X$, we have $f^*(ab) = f^*(a)f^*(b)$. The extension to relative cohomology has the form of a natural pairing

$$H^p(X,A') \times H^q(X,A'') \longrightarrow H^{p+q}(X,A' \cup A'').$$

7.2. Algebraic Properties of the Cup Product The cup product is associative and has a unit $1 \in H^0(X)$ which comes from the cochain with value 1 on each singular 0-simplex. The cup product is graded commutative meaning that for $a \in H^p(X)$, $b \in H^q(X)$ one has

$$ab = (-1)^{pq}ba.$$

This is the same graded commutativity that we have in a graded exterior algebra. It means that for p odd that $2a^2 = 0$, and when there is no 2 torsion, we have

$a^2 = 0$. In particular, a polynomial ring will have to be on an even generator or it will have to be over a ring with $2 = 0$. This is illustrated by the following examples of cohomology rings.

7.3. Example

(1) For the sphere S^q, the cohomology ring $H^*(S^q, k)$ is the exterior algebra $E[a; q]$ on one generator a in degree q with $a^2 = 0$. Here, $H^0(S^q) = k$ and $H^q(S^q) = k.a$ while $H^i(S^q) = 0$ for $i \neq 0, q$.
(2) For the infinite real projective space $P_\infty(\mathbb{R})$, the cohomology over the field $\mathbb{F}_2 = \mathbb{Z}/2$ of two elements is the polynomial algebra $H^*(P_\infty(\mathbb{R}), \mathbb{F}_2) = \mathbb{F}_2[w]$ with generator w in degree 1.
(3) For the infinite complex projective space $P_\infty(\mathbb{C})$, the cohomology over any k is the polynomial algebra $H^*(P_\infty(\mathbb{R}), k) = k[c]$ with generator c in degree 2.
(4) For the infinite quaternionic projective space $P_\infty(\mathbb{H})$, the cohomology over any k is the polynomial algebra $H^*(P_\infty(\mathbb{H}), k) = k[v]$ with generator v in degree 4.

In the next chapter, we will see that example (2) has to do with characteristic classes of real vector bundles, called Stiefel–Whitney classes, and example (3) has to do with characteristic classes of complex vector bundles, called Chern classes. The proof that these are polynomial algebras can be derived from Poincaré duality. For Poincaré duality in Chap. 11, we need another product called the cap product mixing cohomology and homology for a good formulation of the duality map.

7.4. Cap Product The cap product is a natural pairing over k which is defined by $H^i(X) \times H_n(X, A) \to H_{n-i}(X, A)$. For $c \in H^i(X)$ and $u \in H_n(X, A)$ the image under the cap product is denoted by $c \frown u$. For an explicit description of the cap product, we refer to Spanier's book.

The naturality will be explained further in the context of duality. Its relation to the cup product is contained in the next remark.

7.5. Cup–Cap Relation For $u \in H_n(X, A)$ and $a \in H^p(X), b \in H^q(X)$, we have the relation $(a \smile b) \frown u = a \frown (b \frown u)$ in $H_{n-p-q}(X, A)$. Also for $i = n$ and $H_0(X, A) = k$, we have simple evaluation $\langle c, u \rangle = c \frown u$ of a cocycle on a chain.

8 Introduction to Morse Theory

In (1.1), we introduced the notion of filtered space, and Morse theory generates from the following construction filtered manifolds which have a cell decomposition up to homotopy. Hence, there are cases where there is a down to earth calculation of homology.

8.1. Example Let $f : X \to [0, +\infty)$ be a continuous function, and let $0 \le t_0 < t_1 < \ldots < t_n < \ldots$ be an increasing sequence of real numbers converging to infinity. Then, the filtration defined by f and the sequence $(t_n)_{n \in \mathbb{N}}$ is given by

$$X_0 = f^{-1}([0, t_1]) \subset \ldots \subset X_p = f^{-1}([0, t_p]) \subset \ldots \subset X_\infty = f^{-1}([0, t_\infty)).$$

It is difficult, in general, to say anything about the layers (X_p, X_{p-1}) of this filtration without further information about the function. Morse functions are a class of functions where more information is available about such filtrations defined by positive real-valued functions.

8.2. Notation Let $f : M \to \mathbb{R}$ be a smooth real-valued function on a smooth manifold with tangent bundle $T(M)$. The derivative at $x \in M$ is a linear map $T(f)_x : T(M)_x \to \mathbb{R}$, and when $T(f)_x = 0$, the second derivative at $x \in M$ is defined as a bilinear map

$$T^2(f)_x : T(M)_x \times T(M)_x \longrightarrow \mathbb{R}.$$

8.3. Definition Let $f : M \to \mathbb{R}$ be a smooth real-valued function on a smooth manifold M. A critical point x of f is an $x \in M$, where $T(f)_x = 0$, and the associated critical value is $y = f(x) \in \mathbb{R}$. A critical point $x \in M$ is nondegenerate provided the bilinear second derivative $T^2(M)_x$ is a nondegenerate bilinear form. The index of f at critical point is the number of negative eigenvalues of the $T^2(f)$.

If $x \in M$ is a minimum or maximum of f, then x is a critical point, and further if f is nondegenerate at a minimum x (resp. maximum x), then the index of f is 0 (resp. $\dim(M)$).

8.4. Definition A Morse function on a compact, smooth manifold M is a smooth function $f : M \to \mathbb{R}$ with only a finite number of critical points all of which are nondegenerate. A nice Morse function F is one where the critical point of index i has critical value $i \in \mathbb{R}$.

There is a version of this definition for noncompact manifolds and even infinite dimensional ones due to Palais and Smale. This is treated in Schwarz (1993). The two basic references for Morse theory are Milnor (1963) and Milnor (1965). In particular, in these books, one finds a proof of the following theorem.

8.5. Theorem *Every compact smooth manifold M admits a nice Morse function.*

The existence of a Morse function is a partition of unity argument while to change it to a nice Morse function is a much deeper result.

Applying the scheme in (8.1) with the following proposition, we see how a Morse function leads to a cell decomposition of a smooth manifold up to homotopy.

8.6. Proposition *Let $f : M \to \mathbb{R}$ be a smooth real-valued function on a smooth manifold M such that there are only q critical points in the subset $f^{-1}([t - \varepsilon, t + \varepsilon])$ and all are of index i. Then $f^{-1}((-\infty, t + \varepsilon])$ has the homotopy type of $f^{-1}((-\infty, t - \varepsilon])$ with q cells of dimension i adjoined by maps $S^{i-1} \to f^{-1}(t - \varepsilon) \subset f^{-1}((-\infty, t - \varepsilon])$.*

8.7. Remark In fact, a Morse function not only can be used to control the homotopy type of M but also its diffeomorphism type. The attachment of the sphere in the previous proposition can be done up to diffeomorphism by thickening $S^{i-1} \subset S^{i-1} \times D^{n-i}$ and attaching $D^i \times D^{n-i}$, called a handle. This is also explained very well in the books of Milnor cited above.

References

Milnor, J.: Morse theory. Ann. of Math. Stud. **51** (1963)

Milnor, J.: Lectures on the h-Cobordism Theorem. Princeton University Press, Princeton, NJ (1965)

Schwarz, M.: Morse Homology. Birkhäuser, Basel (1993)

Spanier, E.: Algebraic Topology. McGraw-Hill, New York (1963)

Warner, F.: Foundations of Differential Manifolds and Lie Groups. Springer-Verlag, New York (1984) (reprint of the1971 edition.)

Whitehead, G.: Elements of homotopy theory. Springer Graduate Texts, Vol. 61. Springer Verlag, New York (1978)

Chapter 10
Basic Characteristic Classes

The theory of characteristic cohomology classes of bundles especially vector bundles grew up in several contexts like the notion of cohomology of a space. It was natural to try and make calculations with bundles in terms of cohomology for two reasons. With homology and cohomology, there were combinatorial tools for computation. Then, cohomology and bundles each had contravariant properties under continuous mappings. The first definitions of characteristic classes were given by obstructions to existence to cross sections of a bundle or related fibre bundle. Examples of this can be found in the book by Steenrod (1951).

With the understanding of the cohomology of fibre spaces including the Leray spectral sequence, the theory of characteristic classes was developed along more intrinsic lines. It was clear that the separate theories of Chern classes for complex vector bundles and of Stiefel–Whitney classes of real vector bundles had a parallel structure. This became completely clear with Hirzebruch's axiomatization of the Chern classes which carries over immediately to Stiefel–Whitney classes. In the first sections, we carry this out using an approach of Grothendieck which also works in the context of algebraic geometry.

Then, we have the Euler class and Pontrjagin classes which are introduced also by elementary fibre space methods. Pontrjagin classes can be related nicely to Chern classes. Then, it becomes clear that with splitting principles for vector bundles into line bundles that families of characteristic classes can be introduced from power series by their properties on line bundles. From this, we generate important classes in the theory of manifolds, that is, the Todd class, the L-class, and the $\hat{A}$-class, and these classes evaluated on the fundamental homology class of a manifold lead to specific characteristic numbers. This is considered in the next chapter.

Chapter 17 of *Fibre Bundles* (Husemöller 1994) is a reference for this chapter. The topic is also treated in Milnor and Stasheff (1974).

1 Characteristic Classes of Line Bundles

The theory of characteristic classes starts with the observation that the same space can be the classifying space for bundles, so of the form $B(G)$, and for cohomology

D. Husemöller et al.: *Basic Characteristic Classes*, Lect. Notes Phys. **726**, 111–125 (2008)
DOI 10.1007/978-3-540-74956-1_11 © Springer-Verlag Berlin Heidelberg 2008

classes, so of the form $K(\Pi,n)$. There are two cases of this coincidence. The first is related with real vector bundles while the second is related to complex vector bundles.

1.1. The Real Infinite-Dimensional Projective Space $P_\infty(\mathbb{R})$ It classifies three types of objects:

(1) real line bundles over a space X, for $P_\infty(\mathbb{R})$ has a universal real line bundle,
(2) double coverings of a space X, so $P_\infty(\mathbb{R}) = B(\{\pm 1\})$, and
(3) one-dimensional cohomology classes on X with values in the group $\mathbb{Z}/2\mathbb{Z}$ or $\{\pm 1\}$ of two elements, so $P_\infty(\mathbb{R}) = K(\mathbb{Z}/2\mathbb{Z}, 1)$.

A real line bundle L on a space X is classified by a homotopy class of maps $u : X \to P_\infty(\mathbb{R}) = K(\mathbb{Z}/2\mathbb{Z}, 1)$. The related cohomology class in $H^1(X, \mathbb{Z}/2\mathbb{Z}) = [X, K(\mathbb{Z}/2\mathbb{Z}, 1)]$, denoted by $w_1(L)$, is called the first Stiefel–Whitney class of L.

1.2. The Complex Infinite-Dimensional Projective Space $P_\infty(\mathbb{C})$ It classifies three types of objects:

(1) complex line bundles over a space X, for $P_\infty(\mathbb{C})$ has a universal complex line bundle,
(2) principal circle bundles over a space X, so that $P_\infty(\mathbb{C}) = B(S^1)$, and
(3) two-dimensional cohomology classes on X with values in the group $\mathbb{Z}$ of integers, so $P_\infty(\mathbb{C}) = K(\mathbb{Z}, 2)$.

A complex line bundle L on a space X is classified by a homotopy class of maps $u : X \to P_\infty(\mathbb{C}) = K(\mathbb{Z}, 2)$. The related cohomology class in $H^2(X, \mathbb{Z}) = [X, K(\mathbb{Z}, 2)]$, denoted by $c_1(L)$, is called the first Chern class of L.

1.3. Notation Let $\mathrm{Pic}_\mathbb{C}(X)$ (resp. $\mathrm{Pic}_\mathbb{R}(X)$) denote the group of isomorphism classes of complex (resp. real) line bundles on X with group operation induced by the tensor product. The trivial line bundle is the unit, and the dual is the negative.

1.4. Remark Then with the induced line bundle, we define two functors into abelian groups

$$\mathrm{Pic}_\mathbb{C}, \mathrm{Pic}_\mathbb{R} : (\mathrm{top})^{\mathrm{op}} \longrightarrow (\mathrm{ab}).$$

The characteristic classes w_1 and c_1 define isomorphisms of functors

$$w_1 : \mathrm{Pic}_\mathbb{R}() \longrightarrow H^1(, \mathbb{Z}/2\mathbb{Z}) \quad \text{and} \quad c_1 : \mathrm{Pic}_\mathbb{C}() \longrightarrow H^2(, \mathbb{Z})$$

defined on the opposite category of topological spaces to the category of abelian groups. This is the essentially unique situation where the characteristic class completely determines the bundle.

In higher dimensions, there are more characteristic classes of a vector bundle, but they are usually not enough characteristic classes to classify the isomorphism class of the vector bundles. In the next section, we introduce the fibre space results which are used to define general characteristic classes in terms of the characteristic class of line bundles.

2 Projective Bundle Theorem and Splitting Principle

2.1. Definition Let E be a vector bundle over X. Let $q : P(E) \to X$ be the associated bundle of projective spaces, where the fibre $q^{-1}(x) = P(E)_x$ is the projective space of linear forms on the fibre E_x.

Observe that if $E|U$ is trivial, then $q : P(E|U) \to U$ is trivial, and if $E = f^{-1}(E')$ is an induced bundle, then $q : P(E) \to X$ is the induced associated bundle $f^{-1}(P(E'))$ of projective spaces.

2.2. Remark On the associated bundle of projective spaces $q : P(E) \to X$, there is a line bundle $L_E \to P(E)$ with the property that $L_E|P(E)_x$ is the canonical line bundle on the projective space $P(E)_x$ for each $x \in X$.

The next two theorems, the first for complex vector bundles and the second for real vector bundles, are easy consequences of the cohomology spectral sequence for a fibre map q, but they can also be proved for bundles of finite type with an inductive Mayer–Vietoris argument (i.e., by iteratively using (3.10)).

2.3. Theorem *The associated bundle of projective spaces $q : P(E) \to X$ for an n-dimensional complex vector bundle $p : E \to X$ has an injective integral cohomology morphism $q^* : H^*(X) \to H^*(P(E))$ making $H^*(P(E))$ into a free $H^*(X)$-module with basis*

$$1, c_1(L_E), \ldots, c_1(L_E)^{n-1}.$$

2.4. Theorem *The associated bundle of projective spaces $q : P(E) \to X$ for an n-dimensional real vector bundle $p : E \to X$ has an injective mod 2 cohomology morphism $q^* : H^*(X, \mathbb{F}_2) \to H^*(P(E), \mathbb{F}_2)$ making $H^*(P(E), \mathbb{F}_2)$ into a free $H^*(X, \mathbb{F}_2)$-module with basis*

$$1, w_1(L_E), \ldots, w_1(L_E)^{n-1}.$$

The following two theorems are useful for uniqueness results and for doing certain calculations.

2.5. Theorem *For an arbitrary complex vector bundle E over a space B, we have a map $f : B' \to B$ with two properties:*

(1) $f^* : H^*(B, \mathbb{Z}) \to H^*(B', \mathbb{Z})$ is injective and
(2) $f^*(E) = \oplus_i L_i$, where the L_i are line bundles over B'.

2.6. Theorem *For an arbitrary real vector bundle E over a space B we have a map $f : B' \to B$ with two properties:*

(1) $f^* : H^*(B, \mathbb{F}_2) \to H^*(B', \mathbb{F}_2)$ is injective and
(2) $f^*(E) = \oplus_i L_i$, where the L_i are line bundles.

For the proofs of theorems (2.5) and (2.6), we use induction on the dimension of the vector bundle E. For line bundles, there is nothing to prove. Using the associated bundle of projective spaces $q : P(E) \to B$ for the n-dimensional vector bundle $p : E \to B$, we have $q^{-1}(E) = L_E \oplus Q_E$, where L_E is the canonical line bundle on $P(E)$

and Q_E is a quotient which splits off by direct sum. Since $\dim Q_E < \dim E$, we can use induction to obtain a map $g : B' \to P(E)$ with the two properties (1) and (2) of the respective theorems. Then, $f = qg$ is the desired splitting map. Here, we use (2.3) for (2.5) and (2.4) for (2.6).

3 Chern Classes and Stiefel–Whitney Classes of Vector Bundles

3.1. Definition A theory of Chern classes for complex vector bundles assigns to each complex vector bundle E over a space B cohomology classes $c_i(E) \in H^{2i}(B,\mathbb{Z})$ satisfying the axioms:

(1) $c_i(E) = 1$ for $i = 0$ and $c_i(E) = 0$ for $i > \dim(E)$.
(2) For a map $f : B' \to B$ and a vector bundle E over B, we have $c_i(f^{-1}(E)) = f^*(c_i(E))$ in $H^{2i}(B',\mathbb{Z})$
(3) For the Whitney sum, we have the relation

$$c_i(E' \oplus E'') = \sum_{i=j+k} c_j(E')c_k(E'')$$

using the cup product on cohomology, where

$$H^{2j}(B,\mathbb{Z}) \otimes H^{2k}(B,\mathbb{Z}) \to H^{2i}(B,\mathbb{Z})$$

(4) For a line bundle L over B, the first Chern class

$$c_1(L) \in H^2(B,\mathbb{Z})$$

is defined by representing both line bundle isomorphism classes and two-dimensional integral cohomology classes by $K(\mathbb{Z},2) = P_\infty(\mathbb{C})$.

The Chern classes exist and are unique by the projective bundle theorem and the splitting principle.

3.2. Uniqueness of the Chern Classes The first Chern class of a line bundle is unique by axiom (4). For an arbitrary complex vector bundle E over a space B, we have a map $f : B' \to B$ with two properties, namely that $f^* : H^*(B,\mathbb{Z}) \to H^*(B',\mathbb{Z})$ is injective and that $f^*(E) = \oplus_i L_i$, where the L_i are line bundles. Now, the mth Chern class $c_m(E)$ of E must map under an injection f^* to $f^*c_m(E) = c_m(\oplus_i L_i)$ by axiom (2). By axiom (3), the Whitney sum formula, this must be

$$c_m(\bigoplus_i L_i) = \sum_{i(1)<...<i(r)} c_1(L_{i(1)})...c_1(L_{i(r)}),$$

and since only first Chern classes of line bundles appear in the formula, it is well defined by axiom 4. Since f^* is injective on cohomology, the Chern class is uniquely determined by the axioms.

The Grothendieck definition of the Chern classes follows by considering the relation of linear dependence of powers of the first Chern class in the projective bundle theorem for an arbitrary complex vector bundle E over a space B.

3.3. Definition of the Chern Classes The associated bundle of projective spaces $q : P(E) \to X$ for an n-dimensional complex vector bundle $p : E \to X$ has an injective integral cohomology morphism

$$q^* : H^*(X) \longrightarrow H^*(P(E))$$

making $H^*(P(E))$ into a free $H^*(X)$-module with basis

$$1, c_1(L_E), \ldots, c_1(L_E)^{n-1}.$$

In particular, $c_1(L_E)^n$ is a unique linear combination of these basis elements with coefficients which we define to be the Chern classes $c_i(E)$ of E. More precisely, we have the relation

$$c_1(L_E)^n = c_n(E) + c_{n-1}(E)c_1(L_E) + \ldots + c_1(E)c_1(L_E)^{n-1}.$$

3.4. Remark Now, we have to check the axioms (1)–(4) in (3.1). As for axiom (1), it is immediate from the formula, and for axiom (4), the defining relation reduces to $c_1(L_E) = c_1(E)$ for a line bundle E which then $E = L_E$. The functoriality property (2) follows from the functoriality of the associated bundle $q : P(E) \to B$ of projective spaces, the line bundle L_E, and $c_1(L_E)$. For the Whitney sum axiom, we begin with the following space case.

3.5. Proposition *Let $E = L_1 \oplus \ldots \oplus L_n$ be a Whitney sum of complex line bundles over B. Then, we have the relation*

$$1 + c_1(E) + \ldots + c_n(E) = (1 + c_1(L_1)) \ldots (1 + c_1(L_n)) \quad \text{in} \quad H^{ev}(B, \mathbb{Z}).$$

Proof. On $P(E)$, we have two exact sequences

$$0 \longrightarrow L_E \longrightarrow q^{-1}(E) = q^{-1}(L_1) \oplus \ldots \oplus q^{-1}(L_n) \longrightarrow Q_E \longrightarrow 0,$$

and tensoring with the dual $L_E^\vee$, we have

$$0 \to L_E \otimes L_E^\vee \longrightarrow (q^{-1}(L_1) \otimes L_E^\vee) \oplus \ldots \oplus (q^{-1}(L_n) \otimes L_E^\vee) \longrightarrow Q_E \otimes L_E^\vee \to 0.$$

The line bundle $L_E \otimes L_E^\vee$ has an everywhere nontrivial section σ which induces on each of the n summands a nontrivial section σ_i which is everywhere nonzero on an open subset V_i of $P(E)$. Then, the first Chern class $c_1(q^{-1}(L_i) \otimes L_E^\vee) = c_1(q^{-1}(L_i)) - c_1(L_E) = 0$ in $H^2(P(E), V_i; \mathbb{Z})$. The product over $i = 1, \ldots, n$ is zero in $H^2(P(E), \mathbb{Z})$ giving the relation $0 = \prod_{i=1}^n (c_1(q^{-1}(L_i)) - c_1(L_E))$, which compared to $c_1(L_E)^n = c_n(E) + c_{n-1}(E)c_1(L_E) + \ldots + c_1(E)c_1(L_E)^{n-1}$ gives the assertion of the proposition.

3.6. Theorem *The Chern classes for complex vector bundles exist satisfying axioms (1)–(4) of (3.1). They are uniquely determined by the axioms.*

Proof. It remains to check the Whitney sum axiom (3). For two vector bundles E' and E'' over B, we apply the splitting principle two times to obtain a map $f : B' \to B$ inducing a monomorphism in integral cohomology and such that $f^{-1}(E') = L_1 \oplus \ldots \oplus L_m$ and $f^{-1}(E'') = L_{m+1} \oplus \ldots \oplus L_n$. Now, we use the previous proposition for the sum of line bundles $f^{-1}(E') \oplus f^{-1}(E'') = f^{-1}(E' \oplus E'')$. Then, we have $c_i(f^{-1}(E' \oplus E'')) = \sum_{i=j+k} c_j(f^{-1}(E'))c_k(f^{-1}(E''))$ in $H^*(B', \mathbb{Z})$, and this implies the Whitney sum property

$$c_i(E' \oplus E'') = \sum_{i=j+k} c_j(E')c_k(E'')$$

in $H^*(B, \mathbb{Z})$. This proves the theorem.

The Stiefel–Whitney characteristic classes $w_i(E)$ of a real vector bundle E over B are elements of $H^i(B, \mathbb{F}_2)$. They have the following axiomatic characterization which parallels the axioms of the Chern classes.

3.7. Definition A theory of Stiefel–Whitney classes for real vector bundles assigns to each real vector bundle E over a space B cohomology classes $w_i(E) \in H^i(B, \mathbb{F}_2)$ satisfying the axioms:

(1) $w_i(E) = 1$ for $i = 0$ and $w_i(E) = 0$ for $i > \dim(E)$.
(2) For a map $f : B' \to B$ and a vector bundle E over B, we have $w_i(f^{-1}(E)) = f^*(w_i(E))$ in $H^i(B', \mathbb{F}_2)$
(3) For the Whitney sum, we have the relation

$$w_i(E' \oplus E'') = \sum_{i=j+k} w_j(E')w_k(E'')$$

using the cup product on cohomology where

$$H^j(B, \mathbb{F}_2) \otimes H^k(B, \mathbb{F}_2) \to H^i(B, \mathbb{F}_2).$$

(4) For a line bundle L over B, the first Stiefel–Whitney class

$$w_1(L) \in H^1(B, \mathbb{F}_2)$$

is defined by representing both line bundle isomorphism classes and two-dimensional integral cohomology classes by $K(\mathbb{Z}/2, 1) = P_\infty(\mathbb{R})$.

The Stiefel–Whitney classes exist and are unique by an argument completely parallel to that for the Chern classes.

4 Elementary Properties of Characteristic Classes

From the axioms, it is possible to derive some simple stability properties of characteristic classes.

4.1. Notation For a complex vector bundle E, we denote by

$$c(E) = 1 + c_1(E) + \ldots + c_n(E) \in G^{ev}(B, \mathbb{Z}) \subset H^{ev}(B, \mathbb{Z}) = \prod_i H^{2i}(B, \mathbb{Z}),$$

and for a real vector bundle E, we denote by

$$w(E) = 1 + w_1(E) + \ldots + w_n(E) \in G^*(B, \mathbb{F}_2) \subset H^*(B, \mathbb{F}_2) = \prod_i H^i(B, \mathbb{F}_2).$$

Here, G^{ev} or G^* denotes the subset in the direct product with zero component equal to 1. These are commutative monoids of units with the multiplication as monoid operation. The Whitney sum formula becomes

$$c(E' \oplus E'') = c(E')c(E'') \quad \text{or} \quad w(E' \oplus E'') = w(E')w(E'').$$

4.2. Remark For a trivial bundle T over B, we have $c(T) = 1$ or $w(T) = 1$. This is equivalent to $c_i(T) = 0$ or $w_i(T) = 0$ for $i > 0$. A trivial bundle on B is induced by the constant map $B \to *$, to a point $*$ where it is obvious.

4.3. Remark Two bundles E' and E'' are stably equivalent provided there exists trivial bundles T' and T'' with $E' \oplus T'$ and $E'' \oplus T''$ isomorphic. For two stably equivalent complex vector bundles E' and E'', we have $c(E') = c(E'')$, and for two stably equivalent real vector bundles E' and E'', we have $w(E') = w(E'')$. The Whitney sum relation gives a morphism of functors from the K-theory of vector bundles, that is, stable equivalence classes of vector bundles to these multiplicative monoids of cohomology

$$c : K() \longrightarrow G^{ev}(, \mathbb{Z}) \quad \text{and} \quad w : KO() \longrightarrow G^*(, \mathbb{F}_2).$$

4.4. Example For the tangent bundle $T(S^n)$ to the n-sphere S^n, we have $w(T(S^n)) = 1$ because $T(S^n) \oplus N(S^n)$ and also $N(S^n)$, the normal bundle to the sphere, is a trivial real vector bundle. Thus,

$$1 = w(\text{trivial}) = w(T(S^n) \oplus N(S^n)) = w(T(S^n))w(N(S^n))$$
$$= w(T(S^n))w(\text{trivial}) = w(T(S^n)).$$

4.5. Example The real projective space $P_n(\mathbb{R})$ comes from S^n by identifying x with $-x$. The same can be done for the tangent bundle Whitney sum with the normal bundle. By identifying x with $-x$ in the previous example (4.4), we obtain an isomorphism between $T(P_n(\mathbb{R})) \oplus N$ and $(n+1)L$, the $n+1$ Whitney sum of the canonical line bundle L on $P_n(\mathbb{R})$. Since $w(L) = 1 + z$, where $z \in H^1(P_n(\mathbb{R}), \mathbb{F}_2)$ is the canonical generator, the Stiefel–Whitney class of the tangent bundle to the real projective space is $w(T(P_n(\mathbb{R}))) = (1+z)^{n+1}$. For the Chern class of the tangent bundle to complex projective space, the same type of calculation gives $c(T(P_n(\mathbb{C}))) = (1+z)^{n+1}$, where z is a generator of $H^2(P_n(\mathbb{C}), \mathbb{Z})$.

4.6. Remark For $E = E' \oplus E''$, where E' and E'' are of dimension p and q, respectively, we have $c_{p+q}(E) = c_p(E')c_q(E'')$. If E has an everywhere nonzero cross section, then $E = E' \oplus E''$, where E'' is a trivial line bundle. Then, we see that $c_n(E) = 0$ in the complex case or $w_n(E) = 0$ in the real case. Thus, the top characteristic class must be zero for the existence of an everywhere nonzero cross section of the vector bundle. In this sense, we speak of the top characteristic class as an obstruction to the existence of an everywhere nonzero cross section of the bundle. There are interesting cases where one can assert that the vanishing of the top characteristic class implies the existence of a everywhere nonzero cross section.

5 Chern Character and Related Multiplicative Characteristic Classes

The formula for the Chern class of a sum of line bundles,

$$c_m\left(\bigoplus_{i=1} L_i\right) = \sum_{1 \le i(1) < \ldots < i(r) \le n} c_1(L_{i(1)}) \ldots c_1(L_{i(r)}),$$

shows that the mth elementary symmetric function in the first Chern classes of the line bundles in the sum are exactly the Chern class $c_m(\bigoplus_i L_i)$.

5.1. Basics on Elementary Symmetric Functions A symmetric function $f(x_1, \ldots, x_n) \in R[x_1, \ldots, x_n]$ in n variables is a polynomial with the property that $f(x_1, \ldots, x_n) = f(x_{\sigma(1)}, \ldots, x_{\sigma(n)})$ for any permutation σ of n objects. The elementary symmetric function $\sigma_i(x_1, \ldots, x_n)$ of degree i is defined by the following relation $\prod_{1 \le j \le n}(z + x_j) = \sum_{0 \le i \le n} \sigma_i(x_1, \ldots, x_n)z^{n-i}$.

5.2. Fundamental Assertion The subring $R[\sigma_1, \ldots, \sigma_n]$ in the polynomial ring $R[x_1, \ldots, x_n]$ contains all the symmetric functions and the elementary symmetric functions are algebraically independent. See *Fibre Bundles*, 14(1.6), for the classical proof. Now, we use this assertion and the splitting principle of a bundle as a sum of line bundles to define characteristic classes using the fact that Chern classes of sums of line bundles are elementary symmetric functions in the Chern classes of the line bundles.

5.3. The Additive Construction of Symmetric Functions from a Power Series Let $f(t) \in 1 + tR[[t]]$ be a power series with leading term 1, and form the sum $f(x_1 t) + \ldots + f(x_n t)$ with variables $x_1, \ldots, x_n$. Then the sum is symmetric in the variables $x_1, \ldots, x_n$, and hence, it can be written uniquely as

$$f(x_1 t) + \ldots + f(x_n t) = \sum_{0 \le m} g_m(\sigma_1, \ldots, \sigma_n)t^m,$$

where the g_m are polynomial over R in n variables with the elementary symmetric functions substituted by the fundamental assertion in (5.2).

5.4. Definition The Chern character $ch(E)$ of a vector bundle E over B is a formal infinite sum $ch(E) = \sum_m ch_m(E)$ where the summands $ch_m(E) \in H^{2m}(B, \mathbb{Q})$ are defined as follows. For the exponential series $f(t) \in 1 + t\mathbb{Q}[[t]]$ and $n = \dim(E)$

$$f(t) = 1 + t + \ldots + \frac{t^q}{q!} + \ldots,$$

we introduce the related polynomials $g_m(\sigma_1, \ldots, \sigma_n)$ over $\mathbb{Q}$ as in (5.3) and $ch_m(E) = g_m(c_1, \ldots, c_n)$.

Since the Chern character can have infinitely many terms $ch_m(E)$, we introduce the following notation.

5.5. Notation Let $H^{**}(X, R) = \prod_n H^n(X, R)$ be the product cohomology R-module which contains $\bigoplus_n H^n(X, R)$ and each summand $H^n(X, R)$ of degree n. Let

$$H^{ev}(X, R) = \prod_{2n} H^{2n}(X, R)$$

be the corresponding product of the even degree cohomology groups. Since $ch(T) = n$ for the trivial bundle T of dimension n, we see that $ch : K(X) \to H^{ev}(X, \mathbb{Q})$ is defined.

5.6. Proposition *The Chern character*

$$ch : K(X) \longrightarrow H^{ev}(X, \mathbb{Q})$$

is a ring morphism.

Proof. The Chern character is defined to be additive, and for the multiplicative property, it is sufficient to check on line bundles by the splitting principle.

Considerations on the Chern character were basic in Atiyah and Hirzebruch (1961). The following theorem is also proved there.

5.7. Theorem *After tensoring with the rational numbers, we have an isomorphism*

$$ch : K(X) \otimes \mathbb{Q} \longrightarrow H^{**}(X, \mathbb{Q})$$

of functors.

A reference for this is Chap. 19 in *Fibre Bundles* (Husemöller 1994) and Hirzebruch (1962). This results also follows from localizing BU to $BU_{\mathbb{Q}} = \prod_m K(\mathbb{Q}, 2m)$ at the rational numbers. Here, one uses the fact that the rationalization $Y_{\mathbb{Q}}$ of an arbitrary H-space Y always is a product of Eilenberg–MacLane spaces.

5.8. The Multiplicative Construction of Symmetric Functions from a Power Series Let $\phi(t) \in 1 + tR[[t]]$ be a power series with leading term 1, and form

$\phi(x_1 t) \ldots \phi(x_n t)$ with variables $x_1, \ldots, x_n$ and t as a product. Then, the product is symmetric in the variables $x_1, \ldots, x_n$, and hence, it can be written uniquely as

$$\phi(x_1 t) \ldots \phi(x_n t) = \sum_{0 \leq m} \psi_m(\sigma_1, \ldots, \sigma_n) t^n,$$

where the ψ_m are polynomial over R in n variables with the elementary symmetric functions substituted in by the fundamental assertion in (5.2).

5.9. Definition The Todd class $Td(E)$ of a complex vector bundle E over B is a formal infinite sum $Td(E) = \sum_m Td_m(E)$, where the summands $Td_m(E) \in H^{2m}(B, \mathbb{Q})$ are defined as follows. For the series $\phi(t) \in 1 + t\mathbb{Q}[[t]]$ and $n = \dim(E)$

$$\phi(t) = \frac{t}{1 - \exp(-t)},$$

we introduce the related polynomials $\psi_m(\sigma_1, \ldots, \sigma_n)$ over $\mathbb{Q}$ as in (5.8) and $Td_m(E) = \psi_m(c_1, \ldots, c_n)$.

Since the Todd class can have infinitely many terms, we have $Td(E) \in H^{**}(X, \mathbb{Q})$.

5.10. Remark This class is basic for complex manifolds, because a complex manifold X of dimension n has a basic homology class $[X] \in H_{2n}(X, \mathbb{Z})$, and the number $Td(T(X)))[X] = Td_n(T(X))[X]$ is called the Todd genus of the manifold. In the next chapter, we see that the characteristic classes of tangent bundles of manifolds carry information on the structure of the manifold.

5.11. Remark The defining series for the Todd class and Todd genus is related to two other defining series which are used for real manifolds with additional structure, that is,

$$x + \frac{x}{\tanh(x)} = e^x \frac{x}{\sinh(x)} = \frac{2x}{1 - \exp(-2x)}.$$

This identity between three transcendental functions can be seen directly in the following form

$$x\left(1 + \frac{e^x + e^{-x}}{e^x - e^{-x}}\right) = x\frac{2e^x}{e^x - e^{-x}} = \frac{2x}{1 - e^{-2x}}.$$

The two other series mentioned are

$$\frac{\sqrt{z}}{\tanh \sqrt{z}} \quad \text{and} \quad \frac{2\sqrt{z}}{\sinh 2\sqrt{z}}.$$

The first one generating the L-class and the L-genus for oriented manifolds, and the second one generating the $\hat{A}$-class and the $\hat{A}$-genus for oriented manifolds.

Up to now we have only mod 2 characteristic classes for real vector bundles, that is, Stiefel–Whitney classes, and we need integral classes for real vector bundles which have something to do with Chern classes. This leads us to the considerations in the next section where we consider a specific integral class for oriented real vector bundles called the Euler class.

6 Euler Class

Now we come to a theorem which for real vector bundles plays a fundamental role as (2.3) did for complex vector bundles. The best result is for oriented (real) vector bundles.

6.1. Definition A real vector bundle E of dimension n over B is orientable provided E is isomorphic to a fibre bundle $P[\mathbb{R}^n]$, where P is a principal bundle for the group $SO(n)$. In local terms, E has an atlas of charts where the linear transformation changing from one chart to another have strictly positive determinant. From these local charts of the fibre at $b \in B$, we obtain a map $j_b : (\mathbb{R}^n, \mathbb{R}^n - \{0\}) \to (E, E_0)$, where E_0 is $E - \{\text{zero section}\}$ for each $b \in B$. We choose a fixed generator of $H^n(\mathbb{R}^n, \mathbb{R}^n - \{0\})$ with coefficients in $\mathbb{Z}$ for oriented real vector bundles and in the field of two elements $\mathbb{F}_2$ in general. In the latter case though there is no choice.

6.2. Theorem *Let $p : E \to B$ be a real vector bundle of dimension n. Then there exists a unique class $U_E \in H^n(E, E_0)$ such that $j_b(U_E)$ is the fixed generator of $H^n(\mathbb{R}^n, \mathbb{R}^n - \{0\})$, and for $i < n$ all cohomology vanishes: $H^i(E, E_0) = 0$. Moreover, the function $\phi : H^i(B) \to H^{i+n}(E, E_0)$ defined by $\phi(a) = p^*(a) \smile U_E$ is an isomorphism.*

This is again an easy consequence of the spectral sequence of a fibre map or for bundles which are finitely generated, see 5(5.2). It follows from an inductive Mayer–Vietoris sequence argument.

6.3. Definition The class $U_E \in H^n(E, E_0)$ has an image under the natural mapping $H^n(E, E_0) \to H^n(E)$ which corresponds to a class $e(E) \in H^n(B)$, called the Euler class, under the isomorphism $p^* : H^*(B) \to H^*(E)$.

6.4. Theorem *(Gysin sequence) With the cup product with the Euler class $e(E)$, denoted by $\varepsilon(a) = a \smile e(E)$, we have an exact couple*

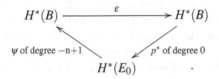

Here, ε has degree n and ψ is the composition of the coboundary $H^*(E_0) \to H^*(E, E_0)$ of degree 1 with the inverse of ϕ in (6.2) of degree $-n$.

The exact couple or Gysin sequence follows from the cohomology exact sequences and (6.2).

6.5. Remark The Euler class $e(E) = \phi^{-1}(U_E^2)$, and so for an odd dimensional vector bundle, $2e(E) = 0$. If $f : B' \to B$ is a map, then $f^*(e(E)) = e(f^{-1}(E))$.

There is a Whitney sum property as a cup product relation for the Euler class.

6.6. Theorem *We have* $e(E' \oplus E'') = e(E')e(E'') \in H^*(B)$.

Proof. We consider $E = E' \oplus E''$ and two subbundles $E_1 \subset E$ and $E_2 \subset E$ with first and second projections denoted by $q_1 : (E, E_1) \to (E', E'_0)$ and $q_2 : (E, E_2) \to (E'', E''_0)$ respectively. Then by considering the restrictions on fibres, we have the cup product relation $U_E = q_1^*(U_{E'})q_2^*(U_{E''})$. The Euler class formula follows from this by the definition (6.3).

7 Thom Space, Thom Class, and Thom Isomorphism

The Thom space is a way to replace the relative class U_E with the Thom class on a space coming from (E, E_0) and the related isomorphism $\phi(a) = p^*(a) \smile U_E$ defined $\phi : H^i(B) \to H^{i+n}(E, E_0)$ by the Thom isomorphism.

7.1. Definition Associated with a vector bundle E having a Riemannian metric, we have the disc bundle $D(E) \subset E$ of all vectors v with $||v|| \leq 1$ and the sphere bundle $S(E) \subset E_0$ of all vectors v with $||v|| = 1$.

Instead of ϕ itself, one also can work with the composition

$$H^i(B) \xrightarrow{\phi} H^{i+n}(E, E_0) \longrightarrow H^{i+n}(D(E), S(E)) \longrightarrow H^{i+n}(D(E)/S(E)),$$

where the first isomorphism is ϕ, the second is a restriction homomorphism which is induced by a homotopy equivalence, and the third is the excision isomorphism for the pair $(D(E), S(E))$.

7.2. Definition The space $D(E)/S(E)$ is the Thom space of the vector bundle E, the image of U_E in $H^n(D(E)/S(E))$ is the Thom class of E, and the composite isomorphism is the Thom isomorphism.

7.3. Remark When E is a vector bundle over a compact space B, the Thom space is isomorphic to the one-point compactification of E. If E is a vector bundle over B and if T is the trivial one-dimensional vector bundle over B, then we have a section $B \to S(E \oplus T)$, and the Thom space is isomorphic to $S(E \oplus T)/\text{im}(B)$. The natural morphism $E \to E \oplus T$ defines an inclusion of the associated projective bundles $P(E) \to P(E \oplus T)$, and the collapsed space $P(E \oplus T)/P(E)$ is another version of the Thom space.

8 Stiefel–Whitney Classes in Terms of Steenrod Operations

A basic reference for this chapter is Epstein and Steenrod (1962).

Recall (6.5) where we established the formula $e(E) = \phi^{-1}(U_E^2)$ for the Euler class. This suggests the possibility of defining more characteristic classes by applying other cohomology operations to U_E.

8.1. Definition For cohomology with coefficients in a group G, a cohomology operation of degree i is a morphism $\theta : H^*(\,,G) \to H^{*+i}(\,,G)$ of functors.

8.2. Theorem *For cohomology over the field $\mathbb{F}_2$ of two elements, there is a unique operation $Sq^i : H^*(\,,\mathbb{F}_2) \to H^{*+i}(\,,\mathbb{F}_2)$ of degree i such that Sq^i commutes with suspension and $Sq^i(x) = x^2$, the cup square, for $x \in H^i(X,\mathbb{F}_2)$.*

The operation Sq^i is called the Steenrod square.

8.3. Theorem *The Steenrod squares satisfy the following properties.*

(1) In degree 0, Sq^0 is the identity, and $Sq^i|H^n(\,,\mathbb{F}_2) = 0$ for $i > n$.
(2) (Cartan formula) For $x,y \in H^*(X \,,\mathbb{F}_2)$, we have

$$Sq^k(xy) = \sum_{k=i+j} Sq^i(x)Sq^j(y).$$

Multiproduct version is

$$Sq^q(x_1 \ldots x_r) = \sum_{i(1)+\ldots+i(r)=q} Sq^{i(1)}(x_1) \ldots Sq^{i(r)}(x_r).$$

(3) (Adem relations) For $0 < a < 2b$, the iterate of squares satisfies

$$Sq^a Sq^b = \sum_{j=0}^{[a/2]} \binom{b-1-j}{a-2j} Sq^{a+b-j} Sq^j.$$

8.4. Steenrod Operations on Low-Dimensional Classes We consider dimensions one and two.

(1) If $x \in H^1(X,\mathbb{F}_2)$, then we have $Sq^i(x^m) = \binom{m}{i} x^{m+i}$.
(2) If $y \in H^2(X,\mathbb{F}_2)$ and if $Sq^1(y) = 0$, then we have $Sq^{2i}(y^m) = \binom{m}{i} y^{m+i}$ and $Sq^{2i+1}(y^m) = 0$.

Proof. We use induction on m, where $m = 0$ is clear. Statement (1) then is obtained as follows.

$$Sq^i(x^m) = Sq^i(x.x^{m-1}) = Sq^0(x).Sq^i(x^{m-1}) + Sq^1(x).Sq^{i-1}(x^{m-1})$$
$$= \left[\binom{m-1}{i} + \binom{m-1}{i-1}\right] x^{m+i} = \binom{m}{i} x^{m+i}.$$

8.5. Theorem *Using the class $U_E \in H^n(D(E)/S(E))$ and the total Steenrod operation $Sq = \sum_{0 \le i} Sq^i$, we have the following formula for the total Stiefel–Whitney class $Sq(U_E) = w(E)U_E$ or $w(E) = \phi^{-1}(Sq(U_E))$.*

Proof. By the splitting principle, we can check a formula by doing it only for $E = L_1 \oplus \ldots \oplus L_n$, a sum of line bundles. Then by (6.6), we have a cup product decomposition of $U_E = U_1 \ldots U_n$ of one-dimensional classes U_i related to L_i. Only

Sq^1 is nontrivial on U_i, and it is $Sq^1(U_i) = U_i^2$. Hence, by the multiproduct version of Cartan's formula, we have the following calculation

$$Sq^r(U_E) = Sq^r(U_1 \ldots U_n) = \Sigma_{i(1)<\ldots<i(r)} U_1 \ldots U_{i(1)}^2 \ldots U_{i(r)}^2 \ldots U_n$$
$$= \Sigma_{i(1)<\ldots<i(r)} U_{i(1)}^2 \ldots U_{i(r)}^2 (U_1 \ldots U_n) = w_r(L_1 \oplus \ldots \oplus L_n)(U_1 \ldots U_n)$$

Using the monomorphism property of the splitting map on cohomology, we obtain $Sq^r(U_E) = w_r(E)U_E$. This proves the theorem.

Later, we will have some use for special Adem relations.

8.6. Special Cases of the Adem Relations

(1) For $a = 1$, we have $1 \le b$, and thus, the corresponding sum consists only the term for $j = 0$, that is, we have

$$Sq^1 Sq^b = \binom{b-1}{1} Sq^{b+1} = \begin{cases} Sq^{b+1} & \text{if } b \text{ is even} \\ 0 & \text{if } b \text{ is odd} \end{cases}$$

with simple cases $Sq^1 Sq^1 = 0$, $Sq^1 Sq^2 = Sq^3$, $Sq^1 Sq^3 = 0$, and $Sq^1 Sq^4 = Sq^5$.

(2) For $a = 2$, we have $2 \le b$, so that the sum consists of only the two terms corresponding to $j = 0$ and $j = 1$. Thus, in this case we have

$$Sq^2 Sq^b = \binom{b-1}{2} Sq^{b+2} + \binom{b-2}{0} Sq^{b+1} Sq^1.$$

This divides into two cases depending on $b \bmod 4$.

$$Sq^2 Sq^b = Sq^{b+1} Sq^1 + \begin{cases} Sq^{b+2} & \text{for } b \equiv 0,3 (\bmod 4) \\ 0 & \text{for } b \equiv 1,2 (\bmod 4) \end{cases}$$

with some simple cases $Sq^2 Sq^2 = Sq^3 Sq^1$, $Sq^2 Sq^3 = Sq^4 Sq^1 + Sq^5$, $Sq^2 Sq^4 = Sq^5 Sq^1 + Sq^6$, $Sq^2 Sq^5 = Sq^6 Sq^1$, and $Sq^2 Sq^7 = Sq^8 Sq^1 + Sq^9$.

Over the integers $\mathbb{Z}$, the binomial coefficient $\binom{n}{i}$ is the coefficient of x^i in the polynomial $(1 + x)^n \in \mathbb{Z}[x]$. Here, they are understood as numbers modulo 2.

8.7. Two Mod 2 Congruences We have the following in the field $\mathbb{F}_2 = \{0,1\}$ of two elements for $c \in \mathbb{Z}$

$$\binom{c}{1} = \begin{cases} 0 & \text{if } c \text{ is even} \\ 1 & \text{if } c \text{ is odd} \end{cases}$$

and

$$\binom{c}{2} = \begin{cases} 0 & \text{if } c \equiv 0,1 (\bmod 4) \\ 1 & \text{if } c \equiv 2,3 (\bmod 4) \end{cases}$$

9 Pontrjagin classes

Let $E^\vee$ denote the dual of a complex vector bundle E over B.

9.1. Proposition *The Chern classes of a complex vector bundle E and its dual $E^\vee$ are related by $c_i(E^\vee) = (-1)^i c_i(E)$. If a complex vector bundle E is isomorphic to its dual $E^\vee$, then*

$$2c_{2i+1}(E) = 0.$$

Proof. For a line bundle L, we have $c_1(L^\vee) = -c_1(L)$, and the result follows from the splitting principle and the Whitney sum formula. The second statement is immediate from the first.

9.2. Remark Every real vector bundle E is isomorphic to its real dual E^*. This is true for real line bundles L since $w_1(L) = w_1(L^*)$ and in general, by the splitting principle. Thus, the complexification $E_{\mathbb{C}}$ of a real vector bundle E is isomorphic to its complex dual $E_{\mathbb{C}}^\vee$, and therefore, $2c_{2i+1}(E_{\mathbb{C}}) = 0$.

9.3. Definition The Pontrjagin class $p_i(E)$ of a real vector bundle E is $p_i(E) = (-1)^i c_{2i}(E_{\mathbb{C}}) \in H^{4i}(B, \mathbb{Z})$. The total Pontrjagin class of a real vector bundle is

$$p(E) = 1 + p_1(E) + \ldots \in \prod_{0 \leq i} H^{4i}(B, \mathbb{Z}).$$

9.4. Remark The Whitney sum formula holds in the modified form

$$2(p(E')p(E'') - p(E' \oplus E'')) = 0$$

That is, the difference is a 2-torsion class.

References

Atiyah, M.F., Hirzebruch, F.: Vector bundles and homogeneous spaces. Proc. Symp. Pure Math. Amer. Math. Soc. 3:7–38 (1961)

Epstein, B.D.A., Steenrod, N.E.: Cohomology operations, Annals of Math. Studies 50, 1962

Hirzebruch, Neue topologische Methoden in der algebrischen Geometrie, 2nd ed. Springer-Verlag, Berlin (1962)

Husemöller, D.: *Fibre Bundles*, 3rd ed. Springer-Verlag, New York (1994)

Milnor, J. and Stasheff, J.: Charactarisic classes. Ann. of Math. Sudies 76, 1974

Steenrod, N.E.: Topology of Fibre Bundles. Princeton Univ. Press, Princeton, 1951

Chapter 11
Characteristic Classes of Manifolds

A topological manifold M of dimension n has a fundamental class denoted by ω_M or $[M] \in H_n(M, \mathbb{Z}/2\mathbb{Z})$, and when it has an orientation, this class is defined in $H_n(M, \mathbb{Z})$ with the same notation. In each case, the cap product

$$( ) \frown [M] : H^i(M) \to H_{n-i}(M)$$

for a closed manifold M is an isomorphism called Poincaré duality. This internal symmetry between homology and cohomology is a fundamental property of manifolds. When the manifold has a boundary ∂M or when it is not compact, then Poincaré duality must be modified, but it is always given by cap product with the fundamental class in $H_n(M^n, \mathbb{Z})$ for the oriented case and in $H_n(M^n, \mathbb{Z}/2)$ for the general case.

A smooth manifold M of dimension n has two complementary real vector bundles: the tangent bundle $T(M)$ of dimension n and the normal bundle $v(M)$ to some embedding of M into Euclidean space. The Whitney sum $T(M) \oplus v(M)$ is a trivial bundle on M which is the restriction to M of the trivial tangent bundle on Euclidean space. For the characteristic classes with the Whitney sum property, it is essentially equivalent to work either the characteristic classes of the tangent bundle or the normal bundle.

Polynomial combinations of characteristic classes of the tangent bundle can be evaluated on the fundamental class $[M]$, and the result is characteristic numbers which can be related to other invariants of the manifold. These numbers are related to homological invariants of the manifold and in many cases to other geometric invariants of the manifold and bundles on the manifold. In the last section, we illustrate this with an explication of the Riemann-Roch-Hirzebruch theorem.

Chapter 18 of *Fibre Bundles* (Husemöller 1994) is a reference for this chapter.

1 Orientation in Euclidean Space and on Manifolds

1.1. Linear Orientation A nonsingular linear map $A : \mathbb{R}^n \to \mathbb{R}^n$ preserves orientation provided $\det(A) > 0$. The subgroup of $GL(n, \mathbb{R})$ of orientation-preserving

D. Husemöller et al.: *Characteristic Classes of Manifolds*, Lect. Notes Phys. **726**, 127–135 (2008)
DOI 10.1007/978-3-540-74956-1_12 © Springer-Verlag Berlin Heidelberg 2008

maps is denoted by $GL^+(n, \mathbb{R})$, and its compact form is the rotation group $SO(n) = O(n) \cap GL^+(n, \mathbb{R})$. Here, what is critical for linear orientation is a basis of the vector space $\mathbb{R}^n$. To extend the concept of orientation to nonlinear maps, we use homology.

1.2. Topological Orientation To define orientation, we must choose generators $\alpha_{n-1} \in H_{n-1}(S^{n-1}) = \mathbb{Z}$, which for any open set U in $\mathbb{R}^n$ and $x \in U \subset \mathbb{R}^n$ corresponds by exactness and excision to a generator $\alpha_x \in H_n(U, U - x) = \mathbb{Z}$. A homeomorphism $f : (U, U - x) \to (V, V - f(x))$ preserves (resp. reverses) orientation at x provided $f_*(\alpha_x) = +\alpha_{f(x)}$ (resp. $-\alpha_{f(x)}$).

There are corresponding elements $\beta_{n-1} \in H^{n-1}(S^{n-1}) = \mathbb{Z}$ and $\beta_x \in H^n(U, U - x) = \mathbb{Z}$, and orientation preserving and reversing is the same in cohomology. In the case of a linear f, the determinant definition and the homological definition give the same result. Again by excision $H_n(M^n, M^n - x) = \mathbb{Z}$, and there are two generators for any manifold M^n of dimension n.

1.3. Orientation for Manifolds An orientation of an n-dimensional manifold M is a system of generators $\omega_x \in H_n(M, M - x)$ indexed by $x \in M$ such that for any open ball $B \subset M$ and $x, y \in B$, the classes ω_x and ω_y correspond under the following inclusion-induced isomorphisms

$$H_n(M, M - x) \longleftarrow H_n(M, M - B) \longrightarrow H_n(M, M - y).$$

1.4. Remark When a manifold M has an orientation, we use integral homology, but without an orientation, we implicitly use homology over $\mathbb{Z}/2\mathbb{Z}$, where there is one nonzero element ω_x in the relevant homology group $\omega_x \in H_n(M, M - x, \mathbb{Z}/2\mathbb{Z})$.

For a given orientation $\omega_x \in H_n(M, M - x)$, we define classes $\omega_K \in H_n(M, M - K)$ using the induced morphisms $r : H_*(M, M - L) \to H_*(M, M - K)$ for compact subsets $K \subset L$ in M.

1.5. Proposition *Let M be an n-dimensional topological manifold. For each compact subset K of M, we have $H_i(M, M - K) = 0$ for $i > n$, and a class $a \in H_n(M, M - K)$ is zero if and only if for all $x \in K$ we have $r_x(a) = 0$, where $r_x : H_n(M, M - K) \to H_n(M, M - x)$ is induced by inclusion.*

1.6. Proposition *Let $(\omega_x)_{x \in M}$ be an orientation of a topological manifold M of dimension n. For each compact subset $K \subset M$, there exists a class $\omega_K \in H_n(M, M - K)$ such that $r_x(\omega_K) = \omega_x$ for all $x \in K$, where again $r_x : H_n(M, M - K) \to H_n(M, M - x)$ is induced by inclusion.*

These two results on orientation in a manifold are proved by using the Mayer–Vietoris sequence for $K' \cup K'' = K$ and $K' \cap K''$ starting with closed balls B where the classes ω_B are immediately defined. Then, we use finite coverings of compact sets by small balls.

We thus have an orientation class ω_M for oriented compact manifolds M. There are appropriate definitions for the general case, which we shall not describe here.

2 Poincaré Duality on Manifolds

Using the orientation classes ω_K constructed in (1.6) for an n-dimensional manifold M, we now consider certain cap products related to pairs (K,V), where $K \subset V$ is compact and $V \subset M$ is open in M.

2.1. Remark The inclusion $j : (V, V - K) \longrightarrow (M, M - K)$ induces isomorphisms in homology and in cohomology by excision

$$j_* : H_*(V, V - K) \to H_*(M, M - K) \text{ and } j^* : H^*(M, M - K) \to H^*(V, V - K).$$

With these notations, we obtain a diagram using the cap product mixing cohomology and homology with values in homology.

2.2. Cap Product Diagram With the above notation, we have the following commutative diagram with the right vertical arrow an isomorphism

$$
\begin{array}{ccc}
H^i(M) \otimes H_n(M, M - K) & \xrightarrow{\text{cap}} & H_{n-i}(M, M - K) \\
{\scriptstyle j^* \otimes (j_*)^{-1}} \downarrow & & \uparrow {\scriptstyle j_*} \\
H^i(V) \otimes H_n(V, V - K) & \xrightarrow{\text{cap}} & H_{n-i}(V, V - K)
\end{array}
$$

Now, we fix one of the variables in the cap product to be an orientation class $\omega_K \in H_n(M, M - K)$.

2.3. Definition The dualizing morphism relative to a pair $K \subset V \subset M$ as above is $D_{K,V} : H^i(V) \to H_{n-i}(M, M - K)$ given by $D_{K,V}(a) = j_*(a \frown j_*^{-1}(\omega_K))$. Then, we take the limit of $V \supset V' \supset K$ giving a second dualizing morphism D_K defined on the direct limit

$$\check{H}^i(K) = \varinjlim_{K \subset V} H^i(V)$$

as

$$D_K : \check{H}^i(K) \longrightarrow H_{n-i}(M, M - K)$$

using the cap product diagram.

It is this second dualizing morphism D_K which leads to Poincaré duality.

2.4. Theorem *Let M be a manifold with orientation. Then, the dualizing morphism*

$$D_K : \check{H}^i(K) \longrightarrow H_{n-i}(M, M - K)$$

is an isomorphism for all compact subsets K of M.

Proof. This is proved by using Mayer–Vietoris for $K' \cup K'' = K$ and $K' \cap K''$ starting with closed balls B, where the classes ω_B are immediately defined. Then, we use finite coverings of compact sets by small balls.

The first corollary is the classical Poincaré duality theorem where $\check{H}^i(K) = H^i(M)$.

2.5. Theorem *Let M be a compact n-dimensional manifold with orientation ω_M. Then, $D_M : H^i(M) \to H_{n-i}(M)$ is an isomorphism $i = 0, \ldots, n$.*

The second corollary is the classical Alexander duality theorem.

2.6. Theorem *Let K be a compact subset of $\mathbb{R}^n$. Then, the morphism $\breve{H}^i(K) \to H_{n-i-1}(\mathbb{R}^n - K, *)$ which is the composition of*

$$D_K : \breve{H}^i(K) \to H_{n-i}(\mathbb{R}^n, \mathbb{R}^n - K)$$

and the boundary morphism

$$H_{n-i}(\mathbb{R}^n, \mathbb{R}^n - K) \to H_{n-i-1}(\mathbb{R}^n - K, *)$$

is an isomorphism.

2.7. Remark If K has a neighborhood base of open sets each of which has K as a deformation retract, then the limit cohomology $\breve{H}^i(K)$ is isomorphic to singular homology $H^i(K)$.

3 Thom Class of the Tangent Bundle and Duality

3.1. Remark Let $T(M)$ be the tangent bundle of a smooth manifold M. Then, M has an orientation $\omega_x \in H_n(M, M - x)$ if and only if $T(M)$ has a class $U_{T(M)} \in H^n(T(M), T(M)_0)$, where $T(M)_0 = T(M) - \{\text{zero section}\}$. In general the classes are $\mathbb{Z}/2$ classes, but M has an orientation (as an integral class) if and only if $T(M)$ has the structure of an $SO(n)$ fibre bundle.

Associated with the orientation of M, we have the dualizing morphism $D_K : \breve{H}^i(K) \to H_{n-i}(M, M - K)$ which we wish to characterize in terms of the class $U_{T(M)}$ transfer to a class U_M on $M \times M$ by using the exponential morphism $\exp_x : T(M)_x \to M$ for a Riemannian metric on M, where $\exp_x(v)$ is the value at $t = 1$ of the unique geodesic through x with tangent vector v for $t = 0$.

3.2. Notation We use $h_t(x, v) = (\exp_x(-tv), \exp_x(v))$ to define a diffeomorphism $h_t : (D(T(M)), D_0(T(M))) \to (N_t(\Delta, M \times M), N_t(\Delta, M \times M) - \Delta)$ for all $t \in [0, 1]$, where $N_t(\Delta, M \times M)$ is a closed manifold neighborhood of Δ in $M \times M$. With this h_t, we define a homotopy of diffeomorphisms onto closed submanifolds $k_t : (D(T(M)), D_0(T(M)), M) \to (M \times M, M \times M - \Delta, \Delta)$ by $k_t(x, v) = (h_t(x, v), (x))$, where x is the projection of v from the tangent bundle $T(M)$ to M.

3.3. Remark Using excision, we have an isomorphism

$$(h_t)^* : H^*(M \times M, M \times M - \Delta) \longrightarrow H^*(D(T(M)), D_0(T(M))).$$

Now, we bring in the basic cohomology class $U_{T(M)}$ in 10(6.2) for the tangent bundle $T(M)$. Again, it is a $\mathbb{Z}/2$-class in the general case and an integral class when $T(M)$, or equivalently M, has an orientation.

3.4. Definition The relative product class U'_M of M is defined by the relation $(h_t)^*(U'_M) = U_{T(M)}$ and the fundamental product class $U_M = j^*(U'_M) \in H^n(M \times M)$, where $j : M \times M \to (M \times M, M \times M - \Delta)$ is the inclusion.

3.5. Remark By evaluating cocycles on cycles, we obtain a pairing $\langle \, , \, \rangle : H^i(X,R) \times H_i(X,R) \to R$. We apply this to $X = M$ and $X = M \times M$ and to $R = \mathbb{Z}$ and $\mathbb{Z}/2$.

In terms of the fundamental product class, we have three pairing formulas which are proved in the *Fibre bundles*, 18(6.3), 18(6.4), and 18(6.5) (Husemöller 1994).

3.6. Proposition *Let M be a closed, connected manifold with orientation class $[M]$ and fundamental product class U_M.*

(1) For $a \in H^p(M)$ and $b \in H^q(M)$, we have $U_M(a \times b) = (-1)^{pq} U_M(b \times a)$.
(2) We have the pairing $\langle U_M, 1 \times [M] \rangle = 1$.
(3) For any class $U \in H^n(M \times M)$ with (1) and (2) as for U_M and $a \in H^p(M), c \in H_p(M)$ we have the pairing

$$\langle a,c \rangle = (-1)^{n+p}\langle U, c \times (a \frown [M]) \rangle,$$

where $D(a) = a \frown [M]$ defined as $D : H^p(M) \to H_{n-p}(M)$ is the dualizing morphism.

3.7. Application to Poincaré Duality Over a Field Let F be a field which is $\mathbb{F}_2 = \mathbb{Z}/2$, the field of two elements for a general closed manifold M. Let $D(a) = a \frown [M]$ defined by $D : H^p(M,F) \to H_{n-p}(M,F)$ be the Poincaré duality isomorphism. The fact that D is an isomorphism can be seen from the pairing for $a \in H^p(M), c \in H_p(M)$

$$\langle a,c \rangle = (-1)^{n+p}\langle U, c \times (a \frown [M]) \rangle$$

for two reasons: Firstly, $\langle a,c \rangle = 0$ for all $c \in H_p(M)$ if and only if $a = 0$, and hence, D is a monomorphism. Secondly, cohomology over a field $H^p(M,F) = \operatorname{Hom}(H_p(M),F) = H_p(M,F)^\vee$ is just the dual of homology so that homology and cohomology in degree p have the same dimension. This combined with the two monomorphisms

$$D : H^p(M,F) \longrightarrow H_{n-p}(M,F) \quad \text{and} \quad D : H^{n-p}(M,F) \longrightarrow H_p(M,F)$$

shows that D is an isomorphism.

4 Euler Class and Euler Characteristic of a Manifold

4.1. Definition Let X be a space with finite total dimensional homology. The Euler characteristic $\chi(X)$ of X is given by $\chi(X) = \sum_i (-1)^i \dim_{\mathbb{Q}}(H_i(X,\mathbb{Q}))$.

The reader with a background with the universal coefficient theorem can show that for any field F, we can calculate

$$\chi(X) = \sum_i (-1)^i \dim_F(H_i(X,F)) \quad \text{or} \quad \chi(X) = \sum_i (-1)^i \dim_F(H^i(X,F)).$$

The Euler characteristic of M are calculated in terms of the Euler class of the tangent bundle in the next theorem which is proved in 18(7.2) of *Fibre Bundles* (Husemöller 1994).

4.2. Theorem *Let M^n be a closed, connected, oriented manifold with $[M]' \in H^n(M,\mathbb{Z})$ a generator with $\langle [M]', [M] \rangle = 1$. Then the Euler class $e(T(M))$ of the tangent bundle is given by*

$$e(T(M)) = \chi(M)[M]'.$$

4.3. Corollary *Let M^n be a closed, connected, orientable manifold with an everywhere nonzero vector field. Then, $\chi(M) = 0$.*

It is a theorem of H. Hopf that the converse is also true. The result is illustrated by $\chi(S^{2n}) = 2$ and $\chi(S^{2n+1}) = 0$, where odd-dimensional spheres have such an everywhere nonzero vector field and even spheres do not.

5 Wu's Formula for the Stiefel–Whitney Classes of a Manifold

The Steenrod squares $Sq = \sum_i Sq^i$ and the Stiefel–Whitney class $w(E)$ of a bundle are related by the form $w(E) = \phi^{-1}(Sq(U_E))$, see 10(8.5). Using Poincaré duality and its relation to U_M, we have the Wu class and its relation to the Stiefel–Whitney classes of the tangent bundle.

5.1. Notation Let $Sq^{tr} : H_*(X) \to H_*(X)$ denote the transpose of the (total) Steenrod square $Sq : H^*(X) \to H^*(X)$. In particular, we have $\langle Sq(a), b \rangle = \langle a, Sq^{tr}(b) \rangle$ for $a \in H^*(X), b \in H_*(X)$.

5.2. Definition Let M be closed manifold with Poincaré duality isomorphism $D : H^i(M) \to H_{n-i}(M)$ and fundamental class $[M]$. The Wu class of M is $v = D^{-1}(Sq^{tr}([M]))$.

The Wu class has the property that

$$\langle a, D(v) \rangle = \langle a, Sq^{tr}([M]) \rangle = \langle Sq(a), [M] \rangle = \langle a, v \frown [M] \rangle = \langle av, [M] \rangle.$$

5.3. Theorem *Let M be a closed smooth manifold. Then, the Stiefel-Whitney class $w(M) = w(T(M))$ of the tangent bundle is given as the Steenrod square of the Wu class $w(M) = Sq(v)$.*

For the proof using (3.6), see 18(8.2) of *Fibre Bundles* (Husemöller 1994).

5.4. Corollary *The Stiefel–Whitney classes of closed manifolds are homotopy invariants of the manifold.*

5.5. Corollary *If $v = \sum_i v_i$, where $v_i \in H^i(M)$, then we have $v_i = 0$ for $2i > \dim(M)$.*

6 Cobordism and Stiefel–Whitney Numbers

6.1. Notation Let M' and M'' be two n-dimensional manifolds with $M = M' \sqcup M''$, the disjoint union. Then, the fundamental class of $M' \sqcup M''$ is $[M' \sqcup M''] = [M'] + [M'']$ with suitable injections for $H_n(M' \sqcup M'') = H_n(M') \oplus H_n(M'')$. In particular, when M' and M'' are oriented, then M' and M'' have the sum orientation. In general, we have the $\mathbb{Z}/2$-orientation class. There is one special case where M' or M'' is empty, and then $M' = M' \sqcup \emptyset$. Also $-M'$ is M' with the orientation class $-[M'] = [-M']$.

6.2. Definition Two manifolds M' and M'' are cobordant provided there exists a manifold W^{n+1} with boundary such that the boundary $\partial W = M' \sqcup M''$. Two oriented manifolds M' and M'' are oriented cobordant provided there exists a manifold W with an orientation $[W] \in H_{n+1}(W, \partial W)$ with $\partial W = M' - M''$ so that the homological boundary $\partial[W] = [M'] - [M'']$. A manifold M is a boundary-provided M, and the empty manifold are cobounding, that is, there exists W with $\partial W = M$.

6.3. Example For any manifold M, we can form $W = M \times [-1, 1]$, and we see that $\partial W = M \sqcup M$ in the unoriented sense. For any manifold M with a fixed point-free involution, $T : M \to M$, that is, $T^2 = $ identity and $T(x) \neq x$ for all $x \in M$, we can identify (x, t) with $(T(x), -t)$ in the product $M \times [-1, 1]$ and obtain a smooth quotient manifold W with $\partial W = M$. Then, relation $\partial W = M$ is an oriented cobordism when M is oriented and T is orientation preserving.

Two spaces with involutions are S^n and the odd-dimensional projective spaces $P_{2n+1}(\mathbb{R})$. The odd-dimensional projective space has a fixed point-free involution coming from the quotient of multiplication by i on the odd-dimensional sphere S^{2n+1} as a subspace of $\mathbb{C}^{n+1}$.

Now, we can consider combinations of characteristic classes which are evaluated on the fundamental class $[M]$ of a manifold M to give the so-called characteristic numbers of the manifold M.

6.4. Definition The Stiefel–Whitney number of a manifold M corresponding to the monomial $w = w_1^{r(1)} \ldots w_n^{r(n)}$ is $\langle w, [M] \rangle \in \mathbb{Z}/2$. The Pontrjagin number of an oriented manifold M with orientation class $[M]$ corresponding to the monomial $p = p_1^{r(1)} \ldots p_n^{r(n)}$ is $\langle p, [M] \rangle \in \mathbb{Z}$. In both cases, the characteristic classes are the characteristic classes of the tangent bundle.

For $n = \dim(M)$, there is one Stiefel–Whitney number for each sequence $r(1), \ldots, r(n)$ with $n = 1.r(1) + 2.r(2) + \ldots + n.r(n)$, and for oriented M, there is one Pontrjagin number for each sequence $r(1), \ldots, r(n)$ with $n = 4.r(1) + 8.r(2) + \ldots + 4n.r(n)$. Now, these numbers are related to the concept of cobordism by the following theorem of Pontrjagin.

6.5. Theorem *Let M^n be a closed manifold which is the boundary $M^n = \partial W^{n+1}$ of a manifold with boundary. Then, all Stiefel–Whitney numbers of M are zero. If M' and M'' are cobordant manifolds, then the corresponding Stiefel–Whitney numbers of M' and M'' are equal.*

Proof. The key remark for the proof is that $T(W)|M = T(M) \oplus L$, where L is the trivial line bundle. Hence, the top combinations of characteristic classes are zero on this restriction. Now use that $[W] \in H_{n+1}(W,M)$ maps to $[M] \in H_n(M)$ under the boundary operator.

For further details, see 18(9.2) of *Fibre Bundles* (Husemöller 1994).

6.6. Remark The corresponding theorem holds for Pontrjagin numbers and oriented cobordism.

6.7. Remark There are also stunning converses to Pontrjagin's result. Thom showed in the 1950s that two closed manifolds are cobordant if their Stiefel–Whitney numbers agree. Little later, Wall proved a corresponding result for oriented cobordism. Two oriented closed manifolds are oriented cobordant if their Stiefel–Whitney numbers and their Pontrjagin numbers agree.

7 Introduction to Characteristic Classes and Riemann–Roch

We have seen that characteristic classes evaluated on the orientation class have a geometric significance for cobordism, but when this topological significance was being worked out by Thom, Hirzebruch was applying characteristic classes to the Riemann–Roch problem. This was explained later in two ways: Firstly, Grothendieck extended the work of Hirzebruch to the algebraic domain formulating and proving a Riemann–Roch theorem for an algebraic morphism. Secondly, Atiyah–Singer extended the Hirzebruch Riemann–Roch theorem to an index theorem for elliptic operators. In fact, the first proof of the index theorem was using the cobordism methods of Hirzebruch. In the last 50 years, there has been an explosion in mathematics around these ideas with many far reaching applications. We will just touch on a modest amount of the first ideas of Hirzebruch.

7.1. Topological Notation Consider a compact complex manifold X of complex dimension n. Then, X is a real compact, oriented manifold of real-dimensional $2n$ with an orientation class $[X] \in H_{2n}(X,\mathbb{Z})$ coming from the complex structure since $U(n) \subset SO(2n)$. The complex tangent bundle of complex dimension n has Chern classes $c_1,\ldots,c_n$, where $c_i = c_i(T(X))$ which contains topological information and information about the complex geometry.

7.2. Geometric Notation Consider a complex analytic vector bundle E on X of dimension r. Then, the $\mathbb{C}$-vector space $\Gamma(X,E)$ of complex analytic sections of E has the basic property that it is finitely generated (since X is assumed to be compact), which makes the question of a formula for its dimension an interesting problem. The complex vector bundle E also has Chern classes $c_1(E),\ldots,c_r(E)$.

7.3. Remark If we consider an exact sequence of complex analytic vector bundles $0 \to E' \to E \to E'' \to 0$, then we have only a left exact sequence for the sections

$$0 \longrightarrow \Gamma(X, E') \longrightarrow \Gamma(X, E) \longrightarrow \Gamma(X, E'').$$

This leads to a sheaf cohomology theory $H^i(X, E)$, where $\Gamma(X, E) = H^0(X, E)$. Again, we have finite dimensionality of these cohomology vector spaces and a vanishing above $2 \dim(E)$. Instead of considering one dimension, we group all the dimensions into a Euler characteristic $\chi(E) = \sum_{0 \le i} (-1)^i \dim_{\mathbb{C}} H^i(X, E)$ which will have an additive property relative to short exact sequences of analytic vector bundles, that is, $\chi(E) = \chi(E') + \chi(E'')$, in the case of the exact sequence $0 \to E' \to E \to E'' \to 0$.

It is this Euler characteristic which is used to extend the classical Riemann–Roch for a curve and a complex line bundle to a general smooth projective algebraic variety X. With the above notations, we have the Riemann–Roch–Hirzebruch formula.

7.4. Theorem *For the Todd class $Td(X)$ of $T(X)$ and the Chern character $ch(E)$, we have*

$$\chi(E) = (ch(E)Td(X))[M]$$

Recall from $10(5.5)$ that the Chern character and from $10(5.8)$ that the Todd class of a vector bundle is defined in terms of Chern classes.

In dimension 1, where X is a closed Riemann surface or the complex points on a smooth algebraic complete algebraic curve, we recover the usual Riemann–Roch theorem.

Corollary 7.5 (Riemann–Roch Theorem) *For a line bundle L on a closed Riemann surface X, we have*

$$\dim_{\mathbb{C}} H^0(X, L) - \dim_{\mathbb{C}} H^1(X, L) = \deg(L) + 1 - g,$$

where g is the genus of X.

The term $\deg(L)$ comes from $ch(L)$ and $1 - g$ from $Td(X)$ in the right-hand side of (7.4). The vector space $H^1(X, L)$ is isomorphic to $H^0(X, K \otimes L^{(-1)\otimes})$ by Serre duality.

Reference

Husemöller, D.: *Fibre Bundles*, 3rd ed. Springer-Verlag, New York (1994)

Chapter 12
Spin Structures

Orientation of a real vector bundle E can be described in terms of the $O(n)$-associated principal bundle. Namely, orientability is equivalent to the property that the structure group of the bundle can be reduced to $SO(n) \subset O(n)$.

The group $Spin(n)$ is a double cover of the rotation group $SO(n)$. A spin structure on an oriented vector bundle is a lifting of the $SO(n)$-structure group of the bundle to the $Spin(n)$-structure group. This all takes place with the associated principal bundles, and it is one place where the vector bundle's associated principal bundle plays an essential role. The group $Spin^c(n) = (Spin(n) \times S^1)/\Gamma$ combines real oriented bundle data by the projection $Spin^c(n) \to Spin(n)/\Gamma' = SO(n)$ for $\Gamma' = \{1, e\}$, $e^2 = 1 \in Spin(n)$, and complex line bundle data by the other projection $Spin^c(n) \to S^1/\Gamma''$ for $\Gamma'' = \{1, -1\} \subset S^1$. A $Spin^c(n)$ structure on an oriented vector bundle is a lifting of the $SO(n)$-structure group of the bundle to the group $Spin^c(n)$.

1 The Groups $Spin(n)$ and $Spin^c(n)$

1.1. Orthogonal and Rotation Groups In the discussion of vector bundles, we have used orthogonal group principal bundles. On the orthogonal group in n-dimensions $O(n)$, we have the determinant surjection with kernel $SO(n)$, the rotation group as kernel in the following kernel exact sequence

$$1 \longrightarrow SO(n) = \ker(\det) \longrightarrow O(n) \xrightarrow{\det} O(1) = \{\pm 1\}.$$

Note that $SO(1) = 1$, $SO(2) = S^1$, and $SO(3) \cong P_3(\mathbb{R})$, the three-dimensional projective space where elements in the three-dimensional rotation group are given by an axis of rotation and an angle $\leq \pi$. So the rotation is an element of the ball $B(0, \pi)$ of radius π in $\mathbb{R}^3$ but with x and $-x$ identified for $||x|| = ||-x|| = \pi$. This is the projective space in three dimensions.

D. Husemöller et al.: *Spin Structures*, Lect. Notes Phys. **726**, 137–145 (2008)
DOI 10.1007/978-3-540-74956-1_13

1.2. Remark The first homotopy groups of $SO(n)$ are given by the following:

$$\pi_0(SO(n)) = 0, \qquad \pi_1(SO(n)) = \begin{cases} \mathbb{Z} & \text{for } n = 2 \\ \mathbb{Z}/2\mathbb{Z} & \text{for } n > 2 \end{cases}$$

$$\pi_2(SO(n)) = 0, \qquad \pi_3(SO(n)) = \begin{cases} 0 & \text{for } n = 2 \\ \mathbb{Z} & \text{for } n > 2 \end{cases}.$$

1.3. Spin Group The $Spin(n)$ group is the universal covering group of $SO(n)$ for $n > 2$. For $n = 2$, we require $\mathrm{Spin}(2) = S^1$.

Since $\pi_1(SO(n)) = \mathbb{Z}/2\mathbb{Z}$ for $n > 2$, the universal covering group

$$\{1, \varepsilon\} \longrightarrow Spin(n) \longrightarrow SO(n)$$

is a twofold covering so that the kernel is $\{1, \varepsilon\}$ with $\varepsilon^2 = 1$. For $n = 2$, we consider formally the twofold covering

$$Spin(2) = S^1 \longrightarrow SO(2) = S^1$$

with kernel $= \{\pm 1\}$.

1.4. Example The group $Spin(3)$ is isomorphic to S^3, the topological group of unit quaternions $q = q_0 + q_1 i + q_2 j + q_3 k \in \mathbb{H}$ so that $||q|| = 1$, where $||q||^2 = q_0^2 + q_1^2 + q_2^2 + q_3^2$. The quaternion conjugation is defined by the formula $\bar{q} = q_0 - q_1 i - q_2 j - q_3 k$ for $q = q_0 + q_1 i + q_2 j + q_3 k$. A quaternion u is a unit quaternion if and only if $u^{-1} = \bar{u}$. The rotation of an element $u \in S^3$ on the purely imaginary quaternions $\mathrm{Im}(\mathbb{H}) = \mathbb{R}i + \mathbb{R}j + \mathbb{R}k$, which is given by the formula $r(u)(q) = uqu^{-1}$, defines a twofold cover

$$r : Spin(3) = S^3 \longrightarrow SO(\mathrm{Im}(\mathbb{H})) = SO(3).$$

There is a parallel discussion of the double cover

$$r : Spin(n) \longrightarrow SO(n)$$

using Clifford algebras instead of the quaternions.

1.5. Clifford Algebra The Clifford algebra $Cl(n)$ is the unital algebra over the real numbers generated by n elements $e_1, \ldots, e_n$ satisfying relations

$$e_i e_j + e_j e_i = -2\delta_{i,j} \qquad 1 \le i, j \le n.$$

In particular, we have the anticommutativity $e_i e_j = -e_j e_i$ for $i \neq j$ and the $e_i^2 = -1$. A basis of the algebra consists of monomials $e_{i(1)} \ldots e_{i(r)}$, where $1 \le i(1) < \ldots < i(r) \le n$. In particular, there are exactly 2^n monomials. Observe that

$$(a_1 e_1 + \ldots + a_n e_n)(b_1 e_1 + \ldots + b_n e_n) =$$
$$-2(a_1 b_1 + \ldots + a_n b_n) + (a_1 b_2 - a_2 b_1)e_1 e_2 + \ldots + (a_{n-1} b_n - a_n b_{n-1})e_{n-1} e_n.$$

In particular, every nonzero $u = a_1 e_1 + \ldots + a_n e_n$ with $||a|| = 1$ is a unit in the algebra, and the conjugation by u on the space $\mathbb{R}e_1 + \ldots + \mathbb{R}e_n$ carries this space into itself as a linear automorphism-preserving distance. The group $Pin(n)$ is defined to be the group generated by these units u and its action on $\mathbb{R}e_1 + \ldots + \mathbb{R}e_n$ by conjugation maps $\phi : Pin(n) \to O(n)$. We calculate $\ker(\phi) = \{\pm 1\}$ as with the quaternions. The spin group $Spin(n)$ is $\phi^{-1}(SO(n))$ and the restriction $\phi : Spin(n) \to SO(n)$ is the universal covering of $SO(n)$, the n-dimensional rotation group.

1.6. $Spin^c$ Group We have the following product diagram around $Spin(n) \times S^1$ with the $\mathbb{Z}/2\mathbb{Z}$ subgroups which leads to a definition of the complex spin group $Spin^c$ as a quotient by the central subgroup Γ of order 2.

$$
\begin{array}{ccccc}
\mathbb{Z}/2\mathbb{Z} = \{1,\varepsilon\} & \xleftarrow{\ \mathrm{pr}_1\ } & \{(1,1),(\varepsilon,-1)\} & \xrightarrow{\ \mathrm{pr}_2\ } & \{\pm 1\} \\
\downarrow & & \downarrow & & \downarrow \\
Spin(n) & \longleftarrow & Spin(n) \times S^1 & \longrightarrow & S^1 = K(\mathbb{Z},1) \\
\downarrow & & \downarrow & & \downarrow{\scriptstyle 2} \\
SO(n) & \longleftarrow & Spin^c(n) & \longrightarrow & S^1 = K(\mathbb{Z},1)
\end{array}
$$

In particular, we have $Spin^c(n) = (Spin(n) \times S^1)/\Gamma$, with $\Gamma = \{(1,1),(\varepsilon,-1)\}$.

2 Orientation and the First Stiefel–Whitney Class

2.1. Remark In order to study the first Stiefel–Whitney class of a vector bundle, we take the classifying space fibre sequence of the exact sequence (1.1) relating $O(n)$ and $SO(n)$ to obtain the fibre sequence

$$
O(n) \xrightarrow{\ \det\ } \{\pm 1\} \longrightarrow BSO(n) \longrightarrow BO(n) \xrightarrow{\ B(\det)\ } B\{\pm 1\} = K(\mathbb{Z}/2\mathbb{Z},1).
$$

2.2. Definition Let E be a real n-dimensional vector bundle over a space X classified by a map $f : X \to BO(n)$. The first Stiefel–Whitney class $w_1(E)$ of E is the homotopy class of the composite $[B(\det)f] \in [X,B\{\pm 1\}] = H^1(X,\mathbb{Z}/2\mathbb{Z})$. An orientation of the real vector bundle E is represented by a lifting of the classifying map f to a map $f' : X \to BSO(n)$, and E is orientable provided f lifts to some f'. A homotopy class $[f']$ of liftings of f is called an orientation.

An orientation can also be defined as a reduction of the structure group from $O(n)$ to $SO(n)$.

2.3. Proposition *With the notations in (2.1), the composite $B(\det)f$ is null homotopy if and only if f has a lifting $f' : X \to BSO(n)$. A real vector bundle E has an orientation if and only if the first Stiefel–Whitney class $w_1(E) = 0$. The various orientations are simply and transitively acted on by the group $[X,\{\pm 1\}] = H^0(X,\{\pm 1\}) = H^0(X,\mathbb{Z}/2\mathbb{Z})$.*

Proof. This follows from the properties of homotopy mapping sets into a fibre sequence.

2.4. Remark There is a universal first Stiefel–Whitney class which is a homotopy class of mappings $w_1 : BO(n) \to K(\mathbb{Z}/2\mathbb{Z}, 1)$ or an element $w_1 \in H^1(BO(n), \mathbb{Z}/2\mathbb{Z})$ for n-dimensional bundles and similarly a homotopy class of mappings $w_1 : BO \to K(\mathbb{Z}/2\mathbb{Z}, 1)$ or an element $w_1 \in H^1(BO, \mathbb{Z}/2\mathbb{Z})$ for stable bundles. The element w_1 generates a polynomial algebra in $H^*(BO, \mathbb{Z}/2\mathbb{Z})$.

3 Spin Structures and the Second Stiefel–Whitney Class

3.1. Remark In order to study the second Stiefel–Whitney class of an oriented vector bundle, we take part of the classifying space fibre sequence of the exact sequence (1.3) relating $SO(n)$ and $Spin(n)$ to obtain the fibre sequence

$$\{1, \varepsilon\} \longrightarrow Spin(n) \longrightarrow SO(n) \xrightarrow{\sigma} B\{1, \varepsilon\} = K(\mathbb{Z}/2\mathbb{Z}, 1).$$

With this sequence, we use the group structure on $B\{1, \varepsilon\}$ and the fact that $SO(n) \to B\{1, \varepsilon\}$ is a group morphism to apply again the fibre sequence of this exact sequence to obtain the fibre sequence

$$K(\mathbb{Z}/2, 1) = B\{1, \varepsilon\} \longrightarrow BSpin(n) \longrightarrow BSO(n) \xrightarrow{B(\sigma)} B^2\{1, \varepsilon\} = K(\mathbb{Z}/2, 2).$$

3.2. Definition Let E be a real oriented n-dimensional vector bundle over a space X classified by a map $f : X \to BSO(n)$. The second Stiefel–Whitney class $w_2(E)$ of E is the homotopy class of the composite $[B(\sigma)f] \in [X, B^2\{1, \varepsilon\}] = H^2(X, \mathbb{Z}/2\mathbb{Z}) = H^2(X, \mathbb{F}_2)$. A spin structure on the real oriented vector bundle E is represented by a lifting of the classifying map f to a map $f' : X \to BSpin(n)$, and E has a spin structure provided f lifts to some f'. A homotopy class $[f']$ of liftings of f is called a spin structure.

A spin structure can also be defined as a lifting of the structure group from $SO(n)$ to $Spin(n)$.

3.3. Proposition *With the notations in (3.1), the composite $B(\sigma)f$ is null homotopy if and only if f has a lifting $f' : X \to BSO(n)$. An oriented real vector bundle E has a spin structure if and only if the second Stiefel–Whitney class $w_2(E) = 0$. The various spin structures of an oriented vector bundle over X are simply and transitively acted on by $H^1(X, \mathbb{Z}/2\mathbb{Z})$.*

Proof. This follows from the properties of homotopy mapping sets into a fibre sequence.

3.4. Remark There is always a universal second Stiefel–Whitney class which is a homotopy class of mappings $w_2 : BSO(n) \to K(\mathbb{Z}/2\mathbb{Z}, 2)$ or an element $w_2 \in$

$H^2(BSO(n), \mathbb{Z}/2\mathbb{Z})$. For stable bundles, we have $w_2 : BSO \to K(\mathbb{Z}/2\mathbb{Z}, 2)$ or an element $w_2 \in H^2(BSO, \mathbb{Z}/2\mathbb{Z})$.

4 *Spin^c* Structures and the Third Integral Stiefel–Whitney Class

4.1. Remark In order to define the third integral Stiefel–Whitney class of an oriented vector bundle, we take the classifying space fibre sequences of the product diagram (1.4) relating $Spin^c(n)$ and $SO(n)$ to obtain the following three vertical fibre sequences

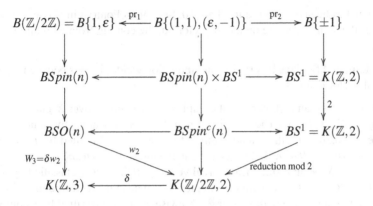

with $\delta =$ Bockstein.

4.2. Definition Let E be a real oriented n-dimensional vector bundle over a space X classified by a map $f : X \to BSO(n)$. The integral third Stiefel–Whitney class $W_3(E)$ of E is the homotopy class of the composite $[\delta w_2 f] \in [X, K(\mathbb{Z}, 3)] = H^3(X, \mathbb{Z})$, where δ is the Bockstein $\delta : K(\mathbb{Z}/2\mathbb{Z}, 2) \to K(\mathbb{Z}, 3)$. A spin^c structure on the real oriented vector bundle E is a complex line bundle L with classifying map $g : X \to B(S^1) = K(\mathbb{Z}, 2)$ such that in $K(\mathbb{Z}/2\mathbb{Z}, 2)$

$$w_2 f = g \bmod 2.$$

With this relation, we have a lifting $f' : X \to B(Spin^c(n))$ which composed with the first projection is f and with the second projection is g. The homotopy class $[f']$ projecting to f is called a spin^c structure on E.

A spin^c structure can also be defined as a lifting of the structure group from $SO(n)$ to $Spin^c(n)$.

4.3. Proposition *With the notations in (4.1), the composite $w_2 f$ is of the form g mod 2 if and only if $\delta w_2 f = W_3 f : X \to K(\mathbb{Z}, 3)$ is null homotopic. An oriented real vector bundle E has a spin^c structure if and only if the third integral Stiefel–Whitney class $W_3(E) = 0$. The group $[X, BS^1] = H^2(X, \mathbb{Z})$ acts simply and transitively on the set of spin^c structures of an oriented vector bundle E over X.*

Proof. This follows from the properties of homotopy mapping sets into a fibre sequence and a fibre product.

4.4. Remark There is always a universal third integral Stiefel–Whitney class which is a homotopy class of mappings $W_3 : BSO(n) \to K(\mathbb{Z}, 3)$ or an element $W_3 \in H^3(BSO(n), \mathbb{Z})$. For stable bundles, there is a corresponding $W_3 \in H^3(BSO, \mathbb{Z})$.

5 Relation Between Characteristic Classes of Real and Complex Vector Bundles

This topic has been considered in Chap. 10, Sect. 9 on Pontrjagin classes. Now, we consider the spin^c-structure on a real vector bundle coming from a complex vector bundle.

5.1. Change of Scalars Restriction and induction yield relations between complex and real vector bundles.

(1) Let $E \to X$ be an n-dimensional complex vector bundle over X classified by a map $f : X \to BU(n)$. The associated real $2n$-dimensional vector bundle $E|\mathbb{R}$ is given by restricting the scalars from $\mathbb{C}$ to $\mathbb{R}$, and it is classified by f composed with $BU(n) \to BO(2n)$ induced by the inclusion $U(n) \subset O(2n)$.

(2) Let $E_{\mathbb{R}} \to X$ be an n-dimensional real vector bundle over X classified by a map $g : X \to BO(n)$. The associated complex n-dimensional vector bundle $\mathbb{C} \otimes_{\mathbb{R}} E_{\mathbb{R}}$ is given by tensoring with $\mathbb{C}$ over $\mathbb{R}$ fibrewise, and it is classified by g composed with $BO(n) \to BU(n)$ induced by the inclusion $O(n) \subset U(n)$.

5.2. Real Characteristic Classes of Complex Vector Bundles Let E be a complex vector bundle on X with restriction to the real vector bundle $E|\mathbb{R}$. Then, we have

$$w_1(E|\mathbb{R}) = 0, \ W_3(E|\mathbb{R}) = 0, \ \text{and} \ w_2(E|\mathbb{R}) = c_1(E) \bmod 2.$$

In particular, a complex vector bundle as a real vector bundle has an orientation and a spin$^c(n)$-structure.

6 Killing Homotopy Groups in a Fibration

6.1. Notation For a pointed space X, we denote by $X\langle n \rangle \to X$ the fibration having the property that $\pi_i(X\langle n \rangle) \to \pi_i(X)$ is an isomorphism for $i \geq n$ and $\pi_i(X\langle n \rangle) = 0$ for $i < n$. Observe that $X\langle 0 \rangle = X, X\langle 1 \rangle$ is the connected component of the base point of X, and

$$X\langle 2 \rangle \longrightarrow X\langle 1 \rangle \subset X$$

is the universal covering of the connected component $X\langle 1 \rangle$. In general, there are factoring fibrations over X given by

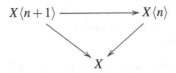

For $n > 0$, the fibre of the fibration $X\langle n+1\rangle \to X\langle n\rangle$ is a $K(\pi_n(X), n-1)$ space. For $n > 1$, this space can be taken to be an abelian group, and the fibration can be made into a principal fibration.

6.2. Definition With these notations, this fibration is classified by a map $c_n : X\langle n\rangle \to K(\pi_n(X), n)$, and we obtain the n-step fibration sequence for the pointed space X

$$K(\pi_n(X), n-1) \longrightarrow X\langle n+1\rangle \longrightarrow X\langle n\rangle \longrightarrow K(\pi_n(X), n).$$

The fibration $X\langle n\rangle \to X$ is called the n-connected covering of X. Note that $X\langle n\rangle$ in general is $n-1$-connected but not necessarily n-connected.

6.3. Remark For any pointed space T, the n-step fibration sequence for X yields the following homotopy mapping sequence

$$[T, K(\pi_n(X), n-1)] \longrightarrow [T, X\langle n+1\rangle] \longrightarrow [T, X\langle n\rangle] \xrightarrow{c} [T, K(\pi_n(X), n)]$$

$$\| \qquad\qquad\qquad\qquad\qquad\qquad\qquad\qquad\qquad\qquad\qquad \|$$

$$H^{n-1}(T, \pi_n(X)) \qquad\qquad\qquad\qquad\qquad\qquad\qquad H^n(T, \pi_n(X))$$

For $\alpha \in [T, X\langle n\rangle]$ and $\beta \in [T, X\langle n+1\rangle]$, we have the following two assertions:

(1) The element $\alpha \in [T, X\langle n\rangle]$ lifts to some $\beta \in [T, X\langle n+1\rangle]$, that is, α is the image of some β if and only if the characteristic class $c(\alpha) = 0$ in $H^n(T, \pi_n(X))$.
(2) Given α which lifts to β, the various liftings β differ by an action of $H^{n-1}(T, \pi_n(X)) = [T, K(\pi_n(X), n-1)]$ on $[T, X\langle n+1\rangle]$.

6.4. Some m-Connected Coverings of BO Now, we consider a stable real vector bundle E on T classified by $f : T \to BO$ and the possibility of lifting it to various connected coverings of BO. The first row is a sequence of identities and the second row is a sequence of fibrations where the fibres are $K(\pi, m)$ spaces with π given by the homotopy of BO as determined by Bott periodicity.

$$
\begin{array}{ccccccc}
T & \xrightarrow{\mathrm{id}} & T & \xrightarrow{\mathrm{id}} & T & \xrightarrow{\mathrm{id}} & T \\
\downarrow{\scriptstyle f_3} & & \downarrow{\scriptstyle f_2} & & \downarrow{\scriptstyle f_1} & & \downarrow{\scriptstyle f} \\
BO\langle 8\rangle & \longrightarrow & BO\langle 4\rangle = BSpin & \longrightarrow & BSO & \longrightarrow & BO
\end{array}
$$

Both homology groups, $H^4(BSpin, \mathbb{Z})$ and $H^4(BSO, \mathbb{Z})$, are isomorphic to $\mathbb{Z}$, and the homomorphism $H^4(BSO, \mathbb{Z}) \to H^4(BSpin, \mathbb{Z})$ induced by the natural map $BSpin \to$

BSO has cokernel isomorphic to $\mathbb{Z}/2\mathbb{Z}$. The image of $p_1 \in H^4(BSO, \mathbb{Z})$ therefore is two times a generator. This generator in $H^4(BSpin, \mathbb{Z})$ is denoted $\frac{1}{2}p_1$.

We have three assertions:

(1) A map f_1 factoring f up to homotopy exists if and only if $w_1(E) = 0$.
(2) When $w_1(E) = 0$ and f_1 is chosen, a map f_2 factoring f_1 up to homotopy exists if and only if $w_2(E) = 0$.
(3) When $w_1(E) = 0, w_2(E) = 0$, and f_2 is chosen, a map f_3 factoring f_2 exists if and only if $\frac{1}{2}p_1(E) = 0$.

6.5. Some m-Connected Coverings of $O(n)$ There are corresponding connected covers for the group $O(n)$. Note that the m-connected cover of a topological group by definition yields a space and not a topological group. However, the sequence of connected covers $O(n)\langle 7 \rangle \to O(n)\langle 3 \rangle \to O(n)\langle 1 \rangle \to O(n)\langle 0 \rangle \to O(n)$ for $n > 4$ can be realized through a sequence of topological groups

$$String(n) \longrightarrow Spin(n) \longrightarrow SO(n) \longrightarrow O(n).$$

The topological group $SO(n)$ is the 0-connected subgroup of $O(n)$ for $n > 1$, $Spin(n)$ is the 3-connected covering of $SO(n)$ and $O(n)$ for $n > 2$, and these are all compact Lie groups. The group $String(n)$ which corresponds to the 7-connected covering of $Spin(n)$, $SO(n)$, and $O(n)$ is defined for $n > 4$ and is called the string group. There are various models for the topological group $String(n)$, but none of them is a compact Lie group since $H^3(String(n), \mathbb{Z}) = 0$. A lifting of an $O(n)$ principal bundle to $SO(n)$ is called an orientation, to $Spin(n)$ is called a spin structure, and to $String(n)$ is called a string structure.

We can interpret (6.3) for the tangent bundle $T(M)$ to a manifold with the following terminology.

6.6. Remark Let M be a smooth manifold with tangent bundle $T(M)$ with characteristic classes of M equal to the classes of $T(M)$.

(a) The manifold M is orientable if and only if the Stiefel–Whitney class $w_1(M) = 0$. The set of orientations of M is simply and transitively acted on by $H^0(M, \mathbb{Z}/2\mathbb{Z})$.
(b) The oriented manifold M is spin if and only if the Stiefel–Whitney class $w_2(M) = 0$. The set of spin structures on M is simply and transitively acted on by $H^1(M, \mathbb{Z}/2\mathbb{Z})$.
(c) The spin manifold M is string if and only if the Pontrjagin class $\frac{1}{2}p_1(M) = 0$. The set of string structures on M is simply and transitively acted on by $H^3(M, \mathbb{Z})$.

6.7. Geometric Significance of Spin Structures On a spin manifold M, we can define a spinor bundle S_M related to the tangent bundle $T(M)$ together with a Dirac operator acting on the sections of the spinor bundle S_M. The index of the Dirac operator can be computed in terms of the $\hat{A}(M)$, the $\hat{A}$-genus of the manifold M. The index is an obstruction to the existence of a metric with positive scalar curvature on M.

6.8. Remark Conjecturally, there is a similar interpretation of the existence of string structures on M in terms of spinors and Dirac operators on the free loop space $\Lambda(M)$ of M. The index of the Dirac operator would be the Witten index, and it is conjecturally an obstruction to the existence of a metric with positive Ricci curvature on M.

Part III
Versions of *K*-Theory and Bott Periodicity

We consider complex and real K-theory from the point of view of the theory of vector bundles on compact spaces, from the point of view of operator algebra, and from the homotopy point of view using classifying spaces. The G-equivariant K-theory is introduced in the context of the Grothendieck construction together with the Atiyah real theory relating complex, self-conjugate, and real versions in terms of one equivariant construction. Bott periodicity and the Thom isomorphism have a rather direct meaning in this context as developed by Atiyah. This is carried out in Chap. 13 and 14. This is the version which is easily adopted to the questions around the index theorem.

In Chap. 15, the homotopy version of K-theory of a space X with $K(X)$ identified with $[X, \mathbb{Z} \times BU]$ is considered. This is the classification space approach which arises out of the classifying space approach to fibre bundles. In fact, BU is an increasing inductive limit or colimit $\varinjlim_n BU(n)$, and the real K-theory $KO(X)$ is isomorphic to $[X, \mathbb{Z} \times BO]$. In this context, Bott periodicity is a series of maps into loop spaces which have certain relation to Clifford algebras. These maps are homotopy equivalences, and their relation to Clifford algebras is pointed out.

In Chap. 16, we prove some of the homotopy properties of general linear groups by elementary considerations.

In Chap. 18, the operator formulation of equivariant K-theory is carried out. For a compact space X, we form the C^*-algebra $C(X)$ of continuous $\mathbb{C}$-valued functions, and the C^*-algebra K-theory $K_0(C(X))$ is isomorphic to $K^0(X)$, the K-theory of vector bundles. For a compact group G acting on a compact space X we form the cross product C^*-algebra $G \ltimes C(X)$, and the C^*-algebra K-theory $K_0(G \ltimes C(X))$ is naturally isomorphic to $K_G^0(X)$, the K-theory of G-equivariant vector bundles on X considered by Atiyah and Segal. In this context, Bott periodicity comes from analysis of Töplitz operators.

Chapter 13
G-Spaces, G-Bundles, and G-Vector Bundles

In this chapter, we introduce the basic notions related to objects being acted on by a group G. Since the objects like spaces and bundles have topologies, we will assume that the group is a topological group. Let G be a topological group which is eventually a compact Lie group. We consider G-spaces X and G-vector bundles with a base G-space. The aim is to develop the theory in a parallel fashion to ordinary bundle theory taking into account the G-action on both the base space and the total space together with actions between fibres. There are some generalities which apply to bundles in general including principal bundles which we outline in the first sections. A G-vector bundle over a point is just a representation of G on a vector space. In particular, we analyze part of the properties of G-vector bundles in terms of representation theory of G. Since the representation theory of compact groups is well understood, we will restrict ourselves to this case for the more precise theory, but compact groups will play a role for topological reasons too. For example, we show that G-homotopic maps always induce isomorphic G-vector bundles when G is compact.

Of special interest is the group G of two elements. The symbol (τ) is a symbol for the group on two elements with generator τ which is always an involution under any action. The first example is that of complex conjugation $\tau(a+ib) = a - ib$, where $\tau : \mathbb{C} \to \mathbb{C}$. We incorporate the concept real structures in K-theory using (τ)-bundles and mix this with the G-equivariant theory, for it plays a basic role in the relation between complex and real structures on vector bundles.

1 Relations Between Spaces and G-Spaces: G-Homotopy

In Chap. 5, Sect. 1, we introduced the basic notions of G-spaces and the G-equivariant maps between G-spaces as background for principal bundles where G acted on total space-preserving fibres. We extend these concepts so that G can act also on the base space. This consists of making certain functors more explicit.

D. Husemöller et al.: *G-Spaces, G-Bundles, and G-Vector Bundles*, Lect. Notes Phys. **726**, 149–161 (2008)
DOI 10.1007/978-3-540-74956-1_14   © Springer-Verlag Berlin Heidelberg 2008

1.1. Notation for the Basic Categories and Functors Let (top) be the category of spaces and maps, and let $(G\backslash\text{top})$ be the category of G-spaces and G-equivariant maps. If G is the trivial group, $(G\backslash\text{top})$ can be identified with (top). For two G-spaces, X and Y, one obtains the space $\text{Map}(X,Y)$ of all continuous maps from X to Y. It carries a natural G-action which is motivated and specified in remark 1.5 and definition 1.6 below.

For a homomorphism $\rho : H \to G$ of topological groups, the restriction functor $\text{Res} : (G\backslash\text{top}) \to (H\backslash\text{top})$ assigns to a G-space X the H-space X with the action $\theta : G \times X \to X$ given by the composition

$$\theta(\rho \times id) : H \times X \to G \times X \to X.$$

The trivial G-space functor $\text{Tr} : (\text{top}) \to (G\backslash\text{top})$ assigns to a space W the G-space $\text{Tr}(W) = W$ as a space with trivial G-action. The stripping (or forgetful) functor $\text{Str} : (G\backslash\text{top}) \to (\text{top})$ assigns to a G-space X the space $\text{Str}(X) = X$ viewed as a space without G-action. The functor Tr can be identified with the restriction functor induced by the homomorphism which maps G to the trivial group while the stripping functor Str can be identified with the restriction functor induced by the inclusion of the trivial group into G.

The functor Res has both a left and a right adjoint.

1.2. Biadjunctions of Res The induction functor $\text{Ind} : (H\backslash\text{top}) \to (G\backslash\text{top})$ assigns to an H-space X the quotient $G \times^H X$ which is obtained by identifying $(g\rho(h^{-1}),x)) = (g,hx)$ for all $g \in G, h \in H$, and $x \in X$. The quotient is regarded as a G-space via the action induced by left multiplication on the first factor. The functor is a left adjoint to the functor Res, that is, for every G-space W, one has

$$\text{Map}_G(G \times^H X, W) = \text{Map}_H(X, \text{Res}(W))$$

The coinduction functor $\text{Coind} : (G\backslash\text{top}) \to (H\backslash\text{top})$ assigns to an H-space Y the H-space $\text{Map}_H(G,Y)$, where G is regarded as the H-space where an element $h \in H$ acts on G by left multiplication with $\rho(h)^{-1}$. That is, the space $\text{Map}_H(G,Y)$ consists of all maps $\phi : G \to Y$ satisfying where an element $h \in H$ acts on G by left multiplication via ρ. More specifically, $\text{Map}_H(G,Y)$ consists of all maps $\phi : G \to Y$ satisfying $t\phi(s) = \phi(\rho(t)s)$ for all $s \in G$ and $t \in H$. The space $\text{Map}_H(G,Y)$ is regarded as a G-space with the action of an element $s \in G$ on a function $\phi \in \text{Map}_H(G,Y)$ is given by $(s.\phi)(s') = \phi(s's)$. The functor Coind is right adjoint to Res, that is, in terms of morphism sets, we have

$$\text{Map}_H(\text{Res}(X),Y) = \text{Map}_G(X, \text{Coind}(Y))$$

From the above, we obtain biadjunctions to the functors Tr and Str.

1.3. Biadjunctions of Tr Let X be a G-space. We have two spaces $\text{Quot}(X) = G\backslash X$ and $\text{Fix}(X) = \text{Map}_G(*,X) \subset X$ with, respectively, the quotient and subspace topologies from X. They correspond to the induction and the coinduction functor for the homomorphism which maps G to the trivial group. We therefore have the following adjunction relations

$$\mathrm{Quot} \dashv \mathrm{Tr} \quad \text{and} \quad \mathrm{Tr} \dashv \mathrm{Fix} : (G \backslash \mathrm{top}) \longrightarrow (\mathrm{top}).$$

In terms of morphism sets we have, for G-spaces W, the natural isomorphisms

$$\mathrm{Map}(G \backslash X, W) = \mathrm{Map}_G(X, \mathrm{Tr}(W))$$

and

$$\mathrm{Map}_G(\mathrm{Tr}(W), Y) = \mathrm{Map}(W, \mathrm{Fix}(Y)).$$

1.4. Biadjunctions of Str Let W be a space. For the inclusion of the trivial group into G, the adjunctions of (1.2) specialize to the adjunction relation

$$G \times (.) \dashv \mathrm{Str} \quad \text{and} \quad \mathrm{Str} \dashv \mathrm{Map}(G, .) : (\mathrm{top}) \longrightarrow (G \backslash \mathrm{top}).$$

In terms of morphism sets, we have the natural isomorphisms

$$\mathrm{Map}_G(G \times W, Y) = \mathrm{Map}(W, \mathrm{Str}(Y))$$

and

$$\mathrm{Map}_G(X, \mathrm{Map}(G, W)) = \mathrm{Map}(\mathrm{Str}(X), W).$$

This adjunction relation depends on the adjunction

$$\mathrm{Map}(X \times T, Y) = \mathrm{Map}(X, \mathrm{Map}(T, Y)),$$

which we discussed in Sect. 1 and 2 of Chap. 6.

The adjunction for spaces

$$c : \mathrm{Map}(X \times T, Y) \to \mathrm{Map}(X, \mathrm{Map}(T, Y))$$

where $c(f)(x)(t) = f(x,t)$ plays a basic role for the concept of homotopy of maps. We extend adjunction to G-spaces in order to consider G-homotopy.

1.5. Remark Let X, T, and Y be G-spaces. The G-structure on $X \times T$ is given by the relation $s(x,t) = (sx, st)$, and a G-map $f : X \times T \to Y$ satisfies the relation $sf(x,t) = f(sx, st)$. In order that $f(x)(t) : X \to \mathrm{Map}(T, Y)$ be a G-morphism, we must have $f(sx)(t) = [s.f(x)](t)$, or in other terms, we require that

$$f(sx, t) = f(sx, ss^{-1}t) = sf(x, s^{-1}t) = [s.f(x)](t),$$

and this leads to the action $G \times \mathrm{Map}(T, Y) \to \mathrm{Map}(T, Y)$, where

$$(s.u)(t) = su(s^{-1}t) \quad \text{for} \quad s \in G, t \in T.$$

1.6. Definition The action $G \times \mathrm{Map}(T, Y) \to \mathrm{Map}(T, Y)$ given by the formula $s.u(t) = (s.u)(t) = su(s^{-1}t)$ for $s \in G$, $t \in T$ makes $\mathrm{Map}(T, Y)$ into a G-space. Now, we can use 6(2.4) in the G-equivariant setting.

1.7. Assertion Let T be a locally compact G-space, and let X and Y be G-spaces. The isomorphism of spaces

$$c : \mathrm{Map}(X \times T, Y) \longrightarrow \mathrm{Map}(X, \mathrm{Map}(T, Y))$$

restricts to an isomorphism for G-spaces morphisms

$$c : \mathrm{Map}_G(X \times T, Y) \longrightarrow \mathrm{Map}_G(X, \mathrm{Map}(T, Y)),$$

where $\mathrm{Map}_G(X, Y) \subset \mathrm{Map}(X, Y)$ has the subspace topology. If X is locally compact instead of T, then we have an isomorphism of G-spaces $c : \mathrm{Map}_G(X \times T, Y) \to \mathrm{Map}_G(T, \mathrm{Map}(X, Y))$, and if further G acts trivially on T, then $\mathrm{Map}_G(T, \mathrm{Map}(X, Y)) = \mathrm{Map}(T, \mathrm{Map}_G(X, Y))$ since $\mathrm{Map}_G(X, Y) = \mathrm{Fix}(\mathrm{Map}(X, Y))$, the G-fixed points on $\mathrm{Map}(X, Y)$. This, we apply to the concept of G-homotopy.

1.8. Definition Let X and Y be two G-spaces. Two G-maps $f', f'' : X \to Y$ are G-homotopic provided there exists a G-map $h : X \times [0, 1] \to Y$ with G acting trivially on $[0, 1]$ such that $h(x, 0) = f'(x)$ and $h(x, 1) = f''(x)$.

1.9. Remark As in the nonequivariant case, the relation of G-homotopy is an equivalence relation.

1.10. Remark The homotopy h can be viewed as a path parametrized by t given by $h_t(x) = h(x, t)$ consisting of G-maps $X \to Y$, and if X is locally compact, then all such paths determine a homotopy from $h_0 = f'$ to $h_1 = f''$.

2 Generalities on G-Bundles

The most general concept of a bundle is that of a map $p : E \to B$, where B is called the base space, E is called the total space, and p is the projection. The inverse images $p^{-1}(b)$, also denoted E_b, with the subspace topology are called the fibres of the bundle.

2.1. Definition Let G be a topological group. A G-bundle is a G-equivariant map $p : E \to B$ between G-spaces, in particular, $p(sx) = sp(x)$ for $s \in G$. The G-space E is called the total space, the G-space B is called the base space, and the G-equivariant map p is called the projection. Note that in Chap. 5 we required a G-bundle to have a trivial action of G on the base space B and the right action on the total space.

The equivariant property of the projection p can be formulated in terms of the commutativity of the diagram where the horizontal maps are the G-space structure or action maps

$$
\begin{array}{ccc}
G \times E & \longrightarrow & E \\
{\scriptstyle G \times p} \downarrow & & \downarrow {\scriptstyle p} \\
G \times B & \longrightarrow & B.
\end{array}
$$

2.2. Definition Let $p' : E' \to B'$ and $p'' : E'' \to B''$ be two G-bundles. A morphism is a pair of G-equivariant maps $u : E' \to E''$ and $f : B' \to B''$ defined, respectively, on the total spaces and base spaces such that $p''u = fp'$. The composition of G-bundle morphisms is defined by composing the maps on the total and base spaces. In the case that $B' = B'' = B$ and f is the identity on B, the morphism $u : E' \to E''$ is called a B-morphism of G-bundles. The composition of two B-morphisms is again a B-morphism of G-bundles.

2.3. Remark Let $p : E \to B$ be a G-bundle, and let $b \in B$ and $s \in G$ with $sb \in B$. The action of s on E restricts to fibres $E_b \to E_{sb}$ as an isomorphism with inverse the action of s^{-1} on the fibres.

2.4. Notation Let $p : E \to B$ be a G-bundle, and let $\Gamma(B,E)$ denote the set of all sections of $p : E \to B$, that is, all maps $\sigma : B \to E$ with $p(\sigma(b)) = b$ for all $b \in B$. We give $\Gamma(B,E)$ the subspace topology from the compact open topology on Map (B,E). It is a G-space with action $(s\sigma)(b) = s(\sigma(s^{-1}b))$ for $b \in B$, $s \in G$, and $\sigma \in \Gamma(B,E)$. A G-equivariant section $\sigma \in \Gamma(B,E)$ is one with $s\sigma = \sigma$ for all $s \in G$, and the subset of G-equivariant sections is denoted by $\Gamma_G(B,E)$.

2.5. Definition Let $p' : E' \to B'$ and $p'' : E'' \to B''$ be two G-bundles. The product G-bundle is $p' \times p'' : E' \times E'' \to B' \times B''$. In the case that $B' = B'' = B$, the fibre product G-bundle is $q : E' \times_B E'' \to B$, the fibre product of bundles with G-action $s(x',x'') = (sx',sx'')$, where the condition $p'(x') = p''(x'')$ implies $p'(sx') = p''(sx'')$.

2.6. Definition Let $p : E \to B$ be a G-bundle, and let $f : B' \to B$ be a G-map. Then, the induced bundle $f^{-1}(E)$ with total space $f^{-1}(E) \subset B' \times E$ defined by the condition $(b',x) \in f^{-1}(E)$ if and only if $f(b') = p(x)$, and projection $q : f^{-1}(E) \to B'$ given by $q(b',x) = b'$ has a G-bundle structure, where $s(b',x) = (sb',sx)$.

2.7. Remark Let $p' : E' \to B'$ and $p'' : E'' \to B''$ be two G-bundles with a morphism consisting of a pair of G-equivariant maps $u : E' \to E''$ and $f : B' \to B''$ defined, respectively, on the total spaces and base spaces such that $p''u = fp'$. Then, there is a factorization $u = u''u'$ of u as a G-bundle morphism $u' : E' \to f^{-1}(E'')$ over B' and $u'' : f^{-1}(E'') \to E''$ as a G-bundle morphism over f. The formulas are $u'(x') = (p'(x'),u(x'))$ and $u''(b',x'') = x''$.

3 Generalities on G-Vector Bundles

The primary interest in this chapter is in (finite dimensional) complex vector bundles, and in the course of our discussion, real and quaternionic vector bundles will arise within the equivariant structure. As usual, G denotes a topological group.

3.1. Definition Let X be a G-space. A G-vector bundle over X is a G-map $p : E \to X$ which is a G-bundle and a vector bundle such that for all $s \in G$, the action of $s : E_b \to E_{sb}$ is a vector space isomorphism.

Observe that if $sb = b$, then s acts as a linear automorphism of the vector space fibre E_b.

3.2. Definition Let $p' : E' \to X'$ and $p'' : E'' \to X''$ be two G-vector bundles, and let $f : X' \to X''$ be a G-map. An f-morphism $u : E' \to E''$ of G-vector bundles is a morphism of G-bundles which also is a morphism of vector bundles. When $X' = X'' = X$ and f is the identity on X, then u is called an X-morphism of G-vector bundles.

3.3. Categories of G-Vector Bundles Let X be a G-space. The categories of real and complex G-vector bundles over X are denoted by $\mathrm{Vect}_{\mathbb{R}}(X, G)$ and $\mathrm{Vect}_{\mathbb{C}}(X, G)$. Tensoring with $\mathbb{C}$ (over $\mathbb{R}$) gives a functor $\mathrm{Vect}_{\mathbb{R}}(X, G) \to \mathrm{Vect}_{\mathbb{C}}(X, G)$, and reducing the scalar multiplication from the complex numbers $\mathbb{C}$ to the real numbers $\mathbb{R}$ gives a functor $\mathrm{Vect}_{\mathbb{C}}(X, G) \to \mathrm{Vect}_{\mathbb{R}}(X, G)$. The composition of the first functor followed by the second is a functor $\mathrm{Vect}_{\mathbb{R}}(X, G) \to \mathrm{Vect}_{\mathbb{R}}(X, G)$ which carries a real G-bundle V to $V \oplus V$ over (X, G). The composition of the second functor followed by the first is a functor $\mathrm{Vect}_{\mathbb{C}}(X, G) \to \mathrm{Vect}_{\mathbb{C}}(X, G)$ which carries a complex G-bundle E to $E \oplus E^*$ over (X, G).

3.4. Remark In (2.4), the G-action on the set of sections $\Gamma(X, E)$ was introduced for a G-vector bundle. The space $\Gamma(X, E)$ of sections is, first of all, a vector space with the usual addition and scalar multiplication given by the formula

$$(a'\sigma' + a''\sigma'')(b) = a'\sigma'(b) + a''\sigma''(b)$$

for $\sigma', \sigma'' \in \Gamma(X, E)$ and scalars a' and a''. Moreover, by (1.4), the action of $s \in G$ induces the identities

$$(s(a'\sigma' + a''\sigma''))(b) = s((a'\sigma' + a''\sigma'')(s^{-1}b)) =$$
$$a's(\sigma'(s^{-1}b) + a''s(\sigma'')(s^{-1}b) = a'(s\sigma')(b) + a''(s\sigma'')(b)$$

for $b \in X, s \in G$, and $\sigma \in \Gamma(X, E)$ showing that $\Gamma(X, E)$ is a G-vector space. If $u : E' \to E''$ is an X-morphism of G-vector bundles, then $\Gamma(u)(\sigma)(b) = u(\sigma(b))$ defines a morphism $\Gamma(u) : \Gamma(X, E') \to \Gamma(X, E'')$ of G-vector spaces, because $s\Gamma(u)(\sigma)(b) = su(\sigma(b)) = su(\sigma(s^{-1}b)) = u(s(\sigma(s^{-1}b))) = \Gamma(u)(s\sigma)(b)$.

Finally, the cross section vector space is a functor from the categories of complex and real G-vector bundles

$$\Gamma(X,) : \mathrm{Vect}_{\mathbb{C}}(X, G) \longrightarrow \mathrm{Rep}_{\mathbb{C}}(G) \text{ and } \Gamma(X,) : \mathrm{Vect}_{\mathbb{R}}(X, G) \longrightarrow \mathrm{Rep}_{\mathbb{R}}(G)$$

to the categories $\mathrm{Rep}_{\mathbb{C}}(G)$ and $\mathrm{Rep}_{\mathbb{R}}(G)$ of, respectively, complex and real (not necessarily finite dimensional) representations of G. The question of continuity of such representations will not be discussed here.

3.5. Definition Let $p' : E' \to X'$ and $p'' : E'' \to X''$ be two G-vector bundles. The product $p' \times p'' : E' \times E'' \to X' \times X''$ is a G-vector bundle. In the case that $X' = X'' = X$, the fibre product $q : E' \times_X E'' \to X$ is a G-vector bundle.

3.6. Definition Let $p : E \to X$ be a G-vector bundle, and let $f : X' \to X$ be a G-map. Then, the induced bundle $f^{-1}(E)$ is a G-vector bundle as can be checked immediately from the total space $f^{-1}(E) \subset X' \times E$ defined by the condition $(b',x) \in f^{-1}(E)$ if and only if $f(b') = p(x)$ and projection $q : f^{-1}(E) \to X'$ given by $q(b',x) = b'$ has a G-bundle structure, where $s(b',x) = (sb',sx)$ and vector bundle structure $a'(b',x_1) + a''(b',x_2) = (b, a'x_1 + a''x_2)$.

As with (2.7), we have the following remark.

3.7. Remark Let $p' : E' \to X'$ and $p'' : E'' \to X''$ be two G-vector bundles with a morphism consisting of a pair of maps $u : E' \to E''$ and $f : X' \to X''$ defined, respectively, on the total spaces and base spaces such that $p''u = fp'$. Then, there is a factorization $u = u''u'$ of u as a G-vector bundle morphism $u' : E' \to f^{-1}(E'')$ over X' and $u'' : f^{-1}(E'') \to E''$, a G-vector bundle morphism over f. This follows immediately from the formulas $u'(x') = (p'(x'), u(x'))$ and $u''(b',x'') = x''$.

3.8. Definition Analogously, as in Chap. 4, Sect. 2, one defines the G-equivariant K-theory ring $K_G(X)$ of a G-space X as the Grothendieck construction applied to the semiring of isomorphism classes of G-vector bundles on X.

4 Special Examples of G-Vector Bundles

We consider three special cases of G-vector bundles.

4.1. Example Let E be a G-vector bundle over a space B which is trivial as a vector bundle. This means that $p : E \to B$ is isomorphic to $pr_B : B \times V \to B$, where V is a complex vector space. For this product G-bundle, we have an action $G \times B \to B$ and the product $B \times V$ with V has a G-action $G \times B \times V \to B \times V$ of the form

$$s(b,v) = (sb, J(s,b)v),$$

where $J : G \times B \to \text{End}(V)$ is a continuous function. For the unit of the action, we must have $J(1,b) = \text{id}_V$ for all $b \in B$, and for the associativity of the action, we must have

$$(stb, J(st,b)v) = st(b,v) = s(tb, J(t,b)v) = (stb, J(s,tb)J(t,b)v),$$

or for all $b \in B$ and $s,t \in G$, the cocycle relation (or chain rule)

$$J(st,b) = J(s,tb)J(t,b).$$

When $s = t = 1$, we have $J(1,b) = J(1,b)J(1,b)$ so that if $J(1,b)$ is invertible, then we have automatically $J(1,b) = 1$. Otherwise, it is only an idempotent.

Conversely, for a continuous function $J : G \times B \to \text{End}(V)$ satisfying $J(1,b) = 1$ and $J(st,b) = J(s,tb)J(t,b)$, the product bundle has a G-bundle structure given by $s(b,v) = (b, J(b,s)v)$. If we change the trivialization by a function $C : B \to GL(V)$, then we have a new G-bundle structure where $J'(b,s) = C(sb)J(b,s)C(b)^{-1}$, for if $\psi(b,v) = (b, C(b)v)$, then the action

$$(sb, J'(b,s)v) = \psi s \psi^{-1}(b,v) = \psi s(b, C(b)^{-1}v) =$$
$$\psi(sb, J(b,s)C(b)^{-1}v) = (sb, C(sb)J(b,s)C(b)^{-1}v).$$

This gives what is called the cobounding relation between J and J', that is, $J'(b,s) = C(sb)J(b,s)C(b)^{-1}$.

4.2. Homogeneous Space Let X be a G-space with a transitive G-action, and choose a point $x \in X$ with stabilizer subgroup G_x. Then, there is a natural continuous bijection $f : G/G_x \to X$ of G-spaces, where $f(tG_x) = tx$. We assume that f is a topological isomorphism and hence, G-space isomorphism. This is the case for G compact and X separated.

4.3. Example Let $p : E \to G/H$ be a G-vector bundle over the homogeneous space G/H. As we have seen in (3.1), the stabilizer subgroup $G_x = H$ of $1.H = x$ acts on the fibre E_x of E over x as a group representation. We can form the quotient $G \times^H E_x$ of the product $G \times E_x$ which projects to $G \times^H x = G/H$ as a vector bundle $q : G \times^H E_x \to G/H$. This is the fibre bundle construction, and we denote this fibre bundle also by $q : G[E_x] \to G/H$. Next, there is a natural morphism $\alpha : G \times^H E_x \to E$ of vector bundles over G/H given by $\alpha(s,v) = sv \in E$. Since $\alpha(sh^{-1}, hv) = \alpha(s,v)$, the formula defines a morphism of vector bundles which is an isomorphism on each fibre. Hence, it is an isomorphism.

Conversely, to each representation V of H, we can associate a natural G-vector bundle $G[V] = G \times^H V \to G \times^H * = G/H$ over G/H.

4.4. Example Let B be a G-space with trivial G action, that is, $sb = b$ for all $s \in G, b \in B$. Then, a G-vector bundle structure on a vector bundle $p : E \to B$ is just a group morphism

$$G \longrightarrow \text{Aut}(E/B).$$

To go further, we assume that G be a compact group. Then, we can analyze the structure of E in terms of a choice of representatives V_i of the various isomorphisms classes of irreducible representations of G. Let I denote the set of these isomorphism classes. Since G is compact, these representatives V_i are finite dimensional. If V is a finite dimensional representation of G, then the natural map $\bigoplus_{i \in I} \text{Hom}_G(V_i, V) \otimes V_i \to V$ is an isomorphism where we use the relation $\text{Hom}_G(V_i, V_j) = \mathbb{C}\delta_{i,j}$. Let E_i denote the product vector bundle $B \times V_i$ over B. Then $\text{Hom}_G(E_i, E_j) = C(B)\delta_{i,j}$, where $C(B)$ is the $\mathbb{C}$-algebra of continuous complex valued functions on B. Thus, the isomorphism for representations extends to the following isomorphism for vector bundles

$$\bigoplus_{i \in I} \text{Hom}_G(E_i, E) \otimes E_i \longrightarrow E.$$

Recall that $\text{End}(V_i) \cong V_i \otimes V_i^*$ and that the completion of $\bigoplus_{i \in I} \text{End}(V_i)$ are isomorphic to $L^2(G)$, the space of L^2-functions on G with respect to Haar measure. This is the content of the Peter–Weyl theorem.

5 Extension and Homotopy Problems for G-Vector Bundles for G a Compact Group

In this section, G is a compact group, and in general, we will consider bundles over compact G-spaces. A G-vector bundle E over a G-space X is, in particular, a vector bundle, hence locally trivial as a vector bundle. This says nothing about the role of the action of G on these sets where the bundle is trivial in general.

5.1. Remark Let G be a compact group, and let E be a G-vector bundle over a separated space B. The restriction $E|(Gx)$ to the compact orbit $Gx \subset B$ of E is a vector bundle, and the orbit is isomorphic to the homogeneous space G/G_x under the isomorphism $G/G_x \to Gx$. By (4.3), we have a natural isomorphism of this restriction $E|Gx \to G[E_x]$ over Gx, where $G[E_x] \to Gx$ is the fibre bundle associated to the principal G_x-bundle $G \to Gx$ with fibre E_x coming from the representation of G_x on E_x.

5.2. Proposition *Let E be a G-vector bundle over a compact G-space X, and let Y be a closed G-subspace of X. The restriction of G-equivariant sections $\Gamma_G(X,E) \to \Gamma_G(Y,E|Y)$ is a surjection.*

Proof. An open set V with $Y \subset V \subset X$ has a section $\sigma'' \in \Gamma(V,E)$ extending $\sigma' \in \Gamma(Y,E|Y)$, that is, $\sigma''|Y = \sigma'$. This follows from the fact that Y is covered by a finite number of open sets over which E is trivial and over which σ' is described by a map into a vector space. These maps into vector spaces extend by the Tietze extension theorem.

5.3. Proposition *Let E' and E'' be two G-vector bundles over a compact G-space X, and let $u : E'|Y \to E''|Y$ be an isomorphism of the restrictions to a closed G-subspace Y of X. There exists a G-invariant neighborhood V of Y in X and an isomorphism $w : E'|V \to E''|V$ extending u.*

Proof. The first step is to apply the previous proposition (5.2) to $u \in \Gamma_G(Y, \mathrm{Hom}(E',E''))$ to obtain $w' \in \Gamma_G(X, \mathrm{Hom}(E',E''))$, a G-morphism $w' : E' \to E''$. Since being an isomorphism is an open condition, there is G-open set V with $Y \subset V$ in X with $w : E'|V \to E''|V$, an isomorphism. This proves the proposition.

5.4. Proposition *Let E be a G-vector bundle on a G-space Y, and let $f', f'' : X \to Y$ be two G-homotopic maps from a compact G-space X. Then, the induced G-bundles $f'^{-1}E$ and $f''^{-1}E$ are isomorphic.*

Proof. Let $h : X \times [0,1] \to Y$ be a homotopy from f' to f'', and let $E' = h^{-1}(E)$ on $X \times [0,1]$ with $E'|X \times \{0\} = f'^{-1}(E)$. By the previous proposition, this equality extends to an isomorphism

$$E'|X \times [0,\varepsilon] \longrightarrow f'^{-1}(E) \times [0,\varepsilon]$$

using the compactness of X. Then, using the compactness of $[0,1]$, we can extend this to an isomorphism $E' \to f'^{-1}(E) \times [0,1]$ which restricts to $X \times \{1\}$ giving an isomorphism $f''^{-1}(E) \to f'^{-1}(E)$. This proves the proposition.

6 Relations Between Complex and Real G-Vector Bundles

Now the group of two elements plays a special role.

6.1. Notation We use (τ) to denote the group with two elements where τ is element of order 2. The group (τ) acts on $\mathbb{C}$ by complex conjugation $\tau(a+ib) = a-ib$, and the fixed field or space is the real numbers $\mathbb{R} = \mathrm{Fix}(\tau)$ and with complex conjugation on $\mathbb{C}^m$ with fixed space $\mathbb{R}^m = \mathrm{Fix}(\tau)$.

The idea that the real object is the fixed object of an action (τ) plays a basic role.

6.2. Remark Complex conjugation (τ) acts on $P_n(\mathbb{C})$, and the fixed space is just $P_n(\mathbb{R})$. When τ preserves a subalgebraic variety of projective space, the real points are the fixed points of the action of τ. To extend this action to the canonical (complex) line bundle $L \to P_n(\mathbb{C})$, we consider the exact sequence

$$0 \longrightarrow E \longrightarrow P_n(\mathbb{C}) \times \mathbb{C}^{n+1} \longrightarrow L \longrightarrow 0$$

of vector bundles over $P_n(\mathbb{C})$, where the n-dimensional subbundle $E \subset P_n(\mathbb{C}) \times \mathbb{C}^{n+1}$ consists of all $(z_i, w_i) \in P_n(\mathbb{C}) \times \mathbb{C}^{n+1}$, where $z_0 w_0 + \ldots + z_n w_n = 0$. Since the equation for E consists of real coefficients, E and the quotient L have the coordinatewise action of τ. But the action of τ is not by a complex linear automorphism, instead, it is conjugate linear on the fibres of the quotient $P_n(\mathbb{C}) \times \mathbb{C}^{n+1} \to L$. Then, the fixed point line bundle over the fixed space under τ is the real canonical line bundle $L_{\mathbb{R}} \to P_n(\mathbb{R})$. This leads to the study of complex G-bundles where the action is conjugate linear on the fibres or more generally, partially linear and partially conjugate linear on the fibres. These ideas were introduced by Atiyah and extend to a G-vector bundle setting where the group (τ) acts also on G. A G-equivariant real theory with K-groups KR_G is formulated as a mixture of Atiyah's real theory with K-groups KR and Segal's G-equivariant theory with K-groups K_G. This theory KR_G was introduced in Atiyah and Segal (1969).

6.3. Remark Let G be compact Lie group with a real structure $\tau : G \to G$. Form the cross product $(\tau)G$, which is $(\tau) \times G$, as a set and group structure $(\tau', s')(\tau'', s'') = (\tau'\tau'', (\tau''s')s'')$ for $\tau', \tau'' \in (\tau)$ and $s', s'' \in G$. If X is a real $(\tau)G$-space, then both G and (τ) act on X with the property $\tau(s.x) = \tau(s).\tau(x)$, or equivalently, the property that the cross product $(\tau)G$ acts on X.

6.4. Definition A (G, τ)-vector bundle E is a complex G-vector bundle over a $(\tau)G$-space with a real $(\tau)G$-vector bundle structure such that the fibre action of τ on the real $(\tau)G$-vector bundle $E_x \to E_{\tau x}$ is $\mathbb{C}$-conjugate linear. A morphism of (G, τ)-vector bundles $E' \to E''$ is a morphism which is a morphism of complex G-vector bundles and real $(\tau)G$-vector bundles. Since a vector space is a vector bundle over a point, this terminology applies also to complex vector spaces.

6.5. Notation Let $\mathrm{Vect}(X, G, \tau)$ denote the category of (G, τ)-vector bundles over the $(\tau)G$-space X. When G is the identity, this category is denoted by $\mathrm{Vect}(X, \tau)$.

The category $\text{Vect}(X,\tau)$ was studied very early by Atiyah (1966), and he called the space (X,τ) a real space by analogy with algebraic geometry where the fixed points of complex conjugation on the complex points form the set of real points.

6.6. Remark The results of the previous sections apply to the category $\text{Vect}(X,G,\tau)$. For example, a continuous G-map $f : X \rightarrow Y$ defines a functor

$$f^{-1} : \text{Vect}(Y,G,\tau) \rightarrow \text{Vect}(X,G,\tau).$$

The extension and homotopy properties $(5.2),(5.3)$, and (5.4) carry over immediately. The (G,τ)-action on the vector space of sections $\Gamma(X,E)$ is a (G,τ)-action on the vector space where again, that is, it is $\mathbb{C}$-linear for $s \in G$ and is $\mathbb{C}$-conjugate linear for $s = \tau$.

6.7. Remark A $(\tau)G$-space X induces a G-space structure on X^τ, the τ-fixed points of X, and $\text{Vect}(X,G,\tau) \rightarrow \text{Vect}(X^\tau,G)$ is the complexification of a real G-equivariant vector bundle. Under complexification, $\text{Vect}_{\mathbb{R}}(X^\tau,G)$ is mapped injectively onto the image of $\text{Vect}(X,G,\tau) \rightarrow \text{Vect}(X^\tau,G)$ as an equivalence of categories.

7 KR_G-Theory

We apply the Grothendieck construction to the semiring of isomorphism classes in the category $\text{Vect}(X,G,\tau)$ of (G,τ)-vector bundles over the $(\tau)G$-space X.

7.1. Definition The (G,τ)-equivariant real K-theory is a functor defined on the opposite category to the category of compact $(\tau)G$-spaces denoted by KR_G. When $G = (id)$, the functor is denoted simply by KR, and when τ is the identity, KR_G is isomorphic to the ordinary equivariant K-theory K_G. The reduced version of KR_G, denoted $\widetilde{KR}_G$, is defined on the opposite category of compact-pointed $(\tau)G$-spaces as the kernel of the restriction to the base point

$$\widetilde{KR}_G(X) = \ker(KR_G(X) \longrightarrow KR_G(*)).$$

For a G-equivariant cofibration $A \subset X$, the relative group $KR_G(X,A)$ is defined as the reduced group $\widetilde{KR}_G(X/A)$.

These contravariant functors are abelian group valued, KR_G is commutative ring valued, and the relative group $KR_G(X,A)$ is a module over the ring $KR_G(X)$.

7.2. Notation Let $R^{p,q} = \mathbb{R}^q \oplus i\mathbb{R}^p$ with the τ action given by complex conjugation, that is, $\tau(yi,x) = (-yi,x)$. Let $B^{p,q}$ and $S^{p,q}$ denote the unit ball and unit sphere in $R^{p,q}$, respectively. Note that $R^{p,p} = \mathbb{C}^p$, and the sphere $S^{p,q}$ has dimension $p+q-1$.

7.3. Definition The (p,q)-suspension groups of $KR_G(X,A)$ are defined as the following relative KR_G-groups

$$KR_G^{p,q}(X,A) = KR_G(X \times B^{p,q}, (X \times S^{p,q}) \cup (A \times B^{p,q})).$$

Since the index p is related to the τ action, the suspension which is free of the $(\tau)G$-action is the index q, one uses this index to build a corresponding $\mathbb{Z}$-graded cohomology theory. For the nonpositive values, one defines the cohomology groups by $KR_G^{-q}(X,A) = KR_G^{0,q}(X,A)$. For defining the corresponding cohomology groups of positive degree, we have to wait until the next chapter where Bott periodicity is discussed.

One crucial property of the functors KR^{-q} is formulated in the next two propositions.

7.4. Proposition *Let (X,A) be a $(\tau)G$-pair. There is a long exact sequence extending to the left for the single-index-suspension groups $KR_G^{-q}(X,A)$ for $q \geq 0$ as follows*

$$\cdots \longrightarrow KR_G^{-1}(X) \longrightarrow KR_G^{-1}(A) \longrightarrow KR_G(X,A) \longrightarrow KR_G(X) \longrightarrow KR_G(A).$$

Using the exact sequence for the triple

$$(X \times B^{p,q}, (X \times S^{p,q}) \cup (A \times B^{p,q}), X \times S^{p,q}),$$

we obtain the following proposition.

7.5. Proposition *Let (X,A) be a $(\tau)G$-pair. There is a long exact sequence extending to the left for the double-index-suspension groups, we have an exact sequence for $q \leq 0$ ending at $q = 0$, and in each index $p \geq 0$*

$$\cdots \to KR_G^{p,-1}(X) \to KR_G^{p,-1}(A) \to KR_G^{p,0}(X,A) \to KR_G^{p,0}(X) \to KR_G^{p,0}(A).$$

Finally, we have the exterior product which is induced by the exterior tensor product.

7.6. Proposition *The ring structure on $KR_G(X)$ extends to the following external product*

$$KR_G^{p',q'}(X',A') \otimes KR_G^{p'',q''}(X'',A'') \longrightarrow KR_G^{p'+p'',q'+q''}(X' \times X'',A)$$

where $A = (A' \times X'') \cup (X' \times A'')$.

There is a graded ring structure on the groups $KR_G^{p,q}(X,A)$ resulting from the exterior product by restricting to the diagonal $\Delta : X \to X \times X$.

7.7. Corollary *The ring structure on $KR_G(X)$ extends to graded ring structure on the groups additive in p and preserving q*

$$KR_G^{p',q}(X,A) \otimes KR_G^{p'',q}(X,A) \longrightarrow KR_G^{p'+p'',q}(X,A).$$

To complete the picture of this functor $KR_G(X)$ and its suspensions, we need Bott periodicity which is taken up in the next chapter.

References

Atiyah, M.F.: Power Operations in K-theory. Quarterly J. Math. Oxford **17**(2): 165–193 (1966)
Atiyah, M.F., Segal, G.G.: Equivariant K-theory and completion. J. Differential Geom. **3**: 1–18 (1969)

Chapter 14
Equivariant K-Theory Functor K_G : Periodicity, Thom Isomorphism, Localization, and Completion

Using the Grothendieck construction in the preceding chapter, we defined the functors K_G for real and complex G-bundles and the Atiyah real KR_G by mapping the semiring of G-vector bundles into its ring envelope. We saw that the basic properties of the equivariant versions of vector bundle theory have close parallels with the usual vector bundle theory, and the same is true for the related relative K-theories. This we carry further in this chapter for the version of topological K-theory that has close relations to index theory.

The simple form of Bott periodicity for complex K-theory is the isomorphism $K(X) \otimes K(S^2) \to K(X \times S^2)$. The 2-sphere S^2 which also can be identified as the one-dimensional complex projective space $P_1(\mathbb{C})$ plays a basic role. We can say that this is the periodicity theorem for a trivial line bundle $X \times \mathbb{C} \to X$ over X. For the case of KR_G periodicity and Thom isomorphism, it is convenient to have a version of the Bott isomorphism for any line bundle $L \to X$ over a space X or space X with involution τ.

We survey two basic results of Atiyah and Segal: the localization theory and the completion theorem for the calculation of K-theory of classifying spaces.

1 Associated Projective Space Bundle to a G-Equivariant Bundle

As usual for a vector space V, the associated projective space $P(V)$ is the space of one-dimensional linear subspaces $W \subset V$. In algebraic geometry, it is usually defined as the space of one-dimensional quotients $V \to U$. The canonical line bundle $L \to P(V)$ is the subbundle $L \subset P(V) \times V$ of the product of (W, x) with $x \in W$ or the quotient construction of $P(V)$, the canonical line bundle $L \to P(V)$ is the quotient line bundle $P(V) \times V \to L$ of $\{U\} \times U$ for $V \to U$ in $P(V)$. If G acts on V, then G acts on $P(V)$ in both cases, and when V is a (G, τ)-vector space, then $(\tau)G$ also acts on $P(V)$ in both cases. This all carries over to vector bundles.

1.1. Definition Let $E \to X$ be a (G, τ)-vector bundle over X with G-action $G \times E \to E$. The associated projective bundle is $p : P(E) \to X$, where the fibre

D. Husemöller et al.: *Equivariant K-Theory Functor K_G : Periodicity, Thom Isomorphism, Localization, and Completion*, Lect. Notes Phys. **726**, 163–173 (2008)
DOI 10.1007/978-3-540-74956-1_15

$P(E)_x = P(E_x)$ is the associated projective space of the vector space fibre E_x. The (G, τ)-bundle structure on $p : P(E) \to X$ comes from the (G, τ)-vector bundle structure on the vector bundle E over X. The canonical line bundle L_E on $P(E)$ is also a (G, τ)-equivariant bundle which is the canonical line bundle on each fibre. We use the quotient version, and this means that the following diagram with L_E and its dual $L_E^\vee$ as (G, τ)-line bundles over $P(E)$ with $(\tau)G$ actions.

$$
\begin{array}{ccccc}
(\tau)G \times L_E^\vee & \longrightarrow & (\tau)G \times p^{-1}(E) & \longrightarrow & (\tau)G \times L_E \\
\downarrow & & \downarrow & & \downarrow \\
L_E^\vee & \longrightarrow & p^{-1}(E) & \longrightarrow & L_E
\end{array}
$$

1.2. Remark For a one-dimensional vector space U and nonzero $u \in U$ we have for any finite-dimensional vector space V, an isomorphism of vector spaces $V \to U \otimes V, v \mapsto u \otimes v$. It induces an isomorphism $P(V) \to P(U \otimes V)$ which is independent of u. Hence, for each line bundle L and vector bundle E over X, we have a well-defined isomorphism of projective space bundles $P(E) \to P(L \otimes E)$.

2 Assertion of the Periodicity Theorem for a Line Bundle

The following theorem was introduced for a finite group G by Atiyah and proved for a compact group by Segal. It is based on a careful analysis of the clutching functions for a bundle over $X \times S^2$ in terms of a Fourier series. We here give a treatment within the real setting.

2.1. Theorem *Let L be a (G, τ)-equivariant line bundle over a $(\tau)G$-space X with corresponding class $a = [L] \in KR_G(X)$. The ring morphism*

$$
\theta : KR_G(X)[t]/(t-1)(ta-1) \longrightarrow KR_G(P(L \oplus \mathbb{C}))
$$

defined by $\theta(t) = [L_{L \oplus \mathbb{C}}]$ is a well-defined isomorphism.

If we apply the theorem to trivial one-dimensional bundles, we obtain the following corollary. Therefore, note that as $(\tau)G$-spaces, we have $S^{1,1} \cong P(\mathbb{C}^2)$.

2.2. Corollary *For a $(\tau)G$-space X, the ring morphism*

$$
\theta : KR_G(X)[t]/(t-1)^2 \longrightarrow KR_G(X \times S^{1,1})
$$

defined by $\theta(t) = [L_{\mathbb{C} \oplus \mathbb{C}}]$ is a well-defined isomorphism.

2.3. Preliminaries to a Sketch of the Proof of (2.1) Choose a metric on L invariant under the involution τ or in general (G, τ)-structure. The unit circle bundle S is then a real space. The section z of $\pi^*(L)$ defined by the inclusion $S \to L$ is a real section together with all its powers z^k. The function z is the clutching function for $(1,L)$. To see that $(1,z,L)$ is the bundle $H^\vee$ defined as the subbundle of $\pi^*(L \oplus 1)$, we note that for each $y \in P(L \oplus 1)_x$, the fibre $H_y^\vee$ is a subspace of $(L \oplus 1)_x$ with $H_\infty^\vee = L_x \oplus 0$ and $H_0^\vee = 0 \oplus 1_x$. In particular, the composition

$$H^\vee \longrightarrow \pi^*(L \oplus 1) \longrightarrow \pi^*(1)$$

induced by the projection $L \oplus 1 \to 1$ defines an isomorphism

$$f_0 : H^\vee|P^0 \longrightarrow \pi_0^*(1),$$

and similarly, the composition $H^\vee \to \pi^*(L \oplus 1) \to \pi^*(L)$ induced by the projection $L \oplus 1 \to L$ defines an isomorphism

$$f_\infty : H^\vee|P^\infty \longrightarrow \pi_0^*(L).$$

Hence, $f = f_0 f_\infty^{-1} : \pi^*(1) \to \pi^*(L)$ is a clutching function for $H^\vee$. For $y \in S_x$, $f(y)$ is the isomorphism whose graph is $H_y^\vee$. Since $H_y^\vee$ is the subspace of $L_x \oplus 1_x$ generated by $y \oplus 1$, where $y \in S_x \subset L_x$, $1 \in \mathbb{C}$, we see that f is our section z. Thus, we have an isomorphism of real bundles $H^{k\otimes} \to (1, z^{-k}, L^{(-k)\otimes})$.

Now, for any clutching function $f \in \Gamma\mathrm{Hom}(\pi^*(E^0), \pi^*(E^\infty))$ which is a real section, there is a Fourier series with coefficients real sections a_k of $\mathrm{Hom}(L^{k\otimes} \otimes E^0, E^\infty)$ given by $\tau(x) = \bar{x}$

$$\bar{a}_k(x) = \overline{a_k(\bar{x})} = -\frac{1}{2\pi i}\overline{\int_{S_{\bar{x}}} f_{\bar{x}}(z_{\bar{x}})^{-k-1} dz_{\bar{x}}} = \frac{1}{2\pi i}\int_{S_x} f_x(z_x)^{-k-1} dz = a_k(x),$$

where τ reverses the orientation of S_x and f and z are real. At a real point $x \in X$, the condition that f_x is real becomes $f_x(e^{-i\theta}) = \overline{f_x(e^{i\theta})}$, which implies that the Fourier coefficients are real.

The Fourier series is approximated with a finite Laurent series, and this in turn is linearized by adding dimensions so that only a function in x added to a function in x times z remains. This is the linear case, and the real structure goes through the analysis.

It remains to analyze the linear case. In this case, we split the linear clutching function p for two vector bundles E^0 and E^∞ into two parts $p = p_+ \oplus p_-$ with homotopies $p(t) = p_+(t) \oplus p_-(t)$, where

$$p_+(t) = a_+ z + t b_+ \quad \text{and} \quad p_-(t) = t a_- z + b_- \quad \text{for} \quad t \in [0,1]$$

leading to an isomorphism between (E^0, p, E^∞) and the direct sum

$$(E_+^0, z, L \otimes E_+^0) \oplus (E_-^0, 1, E_-^0).$$

For this, we use the fact that for a linear operator $T : V \to V$ on a finite-dimensional space, the complex integral

$$Q = \frac{1}{2\pi i} \int_S (z - T)^{-1} dz,$$

where S is a circle not going through an eigenvalue of T is a projection commuting with T. We have a direct sum decomposition $V = V(+) \oplus V(-)$, where $V(+) = QV$ and $V(-) = (1 - Q)V$, and T decomposes as $T = T(+) \oplus T(-)$, where $T(+)$ has all the eigenvalues of T inside S and $T(-)$ has all the eigenvalues of T outside S.

Now, we go to the related topic of introducing the suspension and relative groups.

2.4. Notation Let $\mathbb{R}^{p,q} = \mathbb{R}^q \oplus i\mathbb{R}^p$ with the involution given by $\tau(a, ib) = (a, -ib)$. Let $S^{p,q} \subset B^{p,q} \subset \mathbb{R}^{p,q}$ denote the unit sphere and unit ball contained in the Euclidean space. Note that $\dim(S^{p,q}) = p + q - 1$ and $\mathbb{R}^{p,p} = \mathbb{C}^p$.

2.5. Definition The relative group

$$KR(X, A) = \ker(KR(X/A) \longrightarrow KR(*)),$$

the kernel of the induced map of $* \to X/A$ restriction to the base point. In this definition, (X, A) is a pair with an involution on X carrying A to itself, and hence, it fixes the base point of X/A. Note that $KR(X) = KR(X, \emptyset)$ where $\emptyset$ is the empty set so that $X/\emptyset = X \sqcup \{*\}$.

In relative equivariant K-theory, we use the same construction.

2.6. Definition For a pair (X, A) with involution, the bigraded suspension groups are defined as

$$KR^{p,q}(X, A) = KR(X \times B^{p,q}, X \times S^{p,q} \cup A \times B^{p,q}).$$

The usual graded suspension groups with an involution are denoted by $KR^{-q} = KR^{0,q}$.

2.7. Exact Sequence for a Pair (X, A) We have the following exact triangle

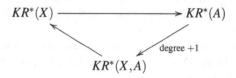

For a triple (X, A, B), it takes the following form

$$
\begin{array}{ccc}
KR^*(X, B) & \longrightarrow & KR^*(A, B) \\
 & & \swarrow \text{ degree } +1 \\
 & KR^*(X, A) &
\end{array}
$$

This, we apply to $(X \times B^{p,0}, (X \times S^{p,0}) \cup (A \times B^{p,0}), X \times S^{p,0})$

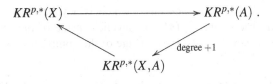

These result from the ring structure on $KR(X)$ and take the form (external products)

$$KR^{p',q'}(X',A') \otimes KR^{p'',q''}(X'',A'') \longrightarrow KR^{p'+p'',q'+q''}(X,A),$$

where $X = X' \times X''$ and $A = (X' \times A'') \cup (A' \times X'')$. Internal products result by restricting to the diagonal.

Now, we return to the periodicity theorem (2.1).

2.8. Notation Let b be the Bott element given by

$$b = [H] - 1 \in KR^{1,1}(*) = KR(B^{1,1}, S^{1,1}) = \widetilde{KR}(P(\mathbb{C}^2)),$$

and let the Bott morphism $\beta : KR^{p,q}(X,A) \to KR^{p+1,q+1}(X,A)$ be given by $\beta(x) = b.x$. Here, H is the canonical line bundle on $P(\mathbb{C})^2$.

2.9. Theorem *For each pair (X,A) with an involution, the Bott morphism β : $KR^{p,q}(X,A) \to KR^{p+1,q+1}(X,A)$ is an isomorphism.*

3 Thom Isomorphism

The Thom isomorphism depends on a construction of elements in $KR_G(X,A)$ given by a complex of vector bundles which is acyclic on A. This generalizes the clutching construction which cannot be used so directly to describe the product structure on $KR_G(X,A)$. The Thom isomorphism is an assertion that a module over $KR_G(X)$ related to a (G,τ)-vector bundle is of rank one generated by the Thom class. The construction of the Thom class is in terms of a complex of exterior algebras.

3.1. Definition The Koszul complex associated to a (G,τ)-vector bundle E and an invariant section σ is

$$\lambda^*\langle E \rangle : 0 \longrightarrow \mathbb{C} \xrightarrow{d} E = \Lambda^1 E \xrightarrow{d} \Lambda^2 E \xrightarrow{d} \dots \xrightarrow{d} \Lambda^i E \xrightarrow{d} \dots,$$

where $d(\xi) = \xi \wedge \sigma(x)$ if $\xi \in \Lambda^i E_x$. Observe that $\lambda^*\langle E \rangle$ is acyclic at all points x, where $\sigma(x) \neq 0$.

3.2. Example Let E be a (G,τ)-vector bundle over X and let $p : E \to X$ denote the projection. Then, the diagonal $\delta : E \to E \times_X E = p^*(E)$ is a natural section of $p^*(E)$, and it vanishes exactly on the image of the zero section. We denote the Koszul complex on E formed from $p^*(E)$ and δ by $\Lambda^*\langle E \rangle$.

3.3. Remark Let M^* be a complex on X which is acyclic outside of a compact set K of X, and observe that $p^*(M^*)$ is acyclic outside of the set $p^*(K)$. Then, $\Lambda^*\langle E \rangle \otimes p^*(M^*)$ is acyclic outside the image of the compact set K under the zero section.

3.4. Definition Let E be a (G, τ)-vector bundle over X. Then, the Thom morphism $\phi_! : KR_G(X) \to KR_G(E)$ carries a class in $KR_G(X)$ defined by a complex M^* of vector bundles acyclic outside a compact set K of X to the complex $\Lambda^*\langle E \rangle \otimes p^*(M^*)$. If X is compact, then $\Lambda^*\langle E \rangle$ is acyclic outside the compact zero section, and the image under the Thom isomorphism of 1 is $\phi_!(1) = \Lambda_E$, where Λ_E is the class of $\Lambda^*\langle E \rangle$ in $KR_G(E)$. It is called Thom class of E. Since Λ_E is related to alternating sum of elements of $\Lambda^*\langle E \rangle$, it also written as $\Lambda_{-1}(E)$.

3.5. Remark Let E be a (G, τ)-vector bundle over X with zero section $\phi : X \to E$. Then, the Thom morphism $\phi_! : KR_G(X) \to KR_G(E)$ is a special example of an induced morphism $f_! : KR_G(X) \to KR_G(Y)$ defined for certain maps $f : X \to Y$, which we will consider in the next section. It has the basic property that $\phi^{-1}\phi_!(\xi) = \xi \lambda_E$, where λ_E is the class of $\lambda^*\langle E \rangle$ in $KR_G(X)$.

3.6. Remark Replacing X and E by, respectively, $X \times \mathbb{R}^q$ and $E \times \mathbb{R}^q$, we obtain the Thom morphism $\phi_! : KR_G^{-q}(X) \to KR_G^{-q}(E)$.

Now, the main theorem is the following which is called the Thom isomorphism theorem.

3.7. Theorem *Let E be any (G, τ)-vector bundle over a locally compact G-space X. Then, the Thom morphism $\phi_! : KR_G(X) \to KR_G(E)$ is an isomorphism.*

3.8. First Reduction If the Thom morphism is an isomorphism for compact $(\tau)G$-spaces X, then it is an isomorphism for locally compact $(\tau)G$-spaces. This is done by observing that it is enough to prove the theorem for relatively compact open $(\tau)G$-subspaces U of X. Then, by the exact sequence for the pair $(\bar{U}, \bar{U} - U)$, one is reduced to the case of a compact base space. Finally, we use the fact that the following diagram has split exact rows,

$$
\begin{array}{ccccccccc}
0 & \longrightarrow & KR_G^{-q}(X) & \longrightarrow & KR_G(X \times S^q) & \longrightarrow & KR_G(X) & \longrightarrow & 0 \\
& & \downarrow{\phi_!} & & \downarrow{\phi_!} & & \downarrow{\phi_!} & & \\
0 & \longrightarrow & KR_G^{-q}(E) & \longrightarrow & KR_G(E \times S^q) & \longrightarrow & KR_G(E) & \longrightarrow & 0.
\end{array}
$$

3.9. Line Bundle Case This is a return to the periodicity theorem in (2.1). Now, we use the fact that E is identified with $P(E \oplus \mathbb{C}) - P(E)$ so that $KR_G(E)$ is the kernel of the restriction morphism

$$
KR_G P(E \oplus \mathbb{C})) \longrightarrow KR_G(P(E)),
$$

which is generated by $H - \pi^*(E)$ or $\mathbb{C} - \pi^*(E) \otimes H^\vee$, where $\pi : P(E \oplus \mathbb{C}) \to X$ is the projection. Restricting to $E \subset P(E \oplus \mathbb{C})$, we obtain the complex Λ_E^* for $H|E$ is trivial. Thus, we obtain that $KR_G(E)$ is the free $KR_G(X)$-module on the one generator Λ_E.

3.10. Sums of Line Bundles Since the compact case is known for line bundles, it follows that locally compact case holds for line bundles. Hence, the general result is true for $(\tau)G$ bundles which are a sum of line bundles because the Thom isomorphism is transitive.

3.11. Proposition *Let $p' : E' \oplus E'' \to E'$ and $p'' : E' \oplus E'' \to E''$ be the projections from the direct sum of $(\tau)G$ bundles over X. Then, we have an isomorphism $\Lambda^*\langle E' \oplus E''\rangle \to (p')^{-1}\Lambda^*\langle E'\rangle \otimes (p'')^{-1}\Lambda^*\langle E''\rangle$, and the following diagram commutes, where ϕ is the zero section*

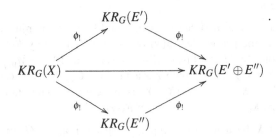

Proof. The first statement is the coproduct-preserving character of the exterior algebra. The second statement uses a natural isomorphism $\Lambda^*\langle E' \oplus E''\rangle \to (p')^{-1}$ $\Lambda^*\langle E'\rangle \otimes \Lambda^*\langle (\pi')^{-1}(E'')\rangle$ for the commutativity of one triangle, where $(p'')^{-1}\Lambda^*$ $\langle E''\rangle$ is isomorphic to $\Lambda^*\langle (\pi'')^{-1}(E'')\rangle$. This proves the proposition.

As a corollary of (3.7) and (3.11), we have the following.

3.12. Corollary *There is a certain element in $KR_G^{-2}(*)$ such that multiplication by this element induces an isomorphism $KR_G^{-q}(X) \to KR_G^{-q-2}(X)$.*

Finally, we mention the final key step which is best seen from the point of view of the index theorem of elliptic differential operators. It is a step reducing the general case to a sum of line bundles.

3.13. Proposition *Let $i : T \to G$ be the inclusion of a maximal torus into a compact connected Lie group. For a locally compact $(\tau)G$-space X, there is a natural morphism of $KR_G^*(X)$-modules $i_! : KR_T^*(X) \to KR_G^*(X)$ such that $i_!(1) = 1$ and $i_! i^*$ is the identity.*

The application to the final step in the proof of the Thom isomorphism can be achieved through an embedding $G \subset U(N)$.

4 Localization Theorem of Atiyah and Segal

The results in this section are contained in Segal (1968) and Atiyah and Segal. The following is a basic property of characters on compact Lie groups relative to subgroups and conjugacy classes.

4.1. Proposition *Let G be a compact Lie group, and let H be a closed subgroup of G, $H \neq G$. If γ is a conjugacy class in G with intersection $\gamma \cap H = \emptyset$, then there exists a character χ on G with $\chi|H = 0$ and $\chi(\gamma) \neq 0$.*

4.2. Remark For each conjugacy class γ, we have a ring morphism $v_\gamma : R(G) \to \mathbb{Z}$ given by substitution $v_\gamma(\chi) = \chi(\gamma)$, where a character $\chi = trace(\rho)$, $\rho \in R(G)$ is always constant on a conjugacy class. The kernel $\mathfrak{p}(\gamma) = \ker(v_\gamma)$ is a prime ideal in $R(G)$, and the function which assigns to the conjugacy class $\gamma \in cl(G)$ the prime ideal $\mathfrak{p}(\gamma)$ is an injection $cl(G) \to \mathrm{Spec}(R(G))$, the space of prime ideals in the commutative ring $R(G)$.

4.3. Notation Let M be a module over $R(G)$, and let γ be a conjugacy class in G. The localization at the prime ideal $\mathfrak{p}(\gamma)$ of $R(G)$ is denoted by $R(G)_\gamma$ and of the module M is denoted by M_γ, that is, $R(G)_\gamma = R(G)_{\mathfrak{p}(\gamma)}$ and $M_\gamma = M_{\mathfrak{p}(\gamma)}$. If $\chi \in R(G)$ is nonzero on γ, that is, $\chi(\gamma) \neq 0$, then χ is a unit in $R(G)_\gamma$. Since $R(G) = K_G(*)$ and the product $K_G(X) \otimes K_G(Y) \to K_G(X \times Y)$ is associative, commutative, and with a unit, it follows that $K_G(X)$ is a module over $R(G)$.

4.4. Notation Let X be a locally compact G-space. For $s \in G$, the subspace of all $x \in X$ with $sx = x$ is denoted by $X(s) \subset X$. If $x \in X(s)$ and $t \in G$, then $tx \in X(tst^{-1})$ since $(tst^{-1})(tx) = tsx = tx$. In other words, $tX(s) = X(tst^{-1})$. Let $X(\gamma) = \bigcup_{s \in \gamma} X(s) = \bigcup_{t \in G} tX(s)$. When $X(s)$ is compact, then $X(\gamma)$ is also compact.

We have that localization and geometry are related by the following basic theorem of Atiyah and Segal.

4.5. Theorem *Let X be a locally compact G-space, and let γ be a conjugacy class in G with inclusion $i : X(\gamma) \to X$.*

(1) If $X(\gamma)$ is empty, then the localization $K_G^*(X)_\gamma = 0$.
(2) In general, the localization of the induced morphism

$$K_G^*(i)_\gamma : K_G^*(X)_\gamma \longrightarrow K_G^*(X(\gamma))_\gamma$$

is an isomorphism.

Proof. Statement (1) implies (2) from the exact triangle in K_G^* cohomology for the induced morphism i_γ^* from inclusion i

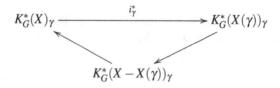

By (1), we have $K_G^*(X - X(\gamma))_\gamma = 0$, and hence, the localization i^* is an isomorphism.

Now, we return to the situation of proposition (4.1) of a closed subgroup H of $G, H \neq G$. If γ is a conjugacy class in G with intersection $\gamma \cap H = \emptyset$, then there exists a character χ on G with $\chi|H = 0$ and $\chi(\gamma) \neq 0$. We localize $R(G) \to R(G)_\gamma$, and $\chi \in R(G)$ becomes a unit in $R(G)_\gamma$, and since $\chi|H = 0$, this unit in $R(G)_\gamma$ annihilates the algebra $R(H)$, its localization $R(H)_\gamma$, and $M_\gamma = 0$ for any $R(H)$-module M.

We apply these remarks to $M = K_G^*(X)$, where X is a compact G-space admitting a surjective G-map $X \to G/H$. The factorization $X \to G/H \to *$ gives a factorization showing that $K_G^*(X)$ is an $R(H)$-module.

$$
\begin{array}{ccc}
R(G) & \longrightarrow & R(H) \\
\downarrow & & \downarrow \\
K_G^*(*) \longrightarrow & K_G^*(G/H) \longrightarrow & K_G^*(X)
\end{array}
$$

Hence, we deduce that if γ is a conjugacy class of G which does not intersect a closed subgroup H of G, then the localization $K_G^*(X)_\gamma = 0$.

Further, if Y is any closed G-subspace of X, then it admits a G-map $Y \to G/H$, and so $K_G^*(Y)_\gamma = 0$ and also $K_G^*(X,Y)_\gamma = 0$ from the exact triangle associated a subspace because localization is an exact functor.

Let X be a locally compact G-space, and let $Y \subset X$ be an orbit with isotropy group H. Since there exists "slices" or local cross sections to $X \to G\backslash X$, we can find a closed G-neighborhood V of Y in X with a G-retraction onto Y. In the case of a G-submanifold Y of X, we can take V to be a closed tubular neighborhood defined by a G-invariant Riemannian metric.

Now in the situation where such neighborhoods V of orbits exist, we can cover any compact G-subspace L of X by a finite number of sets $L_i = V_i \cap L$. Let H_i be the isotropy group connected with V_i with corresponding G-map $L_i \to G/H_i$. Assume that γ is a conjugacy class of elements of G having no fixed points in X so that $\gamma \cap H_i = \emptyset$ for all H_i. Thus, by the above considerations, we have $K_G^*(L_i)_\gamma = 0$. By an induction argument, we have $K_G^*(L)_\gamma = 0$.

Replacing L by any compact G-subset L' and using the exact triangle for (L, L'), we deduce that $K_G^*(L, L')_\gamma = 0$. In particular, if U is an open relatively compact G-subspace of X, we have

$$
K_G^*(U)_\gamma = K_G^*(\bar{U}, \partial \bar{U})_\gamma = 0.
$$

Since $K_G^*(X)_\gamma = \lim_{\substack{\longrightarrow \\ U \subset \subset X}} K_G^*(U)_\gamma$ and localization commutes with direct limits, we have established (1), and this proves the theorem.

5 Equivariant K-Theory Completion Theorem of Atiyah and Segal

The main reference for this section is Atiyah and Segal (1969).

Let G be a compact Lie group with corresponding universal bundle $E(G) \to B(G)$, and let X be a compact $(\tau)G$-space.

5.1. Definition Let V be a (G, τ)-vector bundle on X. Using the fibre bundle construction, we have a τ-vector bundle $E(G)[V] \to E(G)[X]$, and this construction induces a morphism

$$\alpha : KR_G^*(X) \longrightarrow KR^*(E(G)[X])$$

in KR-theory. The base space $E(G)[X]$ is called the Borel construction or homotopy quotient of X by G and is denoted also by X_G.

5.2. Remark Since $KR^*(E(G)[X]) = KR_G^*(E(G) \times X)$, where the action of G is free on $E(G) \times X$, we can interpret α as the morphism induced by the projection $E(G) \times X \to X$ under KR_G^*. There is a difficulty with the definition of $KR^*(E(G)[X])$ since $E(G)$ and $E(G)[X]$ are not compact. As usual in mathematics, we must extend the definition or concept. In particular, α as it stands is not an isomorphism.

5.3. Definition On the Borel construction $E(G)[X]$, we have an ind-structure $(E(G)[X])_n = E_n(G)[X]$ given by the increasing union $E_n(G)$ of the Milnor construction $E(G)$. In K-theory, we have the pro-object

$$KR^*(E(G)[X]) = \{KR^*((E(G)[X])_n)\}_{n \in \mathbb{N}}$$

for a compact group G acting on a compact τ-space.

5.4. Remark On the right side of morphism

$$\alpha : KR_G^*(X) \longrightarrow KR^*(E(G)[X])$$

we have a pro-structure, and now on the left side we define a pro-structure by taking $X = *$, a point where $KR_G(*) = RR(G)$ the character ring of G for G-modules with a τ-structure. Related to the morphism $\alpha : KR_G^*(X) \to KR^*(E(G)[X])$ is the factorization

$$\alpha_n : RR(G) = KR_G^*(*) \longrightarrow KR_G^*(E_n(G)[*]) = KR_G^*(B_n(G)) \xrightarrow{\varepsilon} \mathbb{Z}.$$

5.5. Remark Since $B_n(G)$ is the union of n contractible open subsets U_i, where the ith join coordinate does not vanish, the product of any n elements of the kernel in $KR_G^*(B_n(G))$ is zero. It is also the case then that $I(G)$ is the augmentation ideal, and α_n factors as

$$\alpha_n : KR_G^*(X)/I(G)^n KR_G^*(X) \longrightarrow KR^*((E(G)[X])_n).$$

5.6. Theorem *Let X be a compact G-space such that $KR_G^*(X)$ is a finitely generated RR(G)-module. Then, the morphism α_n defined as*

$$\alpha_n : KR_G(X)/I(G)^n KR_G(X) \longrightarrow KR((E(G)[X])_n)$$

is an isomorphism of pro-rings.

References

Atiyah, M.F., Segal, G.B.: Equivariant K-theory and completion J.Differential Geom. 3: 1–18 (1969)

Atiyah, M.F., Segal, G.B.: The index of elliptic operators II. Ann. of Math. (2) Annals of Mathematics, Second Series 87: 531–545 (1968)

Segal, G.B.: Equivariant K-theory. Publ Math.I.H.E.S. Paris 34: 129–151(1968)

Chapter 15
Bott Periodicity Maps and Clifford Algebras

Before there was K-theory, there was Bott periodicity. It was a result of the analysis of homogeneous spaces, their loop spaces, and classifying spaces of limits of Lie groups. In a parallel study of Clifford algebras, we see a periodicity which has much to do with Bott periodicity. This is pointed out in this chapter.

When topological K-theory was introduced by analogy with Grothendieck's coherence sheaf K-theory, Bott periodicity was there and made possible an entire cohomology theory. Immediately, the vector bundle class approach to K-theory was coupled with the homotopy theory as in the discussion of the universal bundles and classifying spaces. The universal bundles of a group became the classifying space of K-theory by a simple limiting process. These limit spaces is the domain of the first forms of K-theory.

For complex vector bundles, the group $GL(n, \mathbb{C})$ and its compact subgroup $U(n)$ both play a role. The fact that the use of principal bundles over $U(n)$ or over $GL(n, \mathbb{C})$ follows from the existence of a Hermitian metric on vector bundles. There is another approach which we take up in the next chapter by the classical Gram–Schmidt process. We use the compact forms, that is, $U(n)$, for the homotopy classification of K-theory.

After the homotopy analysis of Bott periodicity, we consider the Clifford algebra description. This we do in the context of KR-theory as in the previous chapter.

1 Vector Bundles and Their Principal Bundles and Metrics

In $2(8.2)$, we introduced the notion of a metric on a vector bundle, and in $2(8.3)$ using the metric on the universal bundle, we saw that every vector bundle has a metric.

1.1. Remark When we associate the principal bundle to a complex vector bundle, we can either use changes of coordinates in $GL(n, \mathbb{C})$ preserving only the linear structure or changes of coordinates in $U(n)$ preserving the linear and metric structures. An n-dimensional $\mathbb{C}$-vector bundle E over X is of the form $P[\mathbb{C}^n]$, where P is

D. Husemöller et al.: *Bott Periodicity Maps and Clifford Algebras*, Lect. Notes Phys. **726**, 175–188 (2008)
DOI 10.1007/978-3-540-74956-1_16   © Springer-Verlag Berlin Heidelberg 2008

a $GL(n,\mathbb{C})$-principal bundle or of the form $P'[\mathbb{C}^n]$, where P' is a $U(n)$-principal bundle associated to P by reduction of the structure group from $GL(n,\mathbb{C})$ to the group $U(n)$ of unitary n by n matrices.

1.2. Remark In terms of the classifying spaces, the set of isomorphism classes $\mathrm{Vect}^n(\mathrm{B})$ of n-dimensional complex vector bundles over B can be identified with either $[B,BGL(n,\mathbb{C})]$ or $[B,BU(n)]$. The process of assigning to an n-dimensional $\mathbb{C}$-vector bundle E the Whitney sum with the trivial bundle $E \oplus \mathbb{C}^q$ (or $E \times \mathbb{C}^q$) corresponds to the inclusions $GL(n,\mathbb{C}) \to GL(n+q,\mathbb{C})$ and $U(n) \to U(n+q)$ which induce in terms of either homotopy description the functions corresponding to stabilization of vector bundles

$$[B,BGL(\mathbb{C},n)] \longrightarrow [B,BGL(\mathbb{C},n+q)] \quad \text{and} \quad [B,BU(n)] \longrightarrow [B,BU(n+q)].$$

1.3. Remark Similar assertions hold for real and quaternion vector bundles where stabilization takes, respectively, the form

$$[B,BGL(\mathbb{R},n)] \longrightarrow [B,BGL(\mathbb{R},n+q)] \quad \text{and} \quad [B,BO(n)] \longrightarrow [B,BO(n+q)].$$

$$[B,BGL(\mathbb{H},n)] \longrightarrow [B,BGL(\mathbb{H},n+q)] \quad \text{and} \quad [B,BSp(n)] \to [B,BSp(n+q)].$$

Here, $\mathbb{H}$ is the quaternions and $Sp(n)$ is the symplectic subgroup of quaternionic linear maps preserving the $\mathbb{H}$-metric. In the case of $U(n)$ and $O(n)$, we have the subgroups of determinant 1 matrices.

2 Homotopy Representation of *K*-Theory

K-theory is the result of stabilization using the inclusions of the form $U(n) \to U(n+q)$. Hence, we introduce the following limit groups.

2.1. Definition For each of the five sorts of groups in (1.2) and (1.3), we have the following five direct limit groups

$$O = \bigcup_{n\geq 0} O(n),\ SO = \bigcup_{n\geq 0} SO(n),$$

$$U = \bigcup_{n\geq 0} U(n),\ SU = \bigcup_{n\geq 0} SU(n),\ \text{and}\ Sp = \bigcup_{n\geq 0} Sp(n).$$

We give them the usual direct limit topology, and they are topological groups with respect to this topology.

2.2. Remark The classifying spaces of these groups can be described as limit spaces

$$BO = \bigcup_{n\geq 0} BO(n),\ BSO = \bigcup_{n\geq 0} BSO(n),\ BU = \bigcup_{n\geq 0} BU(n),$$

$$BSU = \bigcup_{n \geq 0} BSU(n), \text{ and } BSp = \bigcup_{n \geq 0} BSp(n).$$

In the case of a compact space, X we have the following relations

$$[X, BO] = \lim_{\substack{\longrightarrow \\ n \geq 0}} [X, BO(n)], \quad [X, BSO] = \lim_{\substack{\longrightarrow \\ n \geq 0}} [X, BSO(n)],$$

$$[X, BU] = \lim_{\substack{\longrightarrow \\ n \geq 0}} [X, BU(n)], \quad [X, BSU] = \lim_{\substack{\longrightarrow \\ n \geq 0}} [X, BSU(n)],$$

$$\text{and } [X, BSp] = \lim_{\substack{\longrightarrow \\ n \geq 0}} [X, BSp(n)].$$

Hence, the stabilized bundle theory, that is, K-theory, has a homotopy interpretation as homotopy classes of maps on a compact space due to the interchanging of increasing unions inside $[X,]$ with inductive limits outside of the sets of homotopy classes of maps. Explicitly, $f : X \rightarrow BU(n)$ and $f' : X \rightarrow BU(n+q)$ represent the same points in $\lim_{\substack{\longrightarrow \\ n \geq 0}} [X, BU(n)]$ if and only if $[f'] = [jf]$ for the natural inclusion $j : BU(n) \rightarrow BU(n+q)$.

2.3. Remark In each of the five sequences of groups, denoted by $G(n) \rightarrow G(n+q)$, we have a natural inclusion

$$G(m) \times G(n) \longrightarrow G(m+n)$$

inducing on the classifying spaces a map $BG(m) \times BG(n) \rightarrow BG(m+n)$. After suitable choices all of which lead to homotopic equivalent maps, we have a map on the limit group $G \times G \rightarrow G$. This observation can be applied to the following cases.

On the limit-classifying spaces, we have multiplications

$$BO \times BO \longrightarrow BO, \quad BSO \times BSO \longrightarrow BSO,$$

$$BU \times BU \longrightarrow BU, \quad BSU \times BSU \longrightarrow BSU, \text{ and } BSp \times BSp \rightarrow BSp.$$

These are called the H-space structures on the five limit-classifying spaces.

As with loop spaces, these are not strictly associative maps, but they are homotopy-associative maps. They induce maps

$$[X, BG]_* \times [X, BG]_* = [X, BG \times BG]_* \longrightarrow [X, BG]_*$$

and

$$[X, \mathbb{Z} \times BG]_* \times [X, \mathbb{Z} \times BG]_* = [X, \mathbb{Z} \times BG \times \mathbb{Z} \times BG]_* \longrightarrow [X, \mathbb{Z} \times BG]_*$$

making $[X, BG]_*$ and $[X, \mathbb{Z} \times BG]_*$ into groups for each of the five cases $G = O, SO, U, SU$, or Sp.

2.4. Remark We have the following formulas for real and complex K-theory of a space where the isomorphisms are given by inducing the universal bundle on some $BU(n) \subset BU$ or $BO(n) \subset BO$.

$$[X, \mathbb{Z} \times BU]_* \longrightarrow K(X) \text{ and } [X, \mathbb{Z} \times BO]_* \longrightarrow KO(X).$$

These isomorphisms should be thought of as the analogue of the Eilenberg-MacLane isomorphism 9(6.2) for K-theory.

Here the base point plays a role in that $[\,,\,]_*$ refers to base-point-preserving maps and homotopies, and the extra factor of $\mathbb{Z}$ takes care of the possibly nonconnected X. Now we return to the suspension extension of the previous chapter, 14(2.6), where we handle the situation more directly without a relative theory.

2.5. Definition We define negative K-groups by

$$K^{-q}(X) = [S^q \wedge X, BU]_* \text{ and } KO^{-q}(X) = [S^q \wedge X, BO]_*.$$

These are functors on the category of pointed spaces and pointed homotopy classes of maps. They have all of the properties of classical cohomology theory contained in Chap. 9, Sect. 3, for example, the suspension property $K^{-q+1}(S(X)) \to K^{-q}(X)$ is an isomorphism. At this point, the K-groups are defined only for negative indices, but using the following adjunction relation, we see the relevance of knowing the loop spaces $\Omega^p(Y) = \mathrm{Map}_*(S^p, Y)$ of the infinite classical groups and their classifying spaces.

2.6. Assertion The isomorphisms of spaces

$$\mathrm{Map}(X \times S^p, Y) \to \mathrm{Map}(X, \mathrm{Map}(S^p, Y))$$

and

$$\mathrm{Map}(X \times S^p, Y) \to \mathrm{Map}(S^p, \mathrm{Map}(X, Y))$$

of 6(2.1) restrict on the right and pass to a quotient on the left to give isomorphisms of pointed spaces

$$\mathrm{Map}_*(X \wedge S^p, Y) \longrightarrow \mathrm{Map}_*(S^p, \mathrm{Map}_*(X, Y))$$

or

$$\mathrm{Map}_*(S^p(X), Y) \longrightarrow \Omega^p(\mathrm{Map}_*(X, Y))$$

and

$$\mathrm{Map}_*(X \wedge S^p, Y) \longrightarrow \mathrm{Map}_*(X, \mathrm{Map}_*(S^p, Y))$$

or

$$\mathrm{Map}_*(S^p(X), Y) \longrightarrow \mathrm{Map}_*(X, \Omega^p(Y)).$$

For homotopy classes, the last isomorphism takes the form

$$[S^p(X), Y]_* \longrightarrow [X, \Omega^p(Y)]_*.$$

2.7. Remark The last result implies that the suspensions for K-theory have another form using loop spaces of BU or BO, that is, we have

$$K^{-q}(X) = [S^q \wedge X, BU]_* = [X, \Omega^q(BU)]_*$$

and

$$KO^{-q}(X) = [S^q \wedge X, BO]_* = [X, \Omega^q(BO)]_*.$$

When this theory was being set up by Atiyah and Hirzebruch (1961),

Bott had already studied the loop spaces of BU and BO and discovered in 1957–1958 their periodicity properties which we consider in the next section. It is a twofold periodicity of BU and an eightfold periodicity of BO.

3 The Bott Maps in the Periodicity Series

The *real periodicity* where V, X, and Y are, respectively, countable dimensional real, complex, and quaternionic vector spaces with complexification $V_{\mathbb{C}}$ and quaternionic scalar extensions $V_{\mathbb{H}}$ and $X_{\mathbb{H}}$.

p	$C_{p,0}$		Periodicity map		$C_{0,q}$	q
0	$\mathbb{R}$		$\frac{U(V_{\mathbb{C}})}{O(V)}$	$\Omega\left(\frac{SU(V_{\mathbb{C}}'\oplus V_{\mathbb{C}}'')}{SO(V'\oplus V'')}\right)$	$\mathbb{R}$	0
				$\searrow \phi_4$		
1	$\mathbb{C}$		$\frac{Sp(X_{\mathbb{H}})}{U(X)}$	$\Omega\left(\frac{Sp(V_{\mathbb{H}})}{U(V_{\mathbb{C}})}\right)$	$\mathbb{R}\oplus\mathbb{R}$	1
				$\searrow \phi_6$		
2	$\mathbb{H}$		$Sp(Y)$	$\Omega(Sp(X_{\mathbb{H}}))$	$\mathbb{R}(2)$	2
				$\searrow \nwarrow$		
3	$\mathbb{H}\oplus\mathbb{H}$	$B(Sp)=\frac{Sp(Y'\oplus Y'')}{Sp(Y')\oplus Sp(Y'')}$		$\Omega(B(Sp(Y)))$	$\mathbb{C}(2)$	3
				$\searrow \phi_1$		
4	$\mathbb{H}(2)$		$\frac{U(Y)}{Sp(Y)}$	$\Omega\left(\frac{Sp(Y'\oplus Y'')}{Sp(Y')\oplus Sp(Y'')}\right)$	$\mathbb{H}(2)$	4
				$\searrow \phi_3$		
5	$\mathbb{C}(4)$		$\frac{SO(X)}{U(X)}$	$\Omega\left(\frac{SO(Y)}{U(Y)}\right)$	$\mathbb{H}(2)\oplus\mathbb{H}(2)$	5
				$\searrow \phi_5$		
6	$\mathbb{R}(8)$		$Spin(X)$	$\Omega(Spin(X))$	$\mathbb{H}(4)$	6
				$\nwarrow \nwarrow$		
7	$\mathbb{R}(8)\oplus\mathbb{R}(8)$	$B(O)=\frac{O(V'\oplus V'')}{O(V')\oplus O(V'')}$		$\Omega B(Spin(X))$	$\mathbb{C}(8)$	7
				$\searrow \phi_2$		
8	$\mathbb{R}(16)$		$\frac{U(V_{\mathbb{R}})}{O(V)}$	$\Omega\left(\frac{SU(V_{\mathbb{C}}'\oplus V_{\mathbb{C}}'')}{SO(V'\oplus V'')}\right)$	$\mathbb{R}(16)$	8
9	$\mathbb{C}(16)$				$\mathbb{R}(16)\oplus\mathbb{R}(16)$	9

The complex periodicity where X is a complex vector space.

p	$C(p)$		Periodicity Map
0	$\mathbb{C}$	$B(U) = \frac{U(X' \oplus X'')}{U(X') \oplus U(X'')}$	$\Omega B(U(X))$
1	$\mathbb{C} \oplus \mathbb{C}$	$U(X)$	$\Omega SU(X' \oplus X'')$
2	$\mathbb{C}(2)$	$B(U) = \frac{U(X' \oplus X'')}{U(X') \oplus U(X'')}$	$\Omega B(U(X))$

3.1. Remark The previous table of mappings are defined and shown to be homotopy equivalences of H-spaces. There are two real cases and one complex case where the homotopy equivalence reduces to a map $G \to \Omega B(G)$. This is a full description of the various elements of Bott periodicity. The verification of the maps being homotopy equivalences is the subject of the *Cartan and Moore seminar, Paris, ENS 1959-1960*. A key point is that the maps are homotopy equivalences if they are shown to induce homology isomorphisms since they are maps of H-spaces.

3.2. Complex Periodicity The first homotopy equivalence is $BU \to \Omega SU$ which extends to $\mathbb{Z} \times BU \to \Omega U$. The second homotopy equivalence is $U \to \Omega BU$, which loops to $\Omega U \to \Omega^2 BU$. The composite is the homotopy equivalence $\mathbb{Z} \times BU \to \Omega^2 BU$. Thus, for any pointed connected space X, we have the sequence of isomorphisms

$$K(X, *) = [X, \mathbb{Z} \times BU]_* \longrightarrow [X, \Omega^2 BU]_* \longrightarrow [S^2(X), BU]_* = K(S^2(X), *),$$

whose composite is known as Bott periodicity for reduced complex K-theory.

3.3. Real Periodicity The eight homotopy equivalences are composed to give $\mathbb{Z} \times BO \to \Omega^8 BO$, and hence, we have the sequence of isomorphisms

$$KO(X, *) = [X, \mathbb{Z} \times BO]_* \longrightarrow [X, \Omega^8 BO]_* \longrightarrow [S^8(X), BO]_* = K(S^8(X), *),$$

whose composite is known as Bott periodicity for reduced real K-theory.

4 $KR_G^*(X)$ and the Representation Ring $RR(G)$

We return to reality, that is to $KR(X)$ and $KR_G(X)$, through the use of involutions.

4.1. Notation Let G be a compact Lie group with an involution τ, for example, the unitary group $U(n)$ with τ complex conjugation of matrices. Let X be a space with an involution τ, and let $E \to X$ be a vector bundle with a real structure, that is, $\tau : E_x \to E_{\tau(x)}$ is a conjugate complex linear isomorphism.

4.2. Definition A (G, τ)-module M is a finite-dimensional G-module over $\mathbb{C}$ together with a conjugate complex linear $\tau : M \to M$ such that $\tau(sx) = \tau(s)\tau(x)$ for

$s \in G, x \in M$. Let $RR(G)$ denote the representation ring of (G, τ)-modules. We have two maps relating it with the complex representation, or character, ring $R(G)$ of G, namely $\sigma : RR(G) \to R(G)$ and $\rho : R(G) \to RR(G)$, where $\sigma(M)$ is M without τ and $\rho(N) = N \oplus \bar{N}$ with $\tau(x', x'') = (x'', x')$, the flip between two factors.

4.3. Remark The simple (G, τ)-modules S divide into three classes depending on $\text{End}_{(G,\tau)}(S)$ which is either $\mathbb{R}, \mathbb{C}$, or $\mathbb{H}$.

(1) $\text{End}_{(G,\tau)}(S) = \mathbb{R}$: $\sigma(E)$ is as simple as a G-module over $\mathbb{C}$,
(2) $\text{End}_{(G,\tau)}(S) = \mathbb{C}$: $S = \rho(N)$, where N and $\bar{N}$ are nonisomorphic, and
(3) $\text{End}_{(G,\tau)}(S) = \mathbb{H}$: $S = \rho(N)$, where N and $\bar{N}$ are isomorphic.

These three types of simple (G, τ)-modules S decompose the simple modules into the disjoint union of three sets which are the basis elements of the three subgroups $R_{\mathbb{R}}(G)$, $R_{\mathbb{C}}(G)$, and $R_{\mathbb{H}}(G)$ giving a direct sum decomposition

$$RR(G) = R_{\mathbb{R}}(G) \oplus R_{\mathbb{C}}(G) \oplus R_{\mathbb{H}}(G).$$

In *Fibre Bundles*, 14(11), we have a characterization of real, complex, and quaternionic representations, see theorem 14(11.4) and the three corollaries which follow (Husemöller 1994).

4.4. Remark If G acts trivially on X, then there is a natural isomorphism $K_G(X) \to R(G) \otimes K(X)$ where a vector bundle E is carried to $\oplus_S S \otimes \text{Hom}_G(S, E)$. This isomorphism carries over to spaces X with involution τ with the same correspondence

$$KR_G(X) \longrightarrow RR(G) \otimes KR(X).$$

Finally, putting together this decomposition with the three fold decomposition of the previous remark, we have for trivial G-spaces with τ-structure a decomposition

$$KR_G(X) \longrightarrow [R_{\mathbb{R}}(G) \otimes K_{\mathbb{R}}(X)] \oplus [R_{\mathbb{C}}(G) \otimes K_{\mathbb{C}}(X)] \oplus [R_{\mathbb{H}}(G) \otimes K_{\mathbb{H}}(X)],$$

where $K_{\mathbb{R}}(X), K_{\mathbb{C}}(X)$, and $K_{\mathbb{H}}(X)$, respectively, are the Grothendieck group of real, complex, and quaternionic vector bundles, respectively, on X.

Note that $K_{\mathbb{C}}(X)$ is just $K(X)$, and it is independent of the involution for a complex vector bundle E decomposes as $E_0 \oplus \tau^*(\bar{E}_0)$, where E_0 is the subvector bundle of E where the two complex structures coincide.

5 Generalities on Clifford Algebras and Their Modules

The following is a summary of some of the results in *Fibre Bundles*, chap. 12 on Clifford algebras (Husemöller 1994). We will be primarily interested in the subject over the real and complex numbers. We use the symbol F for such a field.

5.1. Definition A quadratic form (V, f) over F is an F-vector space V of finite dimension together with a function $f : V \times V \to F$ satisfying

(1) $f(a'x' + a''x'', y) = a'f(x', y) + a''f(x'', y)$ for all $a', a'' \in F$ and $x', x'', y \in V$ and
(2) $f(x, y) = f(y, x)$ for all $x, y \in V$.

The first axiom says that $f(x,y)$ is a linear form in x for fixed $y \in V$, and the second axiom says that also $f(x,y)$ is a linear form in Y for fixed $x \in V$. We also use the terminology of symmetric bilinear form for quadratic form. It is the same over a field with $2 \neq 0$ in F so that we can divide by it.

5.2. Definition The direct sum or Witt sum of two quadratic forms is given by $(V', f') \oplus (V'', f'') = (V' \oplus V'', f' \oplus f'')$, where

$$(f' \oplus f'')((x', x''), (y', y'')) = f'(x', y') + f''(x'', y'').$$

In particular for $v' \in V'$ and $v'' \in V''$, we have $(f' \oplus f'')(v', v'') = 0$, that is, these two vectors are perpendicular in the direct sum.

5.3. Definition Let (V, f) be a quadratic form over F. The Clifford algebra of (V, f) is a pair $(C(f), \theta)$ consisting of an algebra $C(f)$ over F and a linear map $\theta : V \to C(f)$, where $\theta(x)^2 = f(x,x).1$ in the algebra $C(f)$ with the following universal property: For any other F-linear map $\alpha : V \to A$ into an algebra over F with the property that $\alpha(x)^2 = f(x,x).1$, there exists a morphism $u : C(f) \to A$ of algebras with $\alpha = u\theta$. Moreover, u is unique with respect to the property that $\alpha = u\theta$.

5.4. Remark The Clifford algebra of (V, f) exists and is unique up to isomorphism. The universal property gives uniqueness. As for existence, we can start with the tensor algebra $T(V) = \bigoplus_n V^{\otimes n}$ on V and divide by the ideal generated by all $x \otimes x - f(x,x).1$ for $x \in V$. Then $\theta : V \to C(f)$ is the quotient of $V \to T(V)$.

The universal property of the Clifford algebra follows from that of $T(V)$ and the generators of the ideal in $T(V)$ divided out to construct $C(f)$.

5.5. Involution and Even/Odd Grading of the Clifford Algebra There is an involution β of $C(f)$ by applying the universal property of $(C(f), \theta)$ to $-\theta$. This yields a splitting $C(f) = C(f)_0 \oplus C(f)_1$, where $C(f)_i$ is the subspace of $x \in C(f)$ with $\beta(x) = (-1)^i x$.

The example of the Clifford algebra brings us to the study of $\mathbb{Z}/2$-graded or parity-graded objects. For two elements $x, y \in V$ viewed as elements in $C(f)_1$, we have the commutativity relation in $C(f)_0 \subset C(f)$

$$xy + yx = 2f(x,y).1.$$

For these parity-graded objects, either modules or algebras always coming in two parts indexed 0 and 1 or equally well by even and odd we organize the tensor product into a graded object.

5.6. Graded Tensor Product Let $L = L_0 \oplus L_1$ and $M = M_0 \oplus M_1$ be two graded modules. The graded tensor product $L \otimes M = (L \otimes M)_0 \oplus (L \otimes M)_1$, where $(L \otimes M)_0 = (L_0 \otimes M_0) \oplus (L_1 \otimes M_1)$ and $(L \otimes M)_1 = (L_0 \otimes M_1) \oplus (L_1 \otimes M_0)$. The new ingredient is that the involution $\tau : L \otimes M \to M \otimes L$ depends on the grading with a sign $\tau(x \otimes y) = (-1)^{ij} y \otimes x$, where $x \in L_i$ and $y \in M_j$.

This applies to two-graded algebras $A \otimes B$, but with a new definition of multiplication related to the sign for τ, namely

$$(x' \otimes y')(x'' \otimes y'') = (-1)^{ij}(x'x'') \otimes (y'y''),$$

where $x'' \in M_i, y' \in L_j$.

To distinguish the graded tensor product multiplication from the usual multiplication, we will write $A \hat{\otimes} B$ in the graded case.

This leads to two basic isomorphisms.

5.7. Two Basic Isomorphisms We have natural isomorphisms

(1) $C(f' \oplus f'') \to C(f') \hat{\otimes} C(f'')$ as graded algebras and

(2) $C(f) = C(f)_0 \oplus C(f)_1 \xrightarrow{\phi} C(\varepsilon \oplus f)_0$ as ungraded algebras, where ε is the one-dimensional quadratic form (Fe, ε) with $\varepsilon(x, y) = -xy$ for $x, y \in F$.

The formula for ϕ is given by $\phi(x) = e \otimes x$. For $x \in V$, we have $\phi(\theta(x))^2 = (e \otimes \theta(x))(e \otimes \theta(x)) = -e^2 \otimes \theta(x)^2 = f(x,x).1$ showing that ϕ is defined as an algebra morphism from the universal property.

For the first isomorphism, we apply the universal mapping property to $\psi(x', x'') = \theta'(x') \otimes 1 + 1 \otimes \theta''(x'')$ and obtain an algebra morphism $C(f' \oplus f'') \to C(f') \hat{\otimes} C(f'')$ with inverse given tensoring together the algebra morphisms defined by inclusion $C(f') \to C(f' \oplus f'')$ and $C(f'') \to C(f' \oplus f'')$.

5.8. Main Example Let $\mathbb{R}^{q,p}$ denote the quadratic form space for $p + q$-dimensional space over $\mathbb{R}$, that is $\mathbb{R}^{p+q}$, but the quadratic form which is the sum of q-negative squares and p-positive squares. Let $C(q, p)$ be the corresponding Clifford algebra which is

$$C(-x_1 - \ldots - x_q + x_{q+1} + \ldots + x_{q+p}).$$

When we tensor with the complex numbers $\mathbb{C}$, we have the complex Clifford algebra $C(n) = \mathbb{C} \otimes_{\mathbb{R}} C(p, n - p)$.

5.9. Table of Clifford Algebras We denote by $F(n) = M_n(F)$ the n by n matrix algebra over the field $F = \mathbb{R}, \mathbb{C}$, or $\mathbb{H}$. Observe also that $F(n) \otimes_F F(m) = F(mn)$.

n	$C(n, 0)$	$C(0, n)$	$C(n)$
0	$\mathbb{R}$	$\mathbb{R}$	$\mathbb{C}$
1	$\mathbb{C}$	$\mathbb{R} \oplus \mathbb{R}$	$\mathbb{C} \oplus \mathbb{C}$
2	$\mathbb{H}$	$\mathbb{R}(2)$	$\mathbb{C}(2)$
3	$\mathbb{H} \oplus \mathbb{H}$	$\mathbb{C}(2)$	$\mathbb{C}(2) \oplus \mathbb{C}(2)$
4	$\mathbb{H}(2)$	$\mathbb{H}(2)$	$\mathbb{C}(4)$
5	$\mathbb{C}(4)$	$\mathbb{H}(2) \oplus \mathbb{H}(2)$	$\mathbb{C}(4) \oplus \mathbb{C}(4)$
6	$\mathbb{R}(8)$	$\mathbb{H}(4)$	$\mathbb{C}(8)$
7	$\mathbb{R}(8) \oplus \mathbb{R}(8)$	$\mathbb{C}(8)$	$\mathbb{C}(8) \oplus \mathbb{C}(8)$
8	$\mathbb{R}(16)$	$\mathbb{R}(16)$	$\mathbb{C}(16)$

To arrive at this table, we use the isomorphisms $\mathbb{C} \otimes_\mathbb{R} \mathbb{C} = \mathbb{C} \oplus \mathbb{C}$, $\mathbb{C} \otimes_\mathbb{R} \mathbb{H} = \mathbb{C}(2)$, and $\mathbb{H} \otimes_\mathbb{R} \mathbb{H} = \mathbb{R}(4)$.

6 $KR_G^{-q}(*)$ and Modules Over Clifford Algebras

Denote the Clifford algebra $F \otimes_\mathbb{R} C(q,0)$ by just $CF\langle q \rangle$ related to the negative definite form, and view it as a parity or $\mathbb{Z}/2\mathbb{Z}$-graded algebra. Here, $F = \mathbb{R}$, $\mathbb{C}$, or $\mathbb{H}$. Now, we mix the usual theory of Clifford algebras with the reality structure τ on the algebra and a compact Lie group G, in particular, we consider modules over the group algebra $CF\langle q \rangle[G]$.

6.1. Definition A real-graded $CF\langle q \rangle[G]$-module M is a parity or $\mathbb{Z}/2\mathbb{Z}$-graded complex vector space $M = M_0 \oplus M_1$ with the additional structure:

(1) a $\mathbb{C}$-linear action of $CF\langle q \rangle$ making it a graded $CF\langle q \rangle$-module,
(2) an antilinear involution $\tau : M \to M$ of degree zero commuting with $CF\langle q \rangle$, and
(3) a $\mathbb{C}$-linear action of G on M commuting with that of $CF\langle q \rangle$ and such that $\tau(sx) = \tau(s)\tau(x)$ for $s \in G, x \in M$.

6.2. Notation Let $MF\langle q \rangle(G)$ denote the Grothendieck group of such modules. These modules give elements of the G-equivariant K-theory of the spheres using the following clutching construction. We have a natural restriction morphism

$$r : MF\langle q+1 \rangle(G) \longrightarrow MF\langle q \rangle(G)$$

using the inclusion morphism $CF\langle q \rangle \to CF\langle q+1 \rangle$.

6.3. Definition The natural morphism

$$\alpha : MF\langle q \rangle(G) \longrightarrow KF_G(D^q, S^{q-1})$$

assigns to the class of a $CF\langle q \rangle[G]$-module M the pair of (G, τ)-vector bundles $E_i = D^q \times M_i$ for $i = 0, 1$ together with the isomorphism

$$\phi : E_0|S^{q-1} \longrightarrow E_1|S^{q-1}$$

given by $\phi(b,x) = (b, bx)$, where $b \in D^q \subset \mathbb{R}^q \subset CF\langle q \rangle$.

The main calculation of the KR-theory of a point is contained in the following exact sequence where we use $CF\langle q \rangle$ and $MF\langle q \rangle(G)$ for the case $F = \mathbb{R}$.

6.4. Proposition *The sequence*

$$MR\langle q+1 \rangle(G) \xrightarrow{r} MR\langle q \rangle(G) \xrightarrow{\alpha} KR_G^{-q}(*) \longrightarrow 0$$

is exact.

6.5. Notation To study the τ-representation ring $RR(G)$, we use the following decomposition $RR(G) = A(G) \oplus B(G) \oplus C(G)$ in terms of the basis consisting of classes $[E]$ of irreducible G-modules, that is, $[E]$ is in $A(G)$, $B(G)$, or $C(G)$ provided the commuting algebra is, respectively, $\mathbb{R}, \mathbb{C}$, or $\mathbb{H}$. With the decomposition $RR(G) = A(G) \oplus B(G) \oplus C(G)$, we have the following analysis of the morphism α.

6.6. Decomposition of the Morphism α The morphism α splits into a direct sum of three morphisms

$$MR\langle q \rangle(G) = A(G) \otimes M\mathbb{R}(q) \oplus B(G) \otimes M\mathbb{C}(q) \oplus C(G) \otimes M\mathbb{H}(G)$$
$$\downarrow \alpha$$
$$KR_G(*) = A(G) \otimes K\mathbb{R}(*) \oplus B(G) \otimes K\mathbb{C}(*) \oplus C(G) \otimes K\mathbb{H}(*)$$

Here, $K\mathbb{R}(X), K\mathbb{C}(X)$, and $K\mathbb{H}(X)$ are the K-theory based on real, complex, quaternionic vector bundles, respectively.

6.7. Notation We have four morphisms of $RR(G)$-modules.

$$i : R_\mathbb{R}(G) \longrightarrow R_\mathbb{C}(G) \text{ and } \rho : R_\mathbb{C}(G) \longrightarrow R_\mathbb{R}(G)$$

$$j : R_\mathbb{H}(G) \longrightarrow R_\mathbb{C}(G) \text{ and } \eta : R_\mathbb{C}(G) \longrightarrow R_\mathbb{H}(G)$$

given by change of scalars.

6.8. Table of Calculations of $KR_G^{-q}(*)$

q	$CR\langle q \rangle$	$MR\langle q \rangle$	$r : q+1 \mapsto q$	$KR_G^{-q}(*)$
0				$R_\mathbb{R}(G)$
1	$\mathbb{R}$	$R_\mathbb{R}(G)$	ρ	$R_\mathbb{R}(G)/\rho R_\mathbb{C}(G)$
2	$\mathbb{C}$	$R_\mathbb{C}(G) = R(G)$	j	$R_\mathbb{C}(G)/j R_\mathbb{H}(G)$
3	$\mathbb{H}$	$R_\mathbb{H}(G)$	$\text{id} \oplus \text{id}$	0
4	$\mathbb{H} \oplus \mathbb{H}$	$R_\mathbb{H}(G) \oplus R_\mathbb{H}(G)$	$\text{id} \oplus \text{id}$	$R_\mathbb{H}(G)$
5	$\mathbb{H}(2)$	$R_\mathbb{H}(G)$	η	$R_\mathbb{H}(G)/\eta R_\mathbb{C}(G))$
6	$\mathbb{C}(4)$	$R_\mathbb{C}(G)$	i	$R_\mathbb{C}(G)/R_\mathbb{H}(G)$
7	$\mathbb{R}(8)$	$R_\mathbb{R}(G)$	$\text{id} \oplus \text{id}$	0
8	$\mathbb{R}(8) \oplus \mathbb{R}(8)$	$R_\mathbb{R}(G) \oplus R_\mathbb{R}(G)$		

7 Bott Periodicity and Morse Theory

In Chap. 9, Sect. 8, we gave an introduction to Morse theory with the aim of showing how the theory gives cell decompositions of smooth manifolds which yield cell complexes for the calculation of homology. In fact, it was with Morse theory and basic differential geometry that Bott first proved the periodicity theorem for the stable linear groups and their classifying spaces. The basic reference is still Milnor (1963).

The book has four parts, and for the purposes of Chap. 9 in this lecture notes, part I of Milnors book would be sufficient, but for the periodicity theorem, the entire book is needed. In Part II, there is a very readable introduction to some Riemannian geometry, and in Part III, it is applied to the variation of geodesics. This variational theory centers on the index theory of critical points for the energy functional, and the indices are related to a cell decomposition of loop space where the geodesics are now critical points. In the first section of the part IV, there is a general discussion concerning the space of geodesics on homogeneous spaces as a subspace of loop space. For this, we use the space

$$\Omega(SU(2m) : I, -I)$$

of paths from the identity matrix I to the negative $-I$ of the identity matrix, but $\Omega(SU(2m) : I, -I)$ is homotopically equivalent to the $\Omega(SU(2m) : I)$, loop space on $SU(2m)$, by the map which multiplies each element of $\Omega(SU(2m) : I, -I)$ by a fixed path from $-I$ to I.

7.1. Assertion The space of minimal geodesics from I to $-I$ in $SU(2m)$ is homeomorphic to the complex Grassmann manifold $G_m(\mathbb{C}^{2m})$. Every minimal geodesics from I to $-I$ has index $\geq 2m + 2$.

Here, the space of minimal geodesics from I to $-I$ in $SU(2m)$ is considered as a subspace of $\Omega(SU(2m) : I, -I)$, and hence, there is an inclusion map

$$G_m(\mathbb{C}^{2m}) \longrightarrow \Omega(SU(2m) : I, -I),$$

whose connectivity can be measured. This is done by viewing $\Omega(SU(2m) : I, -I)$ as built from Grassmann manifold with cells of high dimension determined by the index of other geodesics.

This assertion is the statement of the two Lemmas 23.1 and 23.2 of Milnor's book (1965, p.128). Its proof is based on calculus of variations application of Morse theory which is explained in Part III and the first part of Part IV of the book. It is a modern account of Marston Morse's theory of calculus of variations in large. This gives the periodicity theorem in the following unstable form.

7.2. Unstable Periodicity Theorem The inclusion map

$$G_m(\mathbb{C}^{2m}) \longrightarrow \Omega(SU(2m) : I, -I)$$

induces an isomorphism of homotopy groups in dimensions $i \leq 2m$

$$\pi_i(G_m(\mathbb{C}^{2m})) \longrightarrow \pi_i(\Omega(SU(2m) : I, -I)) \longrightarrow \pi_{i+1}(SU(2m)).$$

To obtain the homotopy periodicity of the unitary groups, we use the four fibrations and related isomorphisms of homotopy groups coming from the exact homotopy sequence

(1)
$$U(m) \longrightarrow U(m+1) \longrightarrow S^{2m+1}$$

giving the isomorphism

$$\pi_{i-1}(U(m)) \longrightarrow \pi_{i-1}(U(m+1)) \longrightarrow \pi_{i-1}(U(m+2)) \longrightarrow \cdots \longrightarrow \pi_{i-1}(U)$$

for $i \leq m$.

(2)
$$U(m) \longrightarrow U(2m) \longrightarrow U(2m)/U(m)$$

giving the vanishing
$$\pi_i(U(2m)/U(m)) = 0$$

for $i \leq 2m$.

(3)
$$U(m) \longrightarrow U(2m)/U(m) \longrightarrow G_m(\mathbb{C}^{2m})$$

giving the isomorphism

$$\pi_i(G_m(\mathbb{C}^{2m})) \longrightarrow \pi_{i-1}(U(m))$$

for $i \leq 2m$.

(4)
$$SU(m) \longrightarrow U(m) \longrightarrow S^1$$

giving the isomorphism

$$\pi_i(SU(m)) \longrightarrow \pi_i(U(m))$$

for $i \neq 1$.

7.3. Summary For $1 \leq i \leq 2m$, this leads to the sequence of isomorphisms

$$\pi_{i-1}(U) \longrightarrow \pi_{i-1}(U(m)) \longrightarrow \pi_i(G_m(\mathbb{C}^{2m})) \longrightarrow \pi_{i+1}(SU(2m)) \longrightarrow \pi_{i+1}(U)$$

giving the following theorem.

7.4. Periodicity Theorem We have an isomorphism

$$\pi_{i-1}(U) \longrightarrow \pi_{i+1}(U) \quad \text{for all } i > 0.$$

Moreover, $\pi_{2j+1} = \mathbb{Z}$ and $\pi_{2j} = 0$ for $j \geq 0$.

8 The Graded Rings $KU^*(*)$ and $KO^*(*)$

Now, we explain the Bott periodicity theorem in terms of the multiplicative structure of K-theory of a point.

8.1. Assertion We have
$$KU^*(*) = \mathbb{Z}[b][b^{-1}],$$
where b is generator $KU^{-2}(*)$.

8.2. Assertion We have

$$KO^*(*) = \mathbb{Z}[x, y, \eta][x^{-1}]/(2\eta, \eta^3, \eta y, y^2 - 4x),$$

where

(1) η is a generator of $KO^{-1}(*) = \mathbb{Z}/2\mathbb{Z}$
(2) y is a generator of $KO^{-4}(*) = \mathbb{Z}$
(3) x is a generator of $KO^{-8}(*) = \mathbb{Z}$.

References

Atiyah, M.F., Hirzebruch, F.: Vector bundles and homogeneous spaces. Proc.Symp. Pure Math. Amer. Math. Soc.3.7-38 (1961)
Husemöller, D.: Fibre Bundles, 3rd ed. Springer-Verlag New York, (1994)
Milnor, J.: Morse theory. Ann. of Math. Studies 51 (1963)
Milnor, J.: Lectures on the h-Cobordism Theorem. Princeton University Press, Princeton, NJ (1965)

Chapter 16
Gram–Schmidt Process, Iwasawa Decomposition, and Reduction of Structure in Principal Bundles

Using the classical Gram–Schmidt process from the beginning linear algebra, we are able to derive group theory results about the linear groups and reduction of structure group results for vector bundles. By starting with these very elementary considerations, we see that there are applications to both group representation theory as well as to the topology of groups, their classifying spaces, and principal bundles.

This theory also has applications to modular forms which play an important role in systems with an $SL(2,\mathbb{Z})$-symmetry. One example is the Verlinde algebra which arises later in a twisted K-theory calculation.

1 Classical Gram–Schmidt Process

1.1. Notation Let again F denote one of the following three topological fields of scalars: the real numbers $\mathbb{R}$, the complex numbers $\mathbb{C}$, or the quaternions $\mathbb{H}$. An F-Hilbert space H is a vector space H over F with an inner product $(\ |\ ) : H \times H \to F$ which is linear in the first variable, conjugate linear in the second variable, hermitian symmetric, that is, $(y|x) = \overline{(x|y)}$ for all $x, y \in H$, and nondegenerate, that is, $(x|x) > 0$ for all $x \neq 0$ in H.

1.2. Definition For a positive integer n and an F-Hilbert space H, the space $\mathscr{L}_n(H)$ is the open subspace of H^n consisting of linearly independent n-tuples of vectors, the space $\mathscr{O}_n(H)$ is the closed subspace of H^n consisting of n-tuples $y_1, \ldots, y_n$ which are orthogonal, that is, $(y_i|y_j) = 0$ for $i \neq j$, and the space $\mathscr{O}\mathscr{N}_n(H)$ is the closed subspace of n-tuples $z_1, \ldots, z_n$ which are orthonormal, that is, $(z_i\ |\ z_j) = \delta_{i,j}$.

There is a simple pair of operations for the sequence of spaces $\mathscr{O}\mathscr{N}_n(H) \subset \mathscr{O}_n(H) \subset \mathscr{L}_n(H)$ which is retracting to a subspace. These two retractions are called the Gram–Schmidt process.

1.3. Gram–Schmidt Process for $\mathscr{O}_n(H) \subset \mathscr{L}_n(H)$ To $x_1, \ldots, x_n$ in $\mathscr{L}_n(H)$, we associate $y_1, \ldots, y_n$ in $\mathscr{O}_n(H)$ such that the subspace generated by $y_1, \ldots, y_i$ is equal to

D. Husemöller et al.: *Gram–Schmidt Process, Iwasawa Decomposition, and Reduction of Structure in Principal Bundles*, Lect. Notes Phys. **726**, 189–201 (2008)
DOI 10.1007/978-3-540-74956-1_17

the subspace generated by $x_1,\ldots,x_i$ for each $i=1,\ldots,n$. The formula for y_i is given inductively as follows

$$y_1 = x_1 \text{ and } y_i = x_i - \sum_{j=1}^{i-1} y_j(x_i|y_j)/(y_j|y_j).$$

In particular, we can also solve inductively for the x_i in terms of a linear combination of y_j with $1 \le j \le i$ as

$$x_i = \sum_{j=1}^{i} y_j u_{i,j} \tag{1.1}$$

with $u_{i,i} = 1$. If we define $u_{i,j} = 0$ for $j > i$, then we obtain an upper triangular matrix $u = (u_{i,j})$ with ones on the diagonal.

1.4. Gram–Schmidt Process for $\mathcal{O}\mathcal{N}_n(H) \subset \mathcal{O}_n(H)$ To $y_1,\ldots,y_n$ in $\mathcal{O}_n(H)$, we associate $z_1,\ldots,z_n$ in $\mathcal{O}\mathcal{N}_n(H)$ such that the subspace generated by $z_1,\ldots,z_i$ is equal to the subspace generated by $y_1,\ldots,y_i$ for each $i=1,\ldots,n$. The formula for z_i is given by $z_j = y_j/(y_j|y_j)$ for $j=1,\ldots,n$, or equivalently

$$y_j = z_j(y_j|y_j) \text{ for } j=1,\ldots,n. \tag{1.2}$$

2 Definition of Basic Linear Groups

Again F denotes one of the topological fields $\mathbb{R}, \mathbb{C}$, or $\mathbb{H}$.

2.1. Definition Let $M_n(F)$ denote the ring of n by n matrices over the field F with the topology from the matrix element identification with F^{n^2}. The general linear group $GL(n,F)$ is the subgroup of invertible matrices, that is, invertible elements of $M_n(F)$, with the subspace topology.

In the case of F being commutative, that is, $F = \mathbb{R}$ or $\mathbb{C}$, we have a determinant map $\det : M_n(F) \to F$, and for $w \in M_n(F)$, we have $\det(w) \neq 0$ if and only if $w \in GL(n,F)$. Recall that $\det(w'w'') = \det(w')\det(w')$ for two elements $w', w' \in M_n(F)$.

2.2. Definition The special linear group $SL(n,F)$ is the kernel of $\det : GL(n,F) \to GL(1,F)$, where $GL(1,F)$ is the group of invertible one by one matrices, or equivalently, invertible elements $F^* \subset F$.

2.3. Definition The basic Hermitian inner product of F^n is given by the formula $(x|y) = \sum_{i=1}^{n} x_i \bar{y}_i$, where $\bar{y}$ denotes the complex or quaternionic conjugation of y.

2.4. Definition The unitary subgroup $U(n,F)$ of $GL(n,F)$ consists of all $w \in GL(n,F)$ preserving the inner product, that is,

$$(w(x)|w(y)) = (x|y)$$

for all $x,y \in F^n$. The special unitary subgroup is $SU(n,F) = U(n,F) \cap SL(n,F)$.

2.5. Remark The inner product-preserving condition $(w(x)|w(y)) = (x|y)$ is satisfied for all $x, y \in F^n$ if it is satisfied for x, y in a set of generators of F^n, for example, some basis. The inner product-preserving condition $(w(x)|w(y)) = (x|y)$ is satisfied for all $x, y \in F^n$ if it is also satisfied for $x = y$ in F^n where it takes the form that $||w(x)|| = ||x||$ for all $x \in F^n$.

2.6. Notation For the unitary and the special unitary groups, we use the following special notations:

$U(n, \mathbb{R}) = O(n)$ and $SU(n, \mathbb{R}) = SO(n)$ when $F = \mathbb{R}$,

$U(n, \mathbb{C}) = U(n)$ and $SU(n, \mathbb{C}) = SU(n)$ when $F = \mathbb{C}$, and

$U(n, \mathbb{H}) = Sp(n)$.

3 Iwasawa Decomposition for *GL* and *SL*

3.1. Definition An Iwasawa decomposition of a topological group G is a sequence of subgroups $G_1, \ldots, G_m$ such that the multiplication map $G^m \to G$ restricts to $G_1 \times \ldots \times G_m \to G$ as an isomorphism of topological spaces. Such a decomposition is often denoted by simply $G = G_1 \ldots G_m$.

In the case of a Lie group G with a maximal compact subgroup K, we have the original Iwasawa decomposition $G = KAN$, where A is commutative and N is nilpotent. In (4.1), we have an application of the Iwasawa decomposition for classifying spaces of linear groups and their maximal compact subgroups.

See Sect. 7 by B. Krötz to this chapter on general properties of the Iwasawa decomposition.

The Iwasawa decomposition of $GL(n, F)$ and $SL(n, F)$ for $F = \mathbb{R}$ or $F = \mathbb{C}$ comes by considering the right action of $GL(n, F)$ on the space of basis vectors $\mathscr{L}_n(F^n)$ in n-dimensional space F^n.

3.2. Definition The right action $\mathscr{L}_n(F^n) \times GL(n, F) \to \mathscr{L}_n(F^n)$ is given by action $b' = b.w$, where for $b = (x_1, \ldots, x_n)$, $b' = (x'_1 \ldots, x'_1) \in \mathscr{L}_n(F^n)$ and $w = (w_{i,j}) \in GL(n, F)$, the matrix relation $x'_j = \sum_{i=1}^{n} x_i w_{i,j}$ defines the action. This action is just the classical change of basis by an invertible square matrix.

This action has two basic properties which are used to carry out the Iwasawa decomposition.

3.3. Proposition *For $b, b' \in \mathscr{L}_n(F^n)$, there exists a unique $w \in GL(n, F)$ with $b.w = b'$. If two of the following conditions are fulfilled, then the third holds for $b.w = b'$:*

$$b \in \mathscr{O}\mathscr{N}_n(F^n), \ b' \in \mathscr{O}\mathscr{N}_n(F^n) \ and \ w \in U(n, F).$$

Now, we introduce two types of subgroups used in the Iwasawa decomposition.

3.4. Definition For any ring R, the subgroup of upper triangular matrices $N(n, R)$ in $GL(n, R)$ consisting of all matrices $w = (w_{i,j})$, where $w_{i,j} = 0$ for $i > j$ and $w_{i,i} = 1$ for each $i = 1, \ldots, n$.

3.5. Definition Let A be the subgroup of $GL(n, \mathbb{R})$ of matrices $a = (a_{i,j})$ with $a_{i,j} = 0$ for $i \neq j$ and $a_{i,i} > 0$. Let $SA = A \cap SL(n, \mathbb{R})$.

3.6. Remark By extending scalars, there are natural subgroup inclusions $GL(n, \mathbb{R}) \subset GL(n, \mathbb{C})$, $GL(n, \mathbb{C}) \subset GL(n, \mathbb{H})$, and $SL(n, \mathbb{R}) \subset SL(n, \mathbb{C})$.

3.7. Theorem *We have the following five Iwasawa decompositions of the linear groups*

$$GL(n, \mathbb{R}) = O(n).A.N(n, \mathbb{R}) \ and \ SL(n, \mathbb{R}) = SO(n).SA.N(n, \mathbb{R}),$$

$$GL(n, \mathbb{C}) = U(n).A.N(n, \mathbb{C}) \ and \ SL(n, \mathbb{C}) = SU(n).SA.N(n, \mathbb{C}),$$

$$and \ GL(n, \mathbb{H}) = Sp(n).A.N(n, \mathbb{H}).$$

Proof. For $w \in GL(n, F)$, we consider the base point $e \in \mathcal{L}_n(F^n)$ with $e = (e_1, \ldots, e_n)$ and the translate $e.w = (x_1, \ldots, x_n) \in \mathcal{L}_n(F^n)$ by w. Now, we apply the first step of the Gram–Schmidt process (1.3)

$$(*) \qquad x_i = \sum_{j=1} y_j u_{i,j} \ \text{with} \ u_{i,i} = 1,$$

so that $e.w = y.u$ with $u \in N(n, F)$, and then we apply the second step of the Gram–Schmidt process (1.4)

$$(**) \qquad y_j = z_j(y_j | y_j) \ \text{for} \ j = 1, \ldots, n,$$

so that $e.w = y.u = z.a.u$ with $a \in A$. Now $z = (z_1, \ldots, z_n) \in \mathcal{ON}_n(F^n)$ which means that $z = e.k$ with $k \in U(n, F)$. This leads the relations $e.w = y.u = z.a.u = e.(k.a.u)$, and from this, we have the unique decomposition of $w = k.a.u$ for $k \in U(n, F), a \in A$, and $u \in N(n, F)$ which is the Iwasawa decomposition of $GL(n, F)$.

When $F = \mathbb{R}$ or $\mathbb{C}$, we see that $\det(w) = \det(kau) = \det(k) \det(a)$, where $\det(a) > 0$ and $|\det(k)| = 1$. Hence, $w \in SL(n, F)$ if and only if $k \in SU(n, F)$ and $a \in SA$. This establishes the Iwasawa decomposition in these cases.

4 Applications to Structure Group Reduction for Principal Bundles Related to Vector Bundles

The fact that a vector bundle is a fibre bundle with structure group either $GL(n, \mathbb{C})$ or $U(n)$ has already been discussed in the context of a Riemannian metric on the bundle. Using the Iwasawa decomposition, we obtain another proof using homotopy methods with the related classifying spaces.

4.1. Theorem *The following five maps of linear groups and the related five maps of their classifying spaces are homotopy equivalences.*

(1) $U(n) \longrightarrow GL(n, \mathbb{C})$ and $BU(n) \longrightarrow BGL(n, \mathbb{C})$,

(2) $SU(n) \longrightarrow SL(n,\mathbb{C})$ and $BSU(n) \longrightarrow BSL(n,\mathbb{C})$,
(3) $O(n) \longrightarrow GL(n,\mathbb{R})$ and $BO(n) \longrightarrow BGL(n,\mathbb{R})$,
(4) $SO(n) \longrightarrow SL(n,\mathbb{R})$ and $BSO(n) \longrightarrow BSL(n,\mathbb{R})$, and
(5) $Sp(n) \longrightarrow GL(n,\mathbb{H})$ and $BSp(n) \longrightarrow BGL(n,\mathbb{H})$.

Proof. The groups A and SA have a contraction to the identity namely $h_t(a) = a^t$ for a matrix a, where $h_0(a) = 1$ and $h_1(a) = a$. The groups $N(n)$ have a contraction to the identity $h_t(n) = n(t)$, where the matrix element $n(t)_{i,j} = n_{i,j}$ for $i \geq j$ and $t n_{i,j}$ for $i < j$ for a matrix n, where $h_0(n) = 1$ and $h_1(n) = n$. Now, the theorem results by applying these homotopies to the terms in the Iwasawa decomposition.

4.2. Remark By using the spectral theorem in Hilbert space H, we have the same homotopy equivalences

$$U(H) \longrightarrow GL(H) \text{ and } BU(H) \longrightarrow BGL(H).$$

4.3. Remark The homotopy equivalences in (4.1) and (4.2) are compatible with inclusion coming from n to $n+m$ or H to $H \oplus H'$ with the obvious inclusion of groups with the identity added onto the complementary factor.

5 The Special Case of $SL_2(\mathbb{R})$ and the Upper Half Plane

5.1. Notation For $SL(2,\mathbb{R})$, the Iwasawa decomposition takes the form $G = SL(2,\mathbb{R}) = SO(2).A.N = K.A.N$, where the elements of these subgroups are given by the following matrices

(1) $K = SO(2) = \{k(\theta)\}$ for $k(\theta) = \begin{pmatrix} \cos\theta & \sin\theta \\ -\sin\theta & \cos\theta \end{pmatrix}$

(2) $A = \{a(t)\}$ for $a(t) = \begin{pmatrix} t & 0 \\ 0 & t^{-1} \end{pmatrix}$

(3) $N = \{n(v)\}$ for $n(v) = \begin{pmatrix} 1 & v \\ 0 & 1 \end{pmatrix}$.

In order to study the left coset space $SL(2,\mathbb{R})/SO(2) = G/K$ as the group product $A.N = N.A$, we consider the upper half plane $\mathfrak{H}$.

5.2. Definition The upper half plane $\mathfrak{H}$ is the open subset of $z \in \mathbb{C}$ with $\text{Im}(z) > 0$. The action of G on $\mathfrak{H}$ is given by

$$\begin{pmatrix} a & b \\ c & d \end{pmatrix}(z) = \frac{az+b}{cz+d}.$$

If $c = 0$ in an element of G, the image of z is $a^2 z + ab$ since $d = a^{-1}$. Any element $yi + x \in \mathfrak{H}$ is of the form $y = a^2$ and $x = ab$ for the square root of $y = \text{Im}(z) > 0$ and $b = x/a$.

5.3. Proposition *The action of $A.N = N.A$ is transitive on the upper half plane $\mathfrak{H}$, and for $g \in G$, we have $g(i) = i$ if and only if $g \in K$. The maps given by multiplication of matrices $N \times A \times K \to G$ and $K \times A \times N \to G$ are homeomorphisms.*

Proof. The argument for the first assertion is given just before the statement of the proposition. The relation $g(i) = i$ for the two by two matrix $g = \begin{pmatrix} a & b \\ c & d \end{pmatrix}$ is the relation $ai + b = -c + di$ so the fixed point relation $g(i) = i$ reduces to $a = d$ and $b = -c$ with the determinant condition $a^2 + b^2 = 1$. These are exactly the numbers represented by $a = \cos\theta$ and $b = \sin\theta$, that is, $g = k(\theta)$.

Each $g \in G$ when applied to i has the unique form $g(i) = n'a'(i)$ with $a' \in A, n' \in N$. Hence, $(n'a')^{-1}g = k(\theta)$ by the first part giving a unique decomposition and transposing, we have the relation $4g = n'a'k(\theta)$. Giving the topological isomorphism $N \times A \times K \to G$. The other case results by taking inverses reversing the order of multiplication.

5.4. Corollary *The inclusion $K \to G$ is a homotopy equivalence.*

Proof. Every $g \in G$ has a unique representation by $g = k(\theta)a(t)n(v)$, and $g_s = k(\theta)a(st)n(sv)$ is a homotopy from $k(\theta) = g_0$ to $g = g_1$.

5.5. Corollary *The map ψ where $\psi(gK) = g(i)$ defines a topological isomorphism $\psi : G/K \to \mathfrak{H}$.*

Besides the homotopy type implications, this isomorphism $G/K \to \mathfrak{H}$ is used to study discrete subgroups Γ of G in terms of how they act on $\mathfrak{H}$. The first and basic example is $\Gamma = SL(2,\mathbb{Z})$. As an example of how the action on $\mathfrak{H}$ is used, we quote the following two related results.

5.6. Proposition *The group $SL(2,\mathbb{Z})$ is generated by*

$$T = \begin{pmatrix} 1 & 1 \\ 0 & 1 \end{pmatrix} \quad and \quad S = \begin{pmatrix} 0 & 1 \\ -1 & 0 \end{pmatrix}.$$

Every $z \in \mathfrak{H}$ can be transformed by g, a product of S and T and their inverses to $g(z)$ with $|g(z)| \geq 1$ and $|g(z)| \leq 1/2$. Moreover, if $|g(z)| > 1$ and $|g(z)| < 1/2$, then g is unique with this property.

This is proved nicely in Serre (1973).

The operation $S(z) = -1/z$ is called the modular transformation on the upper half plane while $T(z) = z + 1$ is called simple translation.

6 Relation Between $SL_2(\mathbb{R})$ and $SL_2(\mathbb{C})$ with the Lorentz Groups

6.1. Notation Let $M(1,2)$ be the vector space of symmetric real 2×2 matrices, that is, matrices $< t, x, y > = \begin{pmatrix} t+x & y \\ y & t-x \end{pmatrix}$. For $X \in M_2(\mathbb{R})$, we have $X \in M(1,2)$ if and only if $X^t = X$. The space $M(1,2)$ is a vector space over $\mathbb{R}$ of dimension 3, and the determinant

$$\det <t,x,y> = t^2 - x^2 - y^2$$

is the Lorentz metric for one-time coordinate and two-space coordinates.

Let $M(1,3)$ be the vector space of Hermitian complex 2×2 matrices, that is, matrices $<t,x,y,z> = \begin{pmatrix} t+x & y+iz \\ y-iz & t-x \end{pmatrix}$. For $Z \in M_2(\mathbb{C})$, we have $Z \in M(1,3)$ if and only if $Z^* = Z$. The space $M(1,3)$ is a vector space over $\mathbb{R}$ of dimension 4 and the determinant

$$\det <t,x,y,z> = t^2 - x^2 - y^2 - z^2$$

is the Lorentz metric for one-time coordinate and three-space coordinates. Note that $M(1,2) \subset M(1,3)$ is the subspace where $z = 0$.

6.2. The Morphism $\theta : SL_2(\mathbb{R}) \rightarrow O(1,2)$ For $A \in SL_2(\mathbb{R})$, we define $\theta(A)(X) = AXA^{-1}$ and we see that $\theta(A)(X) \in M(1,2)$ for $X \in M(1,2)$. Moreover, $\det(\theta(A)(X)) = \det(X)$ for $X \in M(1,2)$ showing that $\theta(A) \in O(1,2)$, the Lorentz group of type (1,2).

6.3. The Morphism $\theta : SL_2(\mathbb{C}) \rightarrow O(1,3)$ For $A \in SL_2(\mathbb{C})$, we define $\theta(A)(Z) = AZA^{-1}$ and we see that $\theta(A)(Z) \in M(1,3)$ for $Z \in M(1,3)$. Moreover, $\det(\theta(A)(Z)) = \det(Z)$ for $Z \in M(1,3)$ showing that $\theta(A) \in O(1,3)$, the Lorentz group of type (1,3).

6.4. *Remark* We see that $\theta(A'A'') = \theta(A')\theta(A'')$ in both cases, and the two definitions of θ are compatible under restriction to $SL_2(\mathbb{R}) \subset SL_2(\mathbb{C})$ and $M(1,2) \subset M(1,3)$. The kernel $\ker(\theta)$ of θ, that is the matrices A of $SL_2(\mathbb{C})$ with $\theta(A) = $ identity, consists of the two elements $\pm I$, where I is the identity matrix.

6.5. *Remark* Using θ, the Iwasawa decompositions $SL_2(\mathbb{R}) = SO(2).A.N$ and $SL_2(\mathbb{C}) = SU(2).A'.N'$ give Iwasawa decompositions of $O(1,2)$ and $O(1,3)$, respectively. Observe that the image of θ is of finite index in the Lorentz groups $O(1,2)$ and $O(1,3)$. This allows us to study the Lorentz groups using $SL_2(\mathbb{R})$ and $SL_2(\mathbb{C})$.

A Appendix: A Novel Characterization of the Iwasawa Decomposition of a Simple Lie Group *(by B. Krötz)*

This appendix is about (essential) uniqueness of the *Iwasawa (or horospherical) decomposition* $G = KAN$ of a semisimple Lie group G. This means:

A.1. Theorem *Assume that G is a connected Lie group with simple Lie algebra $\mathfrak{g}$. Assume that $G = KL$ for some closed subgroups $K, L < G$ with $K \cap L$ discrete. Then up to order, the Lie algebra $\mathfrak{k}$ of K is maximally compact, and the Lie algebra $\mathfrak{l}$ of L is isomorphic to $\mathfrak{a} + \mathfrak{n}$, the Lie algebra of AN.*

A.2. General Facts on Decompositions of Lie Groups. For a group G, a subgroup $H < G$, and an element $g \in G$, we define $H^g = gHg^{-1}$.

A.3. Lemma *Let G be a group and $H, L < G$ subgroups. Then, the following statements are equivalent:*

(i) $G = HL$ and $H \cap L = \{\mathbf{1}\}$.
(ii) $G = HL^g$ and $H \cap L^g = \{\mathbf{1}\}$ for all $g \in G$.

Proof. Clearly, we only have to show that $(ii) \Rightarrow (i)$. Suppose that $G = HL$ with $H \cap L = \{\mathbf{1}\}$. Then, we can write $g \in G$ as $g = hl$ for some $h \in H$ and $l \in L$. Observe that $L^g = L^h$ and so

$$H \cap L^g = H \cap L^h = H^h \cap L^h = (H \cap L)^h = \{\mathbf{1}\}\,.$$

Moreover, we record

$$HL^g = HL^h = HLh = Gh = G\,.$$

In the sequel, capital Latin letters will denote real Lie groups and the corresponding lower case fractur letters will denote the associated Lie algebra, that is, G is a Lie group with Lie algebra $\mathfrak{g}$.

A.4. Lemma *Let G be a Lie group and $H, L < G$ closed subgroups. Then, the following statements are equivalent:*

(i) $G = HL$ with $H \cap L = \{\mathbf{1}\}$.
(ii) The multiplication map

$$H \times L \to G, \qquad (h, l) \mapsto hl$$

is an analytic diffeomorphism.

Proof. Standard structure theory.

If G is a Lie group with closed subgroups $H, L < G$ such that $G = HL$ with $H \cap L = \{\mathbf{1}\}$, then we refer to (G, H, L) as a *decomposition triple*.

A.5. Lemma *Let (G, H, L) be a decomposition triple. Then:*

$$(\forall g \in G) \qquad \mathfrak{g} = \mathfrak{h} + \mathrm{Ad}(g)\mathfrak{l} \qquad and \qquad \mathfrak{h} \cap \mathrm{Ad}(g)\mathfrak{l} = \{0\}\,. \qquad (A.1)$$

Proof. In view of Lemma A.4, the map $H \times L \to G$, $(h, l) \mapsto hl$ is a diffeomorphism. In particular, the differential at $(\mathbf{1}, \mathbf{1})$ is a diffeomorphism which means that $\mathfrak{g} = \mathfrak{h} + \mathfrak{l}$, $\mathfrak{h} \cap \mathfrak{l} = \{0\}$. As we may replace L by L^g, for example Lemma A.3, the assertion follows.

Question 1 Assume that G is connected. Is it then true that (G, H, L) is a decomposition triple if and only if the algebraic condition (A.1) is satisfied.

A.6. Remark If the Lie algebra $\mathfrak{g}$ splits into a direct sum of subalgebra $\mathfrak{g} = \mathfrak{h} + \mathfrak{l}$, then we cannot conclude in general that $G = HL$ holds. For example, let $\mathfrak{h} = \mathfrak{m} + \mathfrak{a} + \mathfrak{n}$ be a minimal parabolic subalgebra and $\mathfrak{l} = \bar{\mathfrak{n}}$ be the opposite of $\mathfrak{n}$. Then, $HL = MAN\bar{N}$ is the open Bruhat cell in G. A similar example is when $\mathfrak{g} = \mathfrak{sl}(n, \mathbb{R})$ with $\mathfrak{h}$ the upper triangular matrices and $\mathfrak{l} = \mathfrak{so}(p, n - p)$ for $0 < p < n$. In this case, $HL \subset G$ is a proper open subset. Notice that in both examples, condition (A.1) is violated as $\mathfrak{h} \cap \mathrm{Ad}(g)\mathfrak{l} \neq \{0\}$ for appropriate $g \in G$.

A.7. The Case of One Factor Being Maximal Compact. Throughout this section, G denotes a semisimple-connected Lie group with associated Cartan decomposition $\mathfrak{g} = \mathfrak{k} + \mathfrak{p}$. Set $K = \exp \mathfrak{k}$ and note that $\mathrm{Ad}(K)$ is maximal compact subgroup in $\mathrm{Ad}(G)$.

For what follows we have to recall some results of Mostow on maximal solvable subalgebras in $\mathfrak{g}$. Let $\mathfrak{c} \subset \mathfrak{g}$ be a Cartan subalgebra. Replacing $\mathfrak{c}$ by an appropriate $\mathrm{Ad}(G)$-conjugate, we may assume that $\mathfrak{c} = \mathfrak{t}_0 + \mathfrak{a}_0$ with $\mathfrak{t}_0 \subset \mathfrak{k}$ and $\mathfrak{a}_0 \subset \mathfrak{p}$. Write $\Sigma = \Sigma(\mathfrak{a}, \mathfrak{g}) \subset \mathfrak{a}^* \setminus \{0\}$ for the nonzero $\mathrm{ad}\mathfrak{a}_0$-spectrum on $\mathfrak{g}$. For $\alpha \in \Sigma$, write $\mathfrak{g}^\alpha$ for the associated eigenspcae. Call $X \in \mathfrak{a}_0$ *regular* if $\alpha(X) \neq 0$ for all $\alpha \in \Sigma$. Associated to a regular element $X \in \mathfrak{a}$, we associate a nilpotent subalgebra

$$\mathfrak{n}_X = \bigoplus_{\substack{\alpha \in \Sigma \\ \alpha(X) > 0}} \mathfrak{g}^\alpha.$$

If $\mathfrak{a} \subset \mathfrak{p}$ happens to be maximal abelian, then we will write $\mathfrak{n}$ instead of $\mathfrak{n}_X$.

With this notation we have the following

A.8. Theorem *Let $\mathfrak{g}$ be a semisimple Lie algebra. Then, the following assertions hold:*

(i) Every maximal solvable subalgebra $\mathfrak{r}$ of $\mathfrak{g}$ contains a Cartan subalgebra $\mathfrak{c}$ of $\mathfrak{g}$.

(ii) Up to conjugation with an element of $\mathrm{Ad}(G)$, every maximal solvable subalgebra of $\mathfrak{g}$ is of the form

$$\mathfrak{r} = \mathfrak{c} + \mathfrak{n}_X$$

for some regular element $X \in \mathfrak{a}_0$.

Proof. Lemma A.14 and Mostow (1961).

We choose a maximal abelian subspace $\mathfrak{a} \subset \mathfrak{p}$ and write $\Sigma = \Sigma(\mathfrak{g}, \mathfrak{a})$ for the associated root system. For a choice of positive roots, we obtain a unipotent subalgebra $\mathfrak{n}$. Write $\mathfrak{m} = \mathfrak{z}_\mathfrak{k}(\mathfrak{a})$ and fix a Cartan subalgebra $\mathfrak{t} \subset \mathfrak{m}$. Write A, N, T for the analytic subgroups of G corresponding to $\mathfrak{a}, \mathfrak{n}, \mathfrak{t}$. Notice that $\mathfrak{t} + \mathfrak{a} + \mathfrak{n}$ is a maximal solvable subalgebra by Theorem A.8.

A.9. Lemma *Let $L < G$ be a closed subgroup such that $G = KL$ with $K \cap L = \{1\}$. Then, there is an Iwasawa decomposition $G = NAK$ such that*

$$N \subset L \subset TAN \quad \text{and} \quad L \simeq LT/T \simeq AN. \tag{A.2}$$

Conversely, if L is a closed subgroup of G satisfying (A.2), then $G = KL$ with $K \cap L = \{1\}$.

Proof. Our first claim is that L contains no nontrivial compact subgroups. In fact, let $L_K \subset L$ be a compact subgroup. As all maximal compact subgroups of G are conjugate, we find a $g \in G$ such that $L_K^g \subset K$. But $L_K^g \cap K \subset L^g \cap K = \{1\}$ by Lemma A.3. This establishes our claim.

Next, we show that L is solvable. For that, let $L = S_L \times R_L$ be a Levi decomposition, where S is semisimple and R is reductive. If $S \neq \mathbf{1}$, then there is a nontrivial maximal compact subgroup $S_K \subset S$. Hence, $S = \mathbf{1}$ by our previous claim and $L = R_L$ is solvable.

Next, we turn to the specific structure of $\mathfrak{l}$, the Lie algebra of $\mathfrak{l}$. Let $\mathfrak{r} = \mathfrak{c} + \mathfrak{n}_X$ be a maximal solvable subalgebra of $\mathfrak{g}$ which contains $\mathfrak{l}$. As before, we write $\mathfrak{c} = \mathfrak{t}_0 + \mathfrak{a}_0$ for the Cartan subalgebra of $\mathfrak{r}$. We claim that $\mathfrak{a}_0 = \mathfrak{a}$ is maximal abelian in $\mathfrak{p}$. In fact, notice that $\mathfrak{l} \cap \mathfrak{t}_0 = \{0\}$ and so $\mathfrak{l} \hookrightarrow \mathfrak{r}/\mathfrak{t}_0 \simeq \mathfrak{a}_0 + \mathfrak{n}_X$ injects as vector spaces. Hence,

$$\dim \mathfrak{l} = \dim \mathfrak{a} + \dim \mathfrak{n} \leq \dim \mathfrak{a}_0 + \dim \mathfrak{n}_X.$$

But, $\dim \mathfrak{a}_0 \leq \dim \mathfrak{a}$ and $\dim \mathfrak{n}_X \leq \dim \mathfrak{n}$ and therefore $\mathfrak{a} = \mathfrak{a}_0$. Hence, $\mathfrak{r} = \mathfrak{t} + \mathfrak{a} + \mathfrak{n}$. As $\mathfrak{l} \simeq \mathfrak{r}/\mathfrak{t}$ as vector spaces, we thus get hat $L \simeq LT/T \simeq R/T \simeq AN$ as homogeneous spaces. We now show that $N \subset L$ which will follow from $\mathfrak{n} \subset [\mathfrak{l}, \mathfrak{l}]$. For that, choose a regular element $X \in \mathfrak{a}$. By what we know already, we then find an element $Y \in \mathfrak{t}$ such that $X + Y \in \mathfrak{l}$. Notice that $\mathrm{ad}(X + Y)$ is invertible on $\mathfrak{n}$ and hence $\mathfrak{n} \subset [X + Y, \mathfrak{n}]$. Finally, observe that

$$[X + Y, \mathfrak{n}] = [X + Y, \mathfrak{r}] = [X + Y, \mathfrak{l} + \mathfrak{t}] = [X + Y, \mathfrak{l}]$$

which concludes the proof of the first assertion of the lemma.

Finally, the second assertion of the lemma is immediate from the Iwasawa decomposition of G.

A.10. Manifold Decompositions for Decomposition Triples. Throughout this section, G denotes a connected Lie group.

Let (G, H, L) be a decomposition triple and let us fix maximal compact subgroups K_H and K_L of H and L, respectively. We choose a maximal compact subgroup K of G such that $K_H \subset K$. As we are free to replace L by any conjugate L^g, we may assume in addition that $K_L \subset K$.

We then have the following fact, see also Lemma A.4 and Oniščik (1969).

A.11. Lemma *Let (G, H, L) be a decomposition triple. Then, (K, K_H, K_L) is a decomposition triple, that is, the map*

$$K_H \times K_L \to K, \quad (h, l) \mapsto hl$$

is a diffeomorphism.

Before we prove the lemma, we recall a fundamental result of Mostow concerning the topology of a connected Lie group G, cf. Mostow (1952). If $K < G$ is a maximal compact subgroup of G, then there exists a vector space V and a homeomorphism $G \simeq K \times V$. In particular, G is a deformation retract of K and thus $H_\bullet(G, \mathbb{R}) = H_\bullet(K, \mathbb{R})$.

Proof. As $H \cap L = \{1\}$, it follows that $K_H \cap K_L = \{1\}$. Thus, compactness of K_L and K_H implies that the map

$$K_H \times K_L \to K, \quad (h, l) \mapsto hl$$

has closed image. It remains to show that the image is open. This will follow from $\dim K_H + \dim K_L = \dim K$. In fact, $G \simeq H \times L$ implies that G is homeomorphic to $K_H \times K_L \times V_H \times V_L$ for vector spaces V_H and V_L. Thus,

$$H_\bullet(K, \mathbb{R}) = H_\bullet(G, \mathbb{R}) = H_\bullet(K_H \times K_L, \mathbb{R}),$$

and Künneth implies for any $n \in \mathbb{N}_0$ that

$$H_n(K, \mathbb{R}) \simeq \sum_{j=0}^{n} H_j(K_H, \mathbb{R}) \otimes H_{n-j}(K_L, \mathbb{R}).$$

Now, for an orientable connected compact manifold M, we recall that $H_{\dim M}(M, \mathbb{R}) = \mathbb{R}$ and $H_n(M, \mathbb{R}) = \{0\}$ for $n > \dim M$. Next, Lie groups are orientable, and we deduce from the Künneth identity from above that $\dim K_H + \dim K_L = \dim K$. This concludes the proof of the lemma.

Let us write $\mathfrak{k}_\mathfrak{h}$ and $\mathfrak{k}_\mathfrak{l}$ for the Lie algebras of K_H and K_L, respectively. Then, as (K, K_H, K_L) is a decomposition triple, it follows from Lemma A.5 that

$$\mathfrak{k} = \mathfrak{k}_\mathfrak{h} + \mathrm{Ad}(k)\mathfrak{k}_\mathfrak{l} \qquad \text{and} \qquad \mathfrak{h} \cap \mathrm{Ad}(k)\mathfrak{l} = \{0\}.$$

Now, let $\mathfrak{t}_h \subset \mathfrak{k}_\mathfrak{h}$ be a maximal toral subalgebra and extend it to a maximal torus $\mathfrak{t}$, that is, $\mathfrak{t}_\mathfrak{h} \subset \mathfrak{t}$. Now pick a maximal toral subalgebra $\mathfrak{t}_\mathfrak{l}$. Replacing $\mathfrak{l}$ by an appropriate $\mathrm{Ad}(K)$-conjugate, we may assume that $\mathfrak{t}_\mathfrak{l} \subset \mathfrak{t}$ (all maximal toral subalgebras in $\mathfrak{k}$ are conjugate). Finally, write T, T_H, T_L for the corresponding tori in T.

A.12. Lemma *If (K, K_H, K_L) is a decomposition triple for a compact Lie group K, then (T, T_H, T_L) is a decomposition triple for the maximal torus T. In particular*

$$rank\, K = rank\, K_H + rank\, K_L. \tag{A.3}$$

Proof. We already know that $\mathfrak{t}_\mathfrak{h} + \mathfrak{t}_\mathfrak{l} \subset \mathfrak{t}$ with $\mathfrak{t}_\mathfrak{h} \cap \mathfrak{t}_\mathfrak{l} = \{0\}$. It remains to verify that $\mathfrak{t}_\mathfrak{h} + \mathfrak{t}_\mathfrak{l} = \mathfrak{t}$. We argue by contradiction. Let $X \in \mathfrak{t}, X \notin \mathfrak{t}_\mathfrak{h} + \mathfrak{h}_\mathfrak{l}$. As $\mathfrak{k} = \mathfrak{k}_\mathfrak{h} + \mathfrak{k}_\mathfrak{l}$, we can write $X = X_\mathfrak{h} + X_\mathfrak{l}$ for some $X_\mathfrak{h} \in \mathfrak{k}_\mathfrak{h}$ and $X_\mathfrak{l} \in \mathfrak{k}_\mathfrak{l}$.

For a compact Lie algebra $\mathfrak{k}$ with maximal toral subalgebra $\mathfrak{t} \subset \mathfrak{k}$, we recall the direct vector space decomposition $\mathfrak{k} = \mathfrak{t} \oplus [\mathfrak{t}, \mathfrak{k}]$. As $\mathfrak{t}_\mathfrak{h} + \mathfrak{t}_\mathfrak{l} \subset \mathfrak{t}$, we, hence, may assume that $X_\mathfrak{h} \in [\mathfrak{t}_\mathfrak{h}, \mathfrak{k}_\mathfrak{h}]$ and $X_\mathfrak{l} \in [\mathfrak{t}_\mathfrak{l}, \mathfrak{k}_\mathfrak{l}]$. But, then we get

$$X = X_\mathfrak{h} + X_\mathfrak{l} \in [\mathfrak{t}_\mathfrak{h}, \mathfrak{k}_\mathfrak{h}] + [\mathfrak{t}_\mathfrak{l}, \mathfrak{k}_\mathfrak{l}] \subset [\mathfrak{t}, \mathfrak{k}],$$

and therefore, $X \in \mathfrak{t} \cap [\mathfrak{t}, \mathfrak{k}] = \{0\}$, a contradiction.

A.13. Decompositions of Compact Lie Groups. Decompositions of compact Lie groups is an algebraic feature as the following lemma, essentially due to Oniščik, shows.

A.14. Lemma *Let $\mathfrak{k}$ be a compact Lie algebra and $\mathfrak{k}_1, \mathfrak{k}_2 < \mathfrak{k}$ be two subalgebras. Then, the following statements are equivalent:*

(i) $\mathfrak{k} = \mathfrak{k}_1 + \mathfrak{k}_2$ with $\mathfrak{k}_1 \cap \mathfrak{k}_2 = \{0\}$

(ii) Let K, K_1, K_2 be simply connected Lie groups with Lie algebras $\mathfrak{k}, \mathfrak{k}_1$ and $\mathfrak{k}_2$. Write $\iota_i : K_i \to K$, $i = 1, 2$ for the natural homomorphisms sitting over the inclusions $\mathfrak{k}_i \hookrightarrow \mathfrak{k}$. Then the map

$$m : K_1 \times K_2 \to K, \quad (k_1, k_2) \mapsto \iota_1(k_1)\iota_2(k_2)$$

is a homeomorphism.

Proof. The implication $(ii) \Rightarrow (i)$ is clear. We establish $(i) \Rightarrow (ii)$. We need that m is onto and deduce this from Lemma A.11 and Oniščik (1969). Then, K becomes a homogeneous space for the left–right action of $K_1 \times K_2$. The stabilizer of **1** is given by the discrete subgroup $F = \{(k_1, k_2) : \iota_1(k_1) = \iota_2(k_2)^{-1}\}$, that is, $K \simeq K_1 \times K_2 / F$. As K_1 and K_2 are simply connected, we conclude that $\pi_1(K) = F$, and thus, $F = \{\mathbf{1}\}$ as K is simply connected.

We now show the main result of this section.

A.15. Lemma *Let (K, K_1, K_2) be a decomposition triple of a connected compact simple Lie group. Then, $K_1 = \mathbf{1}$ or $K_2 = \mathbf{1}$.*

Before we prove this, a few remarks are in order.

A.16. Remark(a) If K is of exceptional type, then the result can be easily deduced from $\dim K = \dim K_1 + \dim K_2$ and the rank equality *rank K = rank K_1 + rank K_2*, cf. Lemma A.12. For example, if K is of type G_2, Then a nontrivial decomposition $K = K_1 K_2$ must have *rank K_i* = 1, that is, $\mathfrak{k}_i = \mathfrak{su}(2)$. But,

$$14 = \dim K \neq \dim K_1 + \dim K_2 = 6.$$

(b) The assertion of the lemma is not true if we only require $K = K_1 K_2$ and drop $K_1 \cap K_2 = \{\mathbf{1}\}$. For example if K is of type G_2, then $K = K_1 K_2$ with K_i locally $SU(3)$ and $K_1 \cap K_2 = T$ a maximal torus.

Proof. The proof is short but uses a powerful tool, namely the structure of the cohomology ring of the compact group K. See for instance Ozeki (1977) or Koszul (1978).

Putting matters together, this concludes the proof of Theorem 1.

References

Koszul, J.L.: Variante d'un théorème de H. Ozeki: Osaka J. Math. **15**: 547–551 (1978)

Mostow, G.D.: On the L^2-space of a Lie group. Amer. J. Math. **74**: 920–928 (1952)

Mostow, G.D.: On maximal subgroups of real Lie groups. Ann. Math. **74**(3): 503–517 (1961)

Oniščik, A.L.: Decompositions of reductive Lie groups. Mat. Sb. (N.S.) bf **80**(122): 553–599 (1969)

Ozeki, H.: On a transitive transformation group of a compact group manifold. Osaka J. Math, **14**: 519–531 (1977)

Chapter 17
Topological Algebras: G-Equivariance and KK-Theory

In three previous Chaps. 13–15, we have developed the theory of G-equivariant vector bundles over a compact space X for a compact group G. In Chap. 3, vector bundles over a compact space X were related to finitely generated projective modules over the algebra $C(X)$. The K-theory of X was described in terms of projections in the C^*-algebra $C(X)$ tensored with the compact operators in 4(5.14), that is, $K^0(X)$ is naturally isomorphic to C^*-algebra K-theory $K(C(X))$.

In fact, the K-theory $K(A)$ is defined for all C^*-algebras A, and this introduces the possibility of a geometric interpretation of the K-theory for more general classes of topological algebras. The beginnings of this point of view arises when we consider an algebra related to a compact group G acting on a compact space X. This is a noncommutative algebra, called the cross product algebra $G \ltimes A$, where $A = C(X)$, whose K-theory turns out to be the G-equivariant K-theory $K_G(X)$. This theory is outlined in Sects. 1 and 2.

This step leads to a broad extension of the theory involving other topological algebras, called m-algebras and locally convex algebras. The K-theory of these algebras, which includes C^*-algebras, has many formal properties of the K-theory of spaces. We outline aspects of the theory in this chapter in Sects. 3–10.

The index theory of elliptic operators leads to K-theory invariants and suggested a version of K-homology theory with geometric and analytical data evolving "K-cycles." These cycles were first introduced by Atiyah and were related to extension groups $\mathrm{Ext}(B,A)$ by Brown–Douglas–Fillmore. Kasparov has developed a bivariant theory $KK(A,B)$ for C^*-algebras with a G-equivariant version $KK^G(A,B)$. Cuntz has extended the theory to locally convex algebras and a related bivariant theory $k(A,B)$.

We give a short descriptive introduction to these constructions including the smooth case which arises with smooth manifolds naturally, because smooth function spaces are usually only locally convex. Since in one chapter it is only possible to give a sketch of the theory, we will make frequent references to the books denoted by $\langle B \rangle$ or $\langle CMR \rangle$, that is, respectively, Blackadar (1998) and Cuntz et al. (2007).

D. Husemöller et al.: *Topological Algebras: G-Equivariance and KK-Theory*, Lect. Notes Phys. **726**, 203–226 (2008)
DOI 10.1007/978-3-540-74956-1_18   © Springer-Verlag Berlin Heidelberg 2008

1 The Module of Cross Sections for a *G*-Equivariant Vector Bundle

In 3(2.1), we have studied the cross section functor for vector bundles

$$\Gamma : \mathrm{Hom}_X(E', E'') \longrightarrow \mathrm{Hom}_{C(X)}(\Gamma(X, E'), \Gamma(X, E''))$$

with full embedding and equivalence properties under suitable conditions into the category of finitely generated projective modules over the algebra $C(X)$. We wish to extend this picture to modules and spaces with *G*-action. We need the following conventions and definitions.

Convention 1.1 *As in Chap. 3 and 13, X is a compact space and G is a compact group. The C^*-topology on $C(X)$ is also the compact open topology. For a vector bundle E over X, we give $\Gamma(X, E)$, the compact open topology and all automorphism groups occurring get the compact open topology too.*

1.2. Definition For a *G*-space X, we define a multiplicative map $\alpha : G \to \mathrm{Aut}(C(X))$ by the formula $\alpha(s)(f)(x) = f(s^{-1}x)$. A topological algebra A with continuous group homomorphism $\alpha : G \to \mathrm{Aut}(A)$ is called a *G*-algebra.

We extend this definition to cross sections of any vector *G*-vector bundle E over a *G*-space X using the notation $s_\# : E_{s^{-1}x} \to E_x$ which is by definition of a *G*-vector bundle a $\mathbb{C}$-linear isomorphism between fibres of E.

1.3. Definition Let E be a *G*-equivariant vector bundle over a space X. The left *G*-action on $\Gamma(X, E)$ is the continuous group homomorphism $\alpha : G \to \mathrm{Aut}(\Gamma(X, E))$, where $\alpha(s)(\sigma)(x) = s_\# \sigma(s^{-1}x)$.

1.4. Remark The map α is $C(X)$-linear on the vector space $\Gamma(X, E)$ in the sense it satisfies the equivariance formula

$$\alpha(s)(f\sigma) = \alpha(s)(f)\alpha(s)\alpha(\sigma)$$

for the action of $f \in C(X) = \Gamma(X, X \times \mathbb{C})$ on $\sigma \in \Gamma(X, E)$. For this, we use the calculation

$$\alpha(st)(\sigma)(x) = (st)_\# \sigma((st)^{-1}x) = s_\# t_\# \sigma(t^{-1}s^{-1}x)$$

$$= s_\# \alpha(t)(\sigma)(s^{-1}x) = \alpha(s)\alpha(t)(\sigma)(x).$$

1.5. Definition Let A be a topological algebra with *G*-action. A (G, A)-module M is a left A-module together with a linear continuous *G*-action on M such that $s(ax) = s(a)s(x)$, where $s \in G$, $a \in A$, and $x \in M$. Let $_{(G,A)}\mathrm{Mod}$ denote the category of (G, A)-modules and morphisms consisting of A-linear *G*-equivariant maps.

1.6. Example The cross section module $\Gamma(X, E)$ of a *G*-vector bundle E is a $(G, C(X))$-module. The cross section module is a functor

$$\Gamma : \mathrm{Vect}_G(X) \to {}_{(G, C(X))}\mathrm{Mod}.$$

In order to relate two *G*-equivariant *K*-groups for $A = C(X)$, we have the following version of the Serre–Swan theorem using the full subcategory $\text{Vect}_{(G,C(X))}$ of objects in $_{(G,C(X))}\text{Mod}$ which are finitely generated projective $C(X)$-modules.

1.7. Theorem *Let G be a compact group acting on a compact space X. The cross section functor* $\Gamma : \text{Vect}_G(X) \to \text{Vect}_{(G,C(X))}$ *is an equivalence of categories.*

Proof. This theorem is proved by averaging over the compact group the equivalence of categories $\Gamma(X,) : (\text{vect}/X) \to (\text{vect}/C(X))$, see 3(4.3). The functor $\Gamma : \text{Vect}_G(X) \to \text{Vect}_{(G,C(X))}$ is faithful as a restriction of a faithful functor, and

$$\Gamma : \text{Hom}_G(E', E'') \longrightarrow \text{Hom}_{(G,C(X))}(\Gamma(X,E'), \Gamma(X,E''))$$

is a bijection for $f \in \text{Hom}_{(G,C(X))}(\Gamma(X,E'), \Gamma(X,E''))$, there is $v : E' \to E''$ a morphism of vector bundles with $\Gamma(v) = f$. The average u of v will be *G*-equivariant and satisfies $\Gamma(u) = f$, giving a full embedding. To show the restriction of Γ in the assertion is an equivalence, we note the key step is to take an object M in $_{(G,C(X))}\text{Mod}$ and choose a surjective (G,A)-morphism $w : A^n \to M$ together with an A-linear $v : M \to A^n$ which is a section of w. We average v to u and $wv = M$ to $wu = M$. In this way, we construct enough *G*-equivariant vector bundles on X in order that Γ is an equivalence of categories.

1.8. Definition Using the category $\text{Vect}_G(X)$ of *G*-equivariant vector bundles over X, we can form the Grothendieck group $K_G(X)$ to obtain *G*-equivariant *K*-theory $K_G^0(X)$.

1.9. Definition Using the category $\text{Vect}_{(G,A)}$ of finitely generated projective (G,A)-modules, we can form the Grothendieck group $K_G(A)$ to obtain *G*-equivariant *K*-theory of a *G*-algebra A.

As before, the *K*-theory of a space can be calculated in terms of the *K*-theory of algebras, but now in the equivariant setting.

1.10. Corollary *Let G be a compact group acting on a compact space X. The cross section functor* Γ *induces an isomorphism of the K-groups*

$$K_G(X) \xrightarrow{\cong} K^G(C(X)).$$

See $\langle B \rangle$, 11.4, for more details.

2 *G*-Equivariant *K*-Theory and the *K*-Theory of Cross Products

In Chap. 4, Sect. 5, we saw that a C^* property on an algebra could lead to additional properties of the *K*-theory. In this section, we outline how for a compact group G the *G*-equivariant *K*-theory for a C^*-algebra A can be described using the usual *K*-theory of the related crossed product C^*-algebra $G \ltimes A$, where G is a compact group.

2.1. Cross Products of Groups For a topological group G acting on another topological group H by $\alpha : G \to \mathrm{Aut}(H)$, there is a cross product group $G \ltimes H$ and a related topological group extension

$$H \xrightarrow{i} G \ltimes H \xrightarrow{\pi} G,$$

where as a space $G \ltimes H$ is the product space $H \times G$, with $\pi(a,u) = u$ and $i(a) = (a,1)$. The multiplication on $G \ltimes H$ is given by the formula

$$(a',u') \cdot (a'',u'') = (a'\alpha(u')(a''), u'u'') \quad \text{for} \quad (a',u'),(a'',u'') \in G \ltimes H.$$

It is easy to check that $(1,1)$ is the unit, and for the relation

$$(a,u)^{-1} = (\alpha(u^{-1})a^{-1}, u^{-1}),$$

we calculate

$$(a,u) \cdot (\alpha(u^{-1})a^{-1}, u^{-1}) = (a\alpha(u)\alpha(u^{-1})a^{-1}, uu^{-1}) = (1,1).$$

2.2. Example For a field F, its affine group $\mathrm{Aff}(F)$ is

$$F^+ \longrightarrow \mathrm{Aff}(F) = F^* \ltimes F^+ \longrightarrow F^*,$$

where $\mathrm{Aff}(F)$ is a subgroup of the permutation group of F consisting of substitutions $a + ux$ and group law

$$(a' + u'x) \cdot (a'' + u''x) = a' + u'(a'' + u''x) + b' = (a' + u'(a'')) + (u'u'')x.$$

The group $\mathrm{Aff}(F)$ is called the affine group of the line.

There is a matrix realization of $\mathrm{Aff}(F)$ with group law given by

$$\begin{pmatrix} u' & 0 \\ a' & 1 \end{pmatrix} \begin{pmatrix} u'' & 0 \\ a'' & 1 \end{pmatrix} = \begin{pmatrix} u'u'' & 0 \\ a' + u'a'' & 1 \end{pmatrix}.$$

Now, we give the mapping spaces $C(G,A)$ and $L^1(G,A)$ the structure of a G-algebra.

2.3. Definition Let G be a locally compact group, let A be a Banach $*$-algebra, and let $\alpha : G \to \mathrm{Aut}(A)$ be an action of G on A. Let $C_c(G,A)$ be the $*$-algebra of compactly supported continuous functions $G \to A$. The $*$-algebra structure is defined by the convolution multiplication

$$(a*b)(s) = \int a(t)\alpha_t(b)(t^{-1}s)\mathrm{d}t \quad \text{for} \quad s,t \in G, a,b \in C_c(G,\alpha,A)$$

and the $*$-operation

$$(a^*)(s) = \delta(s)^{-1}\alpha_s(a)(s^{-1})^* \quad \text{for} \quad s \in G, a \in C_c(G,A).$$

For the locally compact group G, the modular function is $\delta : G \to \mathbb{R}_+^*$ defined by the formula $d(s^{-1}) = \delta(s)^{-1} ds$. We use the fact that it is morphism of topological groups.

2.4. Definition The cross product $G \ltimes A$ is the enveloping C^*-algebra of $C_c(G,A)$.

2.5. Remark The question of whether there are units or approximate units is the first issue. In fact, we can choose a sequence of elements in $C(G)$ which give an approximate identity in $C(G,A)$ or $L^1(G,A)$.

To assign to a G-equivariant module over A a cross algebra, we follow this procedure.

2.6. Definition Let E be a G-equivariant A-module. Then, E is considered as a right $L^1(G,A)$-module by the following formula

$$x \int a(g)E_g dg = \int \alpha_{g^{-1}}(xa(g)) dg$$

for $x \in E$, $a \in L^1(G,A)$.

2.7. Proposition *Let E be a G-equivariant A-module which is of the form $eM_m(A)$, where e is a projection. The E considered as a right $L^1(G,A)$-module is of the form $pM_n(L^1(G,A))$, where p is a projection.*

For a sketch of a proof, we make the following remarks. We can find a G-equivariant A-module E' such that $E \oplus E'$ is isomorphic $A \otimes W$, where W is finite dimensional with the action of $g \in G$ on $a \otimes w \in A \otimes W$ given by $\alpha_g(a) \otimes g_W(w)$, where g_W is a representation of G on the finite-dimensional W. We use the fact that the G-representation W is of the form $e'L^1(G)$ and put together e and e' to obtain p.

2.8. Theorem *Let G be a compact group acting on a compact space X with cross product algebra $G \ltimes C(X)$. Then, there is an isomorphism between*

$$K^G(C(X)) \text{ and } K(G \ltimes C(X)).$$

See $\langle B \rangle$, 11.7.1, and the appendix (Sect. 10) by S. Echterhoff to this chapter.

The importance of this section is that for a compact G, we do not need a separate G-equivariant theory for operator algebra K-theory as in the case of compact G-spaces.

3 Generalities on Topological Algebras: Stabilization

3.1. Definition Let V be a complex vector space. A seminorm p on V is a function $p : V \to \mathbb{R}$ such that

(a) $p(x) \geq 0$ for all $x \in V$,

(b) $p(cx) = |c|p(x)$ for all $x \in V$, $c \in \mathbb{C}$, and

(c) $p(x+y) \leq p(x) + p(y)$ for all $x, y \in V$.

    If V is an algebra, then p is submultiplicative provided it satisfies also

(d) $p(xy) \leq p(x)p(y)$ for all $x, y \in V$.

A seminorm is a norm provided $p(x) = 0$ implies $x = 0$

3.2. Definition A locally convex vector space V is a topological vector space where the topology is defined by the open balls given by a family of seminorms. A locally convex topological algebra A is a topological algebra whose underlying vector space is locally convex. An m-algebra A is a locally convex topological algebra with the topology given by submultiplicative seminorms.

    The continuity of the multiplication $A \times A \to A$ in the definition of locally convex topological algebras can be formulated in terms of seminorms to the effect that for every seminorm p on A, there exists a seminorm q on A such that $p(xy) \leq q(x)q(y)$ for all $x, y \in A$.

    Let $(C^* \backslash alg), (m \backslash alg),$ and $(lc \backslash alg)$ denote the categories of C^*-algebras, m-algebras, and locally convex algebras, respectively.

3.3. Definition Let A be a locally convex algebra. For each locally compact space X, we denote by $C(X,A)$ the algebra of all continuous functions $f : X \to A$ which vanishes at ∞, that is, for each neighborhood U of the origin in A, there exists a compact subset $K \subset X$ with $f(X - K) \subset U$.

    For a smooth manifold $X \subset \mathbb{R}^n$, we denote by $C^\infty(X,A)$ the algebra of all smooth functions $f : X \to A$ which vanishes at ∞ along with all derivatives.

    The reader should be aware that the notations $C(X), C(X,A)$ as well as $C^\infty(X)$, $C^\infty(X,A)$ are used for different spaces in different contexts.

    Each seminorm p on A gives rise to a seminorm $p(f)$ on $C(X,A)$ denoted by $p(f)$ or $p_X(f) = \max_{x \in X} p(f(x))$. In the case where $X \subset \mathbb{R}^m$ is a smooth submanifold, each seminorm p on A gives rise to an infinite family of seminorms $p_n(f)$ defined by

$$p_n(f) = \max_{x \in X, |J| \leq n} |\partial^J f(x)|,$$

where as usual ∂^J is an iterated derivative of f given by $J = \{j(1), \ldots, j(m)\}$ and $\partial^J = \partial_1^{j(1)} \ldots \partial_m^{j(m)}$. Of course we can extend these definitions to manifolds X which are not necessarily embedded into $\mathbb{R}^n$.

    The following construction is used for describing homotopy and extensions used in K-theory.

3.4. Definition Let I be a finite interval, and let A be a locally convex algebra. Let $A(I)$ or AI denote the subalgebra of continuous functions $f \in C(I,A)$, satisfying $f(t) = 0$ for an open end point t of I otherwise no condition. Let $A^\infty(I)$ or $A^\infty I$ denote the subalgebra of smooth functions $f \in C^\infty(I,A)$ satisfying $f^{(i)}(t) = 0$ for any end point of $t \in I$, where $i > 0$ and $f(t) = 0$ for an open end point t of I.

For $t \in I$, we have a substitution morphism $\varepsilon_t : AI \to A$ and $\varepsilon_t : A^\infty I \to A$, where $\varepsilon_t(f) = f(t) \in A$. This leads to the concept of homotopy and of diffeotopy, that is, smooth homotopy.

3.5. Definition Two morphisms $f', f'' : A \to B$ of locally convex algebras are homotopic (resp. smoothly homotopic) provided there exists a morphism $h : A \to B[0,1]$ (resp. $h : A \to B^\infty[0,1]$) with $\varepsilon_0 h = f'$ and $\varepsilon_1 h = f''$.

If there exists a homotopy with end points not necessarily zero, then there exists a homotopy with zero end points, see $\langle CMR \rangle$, 6.1. Now, it follows that the homotopy and smooth homotopy are equivalence relations.

Stabilizations involve making algebras larger by embedding into matrix algebras and their related completions.

3.6. Remark For an arbitrary algebra, we have embeddings

$$A \longrightarrow M_2(A) \longrightarrow \ldots \longrightarrow M_n(A) \longrightarrow \ldots \longrightarrow M_\infty(A) = \bigcup_{0<n} M_n(A).$$

These embeddings result from putting zero in the bottom and the right entries $M_n(A) \to M_{n+q}(A)$. Note the unit is not preserved. A $*$-structure extends to the matrices by using also the transpose of the matrix, and any seminorm extends by using the operator seminorm for matrices relative to each seminorm on A.

3.7. Definition Let $\mathscr{K}$ denote the C^*-algebra completion of $M_\infty(\mathbb{C})$, and for a C^* algebra A, let $\mathscr{K}(A)$ denote the C^* completion of $M_\infty(A)$.

3.8. Remark The algebra $\mathscr{K}$ can be identified with the closed subalgebra of compact operators in $\mathscr{B}(H)$, the algebra of bounded operators on an infinite dimensional separable Hilbert space H. We use also the notation $\mathscr{K}(H)$ for $\mathscr{K}$ in this case.

3.9. Definition Let $\mathscr{K}^\infty$ denote the m-algebra of all $\mathbb{N} \times \mathbb{N}$ matrices $(a_{i,j})$ for which the following seminorms

$$p_n((a_{i,j})) = \sum_{i,j} |1 + i|^n |1 + j|^n |a_{i,j}|$$

are finite. Such matrices are called rapidly decreasing, and the algebra $\mathscr{K}^\infty$ is called the algebra of all smooth compact operators. The algebra also can be regarded as an algebra of operators acting on an infinite-dimensional separable Hilbert space.

4 Ell(X) and Ext(X) Pairing with K-Theory to $\mathbb{Z}$

We consider two functors $\mathrm{Ell}(X)$ and $\mathrm{Ext}(X)$ which lead to KK-theory through their pairing with K-theory to $\mathbb{Z}$. This pairing suggests that they are related with

K-homology. The first Ell(X) was introduced by Atiyah in the context of elliptic operators on a compact manifold X, and the second Ext(X) was introduced by Brown, Douglas, and Fillmore as a step toward an extension theory for C^*-algebras.

Ell(X) is the set of index data on a manifold X which maps onto the K-homology $K_0(X)$ of the space X. It was introduced by Atiyah and is related to the Fredholm data associated with an elliptic pseudodifferential operator between two vector bundles on a manifold X. It is a natural extension of the work on the Atiyah–Singer index theorem. At this point, we leave the main theme of these notes and only provide a sketch of the ideas.

4.1. Remark Let $D : \Gamma^\infty(M,E') \to \Gamma^\infty(M,E'')$ be a pseudodifferential operator of degree zero between two smooth complex vector bundles on a smooth manifold. If D is elliptic, then D extends to a Fredholm operator

$$F' : H' = L^2(M,E') \to H'' = L^2(M,E'').$$

The index of F' is called the analytic index of D, denoted by $\mathrm{ind}_a(D)$. If ϕ' and ϕ'' are the action of smooth functions $f \in C^\infty(M)$ on H' and H'', respectively, where $\phi' : C^\infty(M) \to \mathcal{B}(H')$ and $\phi'' : C^\infty(M) \to \mathcal{B}(H'')$ are $*$-algebra morphisms, then the difference

$$\phi''(f)F' - F'\phi'(f) \in \mathcal{K}(H',H'')$$

the vector space of compact operators $H' \to H''$. There is also a Fredholm operator $F'' : H'' \to H'$ with $F'F'' - 1 \in \mathcal{K}(H'')$, $F''F' - 1 \in \mathcal{K}(H')$, and $F'' - (F')^* \in \mathcal{K}(H',H'')$.

4.2. Remark We can form the nth direct sum of this situation with the notation $H'\langle n \rangle = (H')^{n\oplus}$, $H''\langle n \rangle = (H'')^{n\oplus}$, and with actions $\phi'\langle n \rangle$, $\phi''\langle n \rangle$, and with a Fredholm operator

$$F'\langle n \rangle = (F')^{n\oplus} : H'\langle n \rangle \longrightarrow H''\langle n \rangle$$

satisfying the compact operator commutator property

$$\phi''\langle n \rangle(f)F'\langle n \rangle - F'\langle n \rangle\phi'\langle n \rangle(f) \in \mathcal{K}(H'\langle n \rangle, H''\langle n \rangle).$$

4.3. Definition Let Ell(X) be the set of triples (H,ϕ,F), where H is the graded Hilbert space $H_0 = H'$ and $H_1 = H''$, $\phi = (\phi',\phi'') : C(X) \to \mathrm{End}(H)_0$ is a $*$-algebra morphism into the graded bounded operators of even degree, and

$$F = \begin{pmatrix} 0 & F' \\ F'' & 0 \end{pmatrix} : H \to H$$

is operator of odd degree such that this data satisfies the following conditions:

(1) $[\phi(f),F] = \phi(f)F - F\phi(f) \in \mathcal{K}(H)$,
(2) $FF^* - 1 \in \mathcal{K}(H)$, and
(3) $F - F^* \in \mathcal{K}(H)$.

Conditions (2) and (3) are the Fredholm conditions. With this grading concept, we put the two Hilbert spaces into one graded Hilbert space and the other data comes in this graded setting.

4.4. Pairing with $K^0(X) = K_0(C(X))$ $K^0(X) = K_0(C(X))$ For the index theorem, the pairing

$$\text{Ell}(X) \times K_0(C(X)) \longrightarrow \mathbb{Z}$$

assigns to a triple (H, ϕ, F) and a projection class $[p] \in K_0(C(X))$, where $p^2 = p = p^* \in M_n(C(X))$, the index of $F'\langle n \rangle | \ker(p\phi\langle n \rangle)$, where $p\phi\langle n \rangle \in C(X)^n$ which acts on $H'\langle n \rangle$ as in (4.2) and $\ker(\phi\langle n \rangle(p)) \subset H'\langle n \rangle$.

This pairing is part of the Kasparov product when $\text{Ell}(X)$ and $K_0(X)$ are interpreted in terms of KK-theory. The next pairing comes from the extension theory developed by Brown, Douglas, and Fillmore.

4.5. Definition The group $\text{Ext}(X)$ is the set of isomorphism classes of extensions of $C(X)$ by $\mathscr{K} = \mathscr{K}(H)$ on a separable H. Such an extension is a type of short exact sequence

$$0 \longrightarrow \mathscr{K} \longrightarrow E \overset{\pi}{\longrightarrow} C(X) \longrightarrow 0,$$

where π is a surjective C^*-algebra morphism with kernel isomorphic to $\mathscr{K}$.

In general, we can replace $C(X)$ by any C^* algebra B, and with the usual notation for extensions, we have $\text{Ext}(X) = \text{Ext}(C(X), \mathscr{K})$

4.6. Remark There is a basic extension with kernel $\mathscr{K} = \mathscr{K}(H)$ namely as an ideal in $\mathscr{B}(H)$, the C^*-algebra $\mathscr{K}(H)$ is an extension over the quotient algebra $\mathscr{A}(H) = \mathscr{B}(H)/\mathscr{K}(H)$ called the Calkin algebra. The elements in $\text{Ext}(X)$ can be shown to be in bijective correspondence with $*$-morphisms $\tau : C(X) \to \mathscr{A}(H) = \mathscr{A}$ up to unitary equivalence in $\mathscr{A}$ under certain circumstances.

4.7. Remark In terms of this classification we can explain the abelian semigroup on Ext by forming for $\tau' : C(X) \to \mathscr{A}(H')$ and $\tau'' : C(X) \to \mathscr{A}(H'')$ the sum $\tau' \oplus \tau'' : C(X) \to \mathscr{A}(H' \oplus H'')$. In fact, this semigroup $\text{Ext}(X)$ usually is a group.

4.8. Example For $X \subset \mathbb{C}$, the elements of $\text{Ext}(X)$ are given by normal elements of $\mathscr{A}$ with spectrum X. Another formulation of this is to take $T \in \mathscr{B}(H)$ with $[T, T^*] = TT^* - T^*T$ in $\mathscr{K}(H)$ such that the essential spectrum of T is X. This element is trivial in $\text{Ext}(X)$ if and only if $T = N + K$ for some normal $N \in \mathscr{B}(H)$ and $K \in \mathscr{K}(H)$.

4.9. Functorial Properties A continuous $f : X \to Y$ induces a $*$-homomorphism $C(f) : C(Y) \to C(X)$, and this in turn induces by a fibre product construction

$$f_*(0 \to \mathscr{K} \to E' \to C(X) \to 0) = (0 \to \mathscr{K} \to E'' \to C(Y) \to 0),$$

where $E'' = E' \times_{C(X)} C(Y)$, and hence, we have $f_* : \text{Ext}(X) \to \text{Ext}(Y)$ the induced morphism. If $g : Y \to Z$ is a second map, then $(gf)_* = g_* f_*$.

4.10. Pairing with $K^1(X)$ We now pair the group $\text{Ext}(X)$ with $K^1(X)$ by defining a group homomorphism $\gamma : \text{Ext}(X) \to \text{Hom}(K^1(X), \mathbb{Z})$. An element of $K^1(X)$ is represented by a homotopy class $[g]$, where $g : X \to GL(n, \mathbb{C})$ is a map or equivalently $g \in Gl(n, C(X))$. An element $[\tau] \in \text{Ext}(X)$ is represented by a map $\tau : C(X) \to \mathscr{A}(H)$ and hence, $\tau(g) \in Gl(n, \mathscr{A}(H))$. Choose an entrywise lifting $\tau(g)' \in M_n(\mathscr{B}(H))$ of the matrix $\tau(g)$. This lifting $\tau(g)'$ is a Fredholm operator on H^n. Then, we define $\gamma[\tau] = \text{ind}(\tau(g)')$. The morphism $\gamma : \text{Ext}(X) \to \text{Hom}(K^1(X), \mathbb{Z})$ is equivalent to a pairing

$$\text{Ext}(X) \times K^1(X) \longrightarrow \mathbb{Z}.$$

5 Extensions: Universal Examples

Extensions are classified by assuming a splitting property of the quotient morphism. In the next definition, we consider examples.

5.1. Definition Let

$$(E) : 0 \longrightarrow A \longrightarrow E \underset{\longleftarrow}{\overset{\pi}{\rightrightarrows}} B \longrightarrow 0$$

be an extension of locally convex algebras. The extension (E) is

(1) linearly split provided there exists a continuous linear map $s : B \to E$ with $\pi s(y) = y$ for all $y \in B$,
(2) algebraically split provided there exists an algebra morphism $s : B \to E$ which is an algebra splitting.
   A C^*-algebra extension (E) is:
(3) positively split provided there exists a continuous linear splitting $s : B \to E$ such that $s(B^+) \subset E^+$, that is, s preserves positive elements,
(3) *completely positively split provided any finite matrix extension of the extension $M_n(\pi) : M_n(E) \to M_n(B)$ is positively split,

5.2. Definition Let $\text{Ext}(B, A)$ denote the set of isomorphism classes of extensions

$$(E) : 0 \longrightarrow A \longrightarrow E \underset{\longleftarrow}{\overset{\pi}{\rightrightarrows}} B \longrightarrow 0$$

of B by A for two given locally convex algebras A and B.

This set $\text{Ext}(B, A)$ is a functor of two variables covariant in B and contravariant in A.

5.3. Functoriality of Ext(B, A) Let $u : A \to A'$ and $v : B' \to B$ be two morphisms of algebras, and consider an extension

$$(E) : 0 \longrightarrow A \longrightarrow E \longrightarrow B \longrightarrow 0 .$$

We form the following extension using a cartesian square over B where $E' = E \times_B B'$ and cocartesian square from A where $E'' = E *_A A'$

$$
\begin{array}{ccccccccc}
v^*(E): & 0 & \longrightarrow & A & \longrightarrow & E' & \longrightarrow & B' & \longrightarrow 0 \\
 & & & \uparrow & & \downarrow & & \downarrow{\scriptstyle v} & \\
(E): & 0 & \longrightarrow & A & \longrightarrow & E & \longrightarrow & B & \longrightarrow 0 \\
 & & & \downarrow{\scriptstyle u} & & \downarrow & & \uparrow & \\
u_*(E): & 0 & \longrightarrow & A' & \longrightarrow & E' & \longrightarrow & B & \longrightarrow 0.
\end{array}
$$

The above construction depends on the existence of finite limits in the category to form E' and finite colimits to form E''.

5.4. Remark In (5.3) with further morphisms $u' : A' \to A''$ and $v' : B'' \to B'$, we have the relations $(u'u)_* = (u')_* u_*$ and $(vv')^* = (v')^* v^*$. The categories of sequences with splittings have to be checked to see whether the constructions preserve the relative splittings. This is the case for (5.1)(1) and (2). Note that $v^* : \mathrm{Ext}(B,A) \to \mathrm{Ext}(B',A)$ and $u_* : \mathrm{Ext}(B,A) \to \mathrm{Ext}(B,A')$ make Ext a functor in each of the variables.

5.5. Sum of Extensions Let

$$
(E') : 0 \longrightarrow A \longrightarrow E' \underset{\longleftarrow}{\overset{\pi'}{\longrightarrow}} B \longrightarrow 0
$$

and

$$
(E'') : 0 \longrightarrow A \longrightarrow E'' \underset{\longleftarrow}{\overset{\pi''}{\longrightarrow}} B \longrightarrow 0
$$

be two extensions. The M_2-sum has the form

$$
(E/2) : 0 \longrightarrow M_2(A) \longrightarrow D \underset{\longleftarrow}{\overset{\pi}{\longrightarrow}} B \longrightarrow 0,
$$

where D is the algebra of matrices of the form $\begin{pmatrix} x' & b \\ c & x'' \end{pmatrix}$, and $b, c \in A$, and $(x', x'') \in E' \times_B E''$, and $\pi \begin{pmatrix} x' & b \\ c & x'' \end{pmatrix} = \pi(x') = \pi(x'')$. A section s' of π' and s'' of π'' combine to a section s of π given $s(b) = \begin{pmatrix} s'(b) & 0 \\ 0 & s''(b) \end{pmatrix}$ for $b \in B$.

This sum is used in E-theory, where A and $M_2(A)$ are isomorphic.

5.6. Remark In general, the kernel A of the projection $\pi : E \to B$ in an extension is an algebra without unit. There is a universal construction of an embedding $\theta : A \to M(A)$ into an algebra with unit called the multiplier algebra, and an extension

$$
0 \longrightarrow \theta(A) \longrightarrow M(A) \longrightarrow Q(A) \longrightarrow 0
$$

with the universal mapping property for an extension (E) maps into the multiplier algebra extension

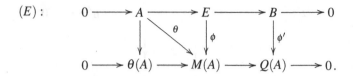

Let $L(A_A)$ and $L(_A A)$ denote the subalgebras of $B(A)$ of right A-linear and left A-linear endomorphisms of the linear space A or the C^*-algebra containing A as an essential ideal, see $\langle B \rangle$, 12.1.

5.7. Construction The multiplier algebra $M(A)$ of locally convex algebra A is the subalgebra of all $(u, v) \in L(A_A) \times L(_A A)$ such that $xu(y) = v(x)y$ for $x, y \in A$.

For (u', v'), $(u'', v'') \in M(A)$, we have $(u', v')(u'', v'') = (u'u'', v'v'')$ for $xu'(u''(y))$ $= v'(x)u''(y) = v''(v'(x))y$. Also $c(u, v) = (cu, cv)$.

To define $\theta : A \to M(A)$, we use the multiplications $l_a(x) = ax$ and $r_a(x) = xa$, and then $\theta(a) = (l_a, r_a)$. The assertion $\theta(A) \in M(A)$ is the relation $x(ay) = (xa)y$ or $xu(y) = v(x)y$. Note also that $l_{a'} l_{a''} = l_{a'a''}$ and $r_{a'} l_{a''} = r_{a''a'}$. For the definition of $\phi : E \to M(A)$, we observe that for $e \in E$ the multiplications $l_e(A) \subset A$ and $r_e(A) \subset A$ so that $\phi(e) = (l_e, r_e)$ is defined since $x(ey) = (xe)y$ in E.

5.8. Remark The $\ker(\theta)$ is the ideal of all $x \in A$ with $xA = 0$ and $Ax = 0$. If A has a unit 1, then $xy(y) = v(x)y$ for $x = 1$ is $u(y) = 1u(y) = v(1)y$ and $y = 1$ is $v(x) = v(x)1 = xu(1)$, but $x = y = 1$ gives a single element $u(1) = v(1) = w \in A$. Hence, θ is surjective because $\theta(w) = (u, v)$. Hence, $\theta : A \to M(A)$ is an isomorphism if A has a unit.

This is an example of an extension by A with the universal property that any extension (E) by A maps uniquely to this multiplier extension.

For an algebra B, there is a universal tensor algebra extension with linear splitting which maps to all extension (E) over B.

5.9. Construction The tensor algebra $T(B)$ extension over a locally convex algebra B is the completion of the algebraic tensor algebra $\bigoplus_{1 \le n} B^{n \otimes}$ with projection $\pi :$ $T(B) \to B$ defined to have the property that $\pi(y_1 \otimes \ldots \otimes y_m) = y_1 \ldots y_m \in B$. The extension is denoted by

$$(T) : 0 \longrightarrow J(B) \longrightarrow T(B) \underset{s}{\overset{\pi}{\underset{\longleftarrow}{\longrightarrow}}} B \longrightarrow 0.$$

Before we consider the locally convex structure on $T(B)$ and $J(B)$, we describe the universal mapping property of this extension.

5.10. Universal Property For each morphism $f : B \to B'$ of algebras and each extension

$$(E') : 0 \longrightarrow A' \longrightarrow E' \underset{s'}{\overset{\pi'}{\rightleftarrows}} B' \longrightarrow 0$$

with a linear splitting s', there exists a unique morphism of extensions over f

$$
\begin{array}{ccccccccc}
(T) : 0 & \longrightarrow & J(B) & \longrightarrow & T(B) & \underset{s}{\overset{\pi}{\rightleftarrows}} & B & \longrightarrow & 0 \\
& & \downarrow{\scriptstyle g'} & & \downarrow{\scriptstyle g} & & \downarrow{\scriptstyle f} & & \\
(E') : 0 & \longrightarrow & A' & \longrightarrow & E' & \underset{s'}{\overset{\pi'}{\rightleftarrows}} & B' & \longrightarrow & 0.
\end{array}
$$

with $g(y_1 \otimes \ldots \otimes y_m) = s'(y_1) \ldots s'(y_m)$ and $g' = g|J(B)$.

The universal property of the tensor algebra $T(V)$ on a vector space or a module is a basic tool in mathematics.

5.11. Remark There are two versions of $T(V) = \bigoplus_{0 \le n} V^{n\otimes}$ and $T(V) = \bigoplus_{0 < n} V^{n\otimes}$, where in the first case $V^{0\otimes} =$ scalars, say $\mathbb{C}$, for the module V and $T(V)$ has a unit, and in the second case, which we use above, there is no unit in the algebra $T(V)$. In both cases, the multiplication is the coproduct of morphisms $V^{p\otimes} \otimes V^{q\otimes} \to V^{(p+q)\otimes}$ where associativity is taken for grant. The universal property relative to the functor v which assigns to an algebra A the underlying module $v(A)$, and it is a natural isomorphism

$$\mathrm{Hom}_{(\mathrm{alg}/\mathrm{k})}(T(V), A) = \mathrm{Hom}_{(\mathrm{k})}(V, v(A))$$

saying that T is a left adjoint functor to v.

5.12. Remark In the context of locally convex vector spaces, we want the structure of a locally convex algebra on $T(V)$ for each locally convex space V such that if $v(A)$ denotes the underlying locally convex vector space of A, we have the same natural isomorphism

$$\mathrm{Hom}_{(lc\backslash\mathrm{alg}/\mathrm{k})}(T(V), A) = \mathrm{Hom}_{(lc\backslash\mathrm{k})}(V, v(A)),$$

abcd where the morphisms are continuous in the locally convex topology.

One reference is Valqui in K-theory (2001). It solves this problem in the topology from seminorms by putting the seminorms on the nonassociative version of $T(A)$ which is a sum indexed by binary trees. The resulting seminorms make multiplication jointly continuous and are universal for general nonassociative algebras and in the quotient tensor algebra $T(A)$.

For further explanations of extensions, the reader should consult $\langle B \rangle$, Chap. 7.

6 Basic Examples of Extensions for K-Theory

In this section, we consider the basic examples of extensions of locally convex algebras used in the formulation of KK-theory and kk-theory exactness properties. In many cases, there is a continuous and smooth version of the extension. See $\langle B \rangle$, 8.2, for these definitions in the case of C^*-algebras.

6.1. Definition Let B be a locally convex algebra. The continuous cone $C(B)$ and suspension $S(B)$ extension is

$$0 \longrightarrow S(B) = B(0,1) \longrightarrow C(B) = B(0,1] \xrightarrow{q} B \longrightarrow 0,$$

where the projection $q(f) = f(1)$. The smooth cone $C^\infty(B)$ and smooth suspension $S^\infty(B)$ extension is

$$0 \longrightarrow S^\infty(B) = B^\infty(0,1) \longrightarrow C^\infty(B) = B^\infty(0,1] \xrightarrow{q} B \longrightarrow 0.$$

6.2. Remark There is a continuous homotopy $h_t : C(B) \to C(B)$ and a smooth homotopy $h_t : C^\infty(B) \to C^\infty(B)$ both given by the formula $h_t(f)(s) = f(ts)$ form the base point to the identity $h_1(f) = f$, or in other words, the cones are contractible in the continuous and smooth senses.

6.3. Definition Let $u : A \to B$ be a morphism of algebras. The mapping cylinder of u is the fibre product $Z(u) = A \times_B B[0,1]$, and the mapping cone of u is the fibre product $C(u) = A \times_B B(0,1]$, where $q : B[0,1] \to B$ and $q : B(0,1] \to B$ are given by $q(f) = f(1)$. We have the smooth versions the mapping cylinder and mapping cone $Z^\infty(u) = A \times_B B^\infty[0,1]$ and $C^\infty(u) = A \times_B B^\infty(0,1]$ also.

6.4. Homotopy Properties of the Mapping Cylinder We have two morphisms of locally convex algebras $\xi = \xi(u) : A \to Z(u)$ given by $\xi(x) = x, f(a)(1)$ and $\eta = \eta(u) : Z(u) \to A$ given by $\eta(x,f) = x$. The composite $\eta\xi = \mathrm{id}_A$ and the composite $\xi\eta(x,f) = x$ is homotopic to $\mathrm{id}_{Z(u)}$ with $k_s(x,f)(t) = (x, f((1-s)t+s))$, where $k_0 = \mathrm{id}_{Z(u)}$ and $k_1 = \xi\eta$. The smooth version of the relation between A and $Z^\infty(u)$ holds by analogy where the homotopy k_s is a smooth homotopy.

6.5. Extensions Defined by the Mapping Cylinder and Cone The projection $p : Z(u) \to B$ given by $p(x,f) = f(0)$ defines a natural extension

$$0 \longrightarrow C(u) \longrightarrow Z(u) \xrightarrow{p} B \longrightarrow 0.$$

Furthermore, if u has a linear section $\sigma : B \to A$ of u, then p has a linear section $\sigma'(y) = (\sigma(y), y(t) = y)$ for all t. Combining (6.4) and (6.5), we have an extension up to homotopy of special importance

$$[u] : C(u) \xrightarrow{a(u)} A \xrightarrow{u} B.$$

The kernel of the surjective algebra morphism $a(u) : C(u) \to A$ is just $C(u)/A = S(B)$, and we have the following basic mapping cone extension for $u : A \to B$.

$$0 \longrightarrow S(B) \xrightarrow{b(u)} C(u) \xrightarrow{a(u)} A \longrightarrow 0.$$

Now, we compare an arbitrary extension to the mapping cone $[u]$ homotopy extension.

6.6. Remark Let $0 \to I \to A \xrightarrow{u} B \to 0$ be an extension of locally convex algebras. There is a comparison between the kernel I and the mapping cone $C(u)$ with $\varepsilon : I \to C(u)$ given by the formula $\varepsilon(x) = (x,0)$ with the following commutative diagram

$$
\begin{array}{ccccc}
I & \longrightarrow & A & \longrightarrow & B \\
\downarrow{\scriptstyle \varepsilon} & & \| & & \\
\end{array}
$$

$$
0 \longrightarrow S(B) \xrightarrow{b(u)} C(u) \xrightarrow{a(u)} A \longrightarrow 0.
$$

The next extension plays a basic role in the definition of kk-theory.

6.7. Definition The universal two-way extension for an algebra A is defined by starting with $Q(A) = A * A$ or $A \coprod A$ called the coproduct (or free product) of A with itself with $\varepsilon', \varepsilon'' : A \to Q(A)$ the two inclusions. There is a codiagonal $\nabla : Q(A) \to A$ which has the defining property $\nabla \varepsilon'$ and $\nabla \varepsilon''$ equal the identity on A. There is a switch morphism $\tau : Q(A) \to Q(A)$ having the defining property $\tau \varepsilon' = \varepsilon''$ and $\tau \varepsilon'' = \varepsilon'$. With the kernel $q(A) = \ker(\nabla)$, we have the two-way extension

$$
0 \longrightarrow q(A) \longrightarrow Q(A) \xrightarrow{\nabla} A \to 0
$$

with two algebra splittings $\varepsilon', \varepsilon'' : A \to Q(A)$.

6.8. Universal Property of the Two-Way Extension Consider an extension with two algebra splittings $\sigma', \sigma'' : A \to E$ over $\pi : E \to A$ by $K \to E$. Then, we have the following universal property of the two-way extension

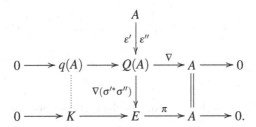

The following extension is used to establish the Bott periodicity for K-theory in the context of operator algebras.

6.9. Töplitz Extensions The algebraic Töplitz algebra $T_{\mathrm{alg}}(v)$ is the $*$-algebra generated by one element v with one relation namely $v^* v = 1$. It has a quotient algebra $C_{\mathrm{alg}}(q)$ of the morphism carrying v to q with two relations namely $q^* q = 1 = qq^*$, and the kernel of $\pi : T_{\mathrm{alg}}(v) \to C_{\mathrm{alg}}(q)$ of the morphism carrying v to q is generated by $e = 1 - vv^*$ which is a projection $e = e^* = e^2$ satisfying $ev = 0$ and $v^* e = 0$. We define an isomorphism $h : (e) \to M_\infty$ where $h(v^i e(v^*)^j) = E_{i,j}$, the matrix with one in the (i,j) entry and zero otherwise satisfying the relations

$$E_{i,j}E_{k,l} = \delta_{j,k}E_{i,l}.$$

The algebraic Töplitz extension itself is

$$0 \longrightarrow M_\infty = (1 - vv^*) \longrightarrow T_{\mathrm{alg}}(v) \longrightarrow C_{\mathrm{alg}}(q) \longrightarrow 0,$$

where $M_\infty = \bigcup_{n \geq 0} M_n$.

The algebraic Töplitz extension has a C^*-completion

$$0 \longrightarrow \mathscr{K} \longrightarrow T(v) \longrightarrow C(T) \longrightarrow 0,$$

where $\mathscr{K}$ is the algebra of compact operators and $C(T)$ is the algebra of continuous functions on the circle $T = U(1)$.

The smooth Töplitz extension is the locally convex algebra extension

$$0 \longrightarrow \mathscr{K}^\infty \longrightarrow T^\infty(v) \longrightarrow C^\infty(T) \longrightarrow 0,$$

where $C^\infty(T)$ is the algebra of smooth functions on the circle $T = U(1)$. In terms of Fourier coefficients $\sum_{k \in \mathbb{Z}} a_k q^k$ where the seminorms $q_n((a_k)) = \sum_{k \in \mathbb{Z}} |1 + k|^n |a_k|$ are finite. These seminorms are submultiplicative. The algebra $\mathscr{K}^\infty$ is described as in (3.9).

6.10. Definition Let A be a locally convex algebra. The dual cone $C^\vee(A)$ and dual suspension $S^\vee(A)$ are defined as kernel in the exact sequence coming from the smooth Töplitz algebra as the kernel of the mapping carrying v and q to $1 \in A$

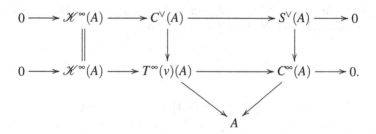

Finally, we have an algebraic version of the first cone and suspension construction as follows.

6.11. Definition Let A be locally convex algebra and form the locally convex algebra $A[t] = A \hat{\otimes} \mathbb{C}[t]$, where $\mathbb{C}[t]$ has the fine topology. Let $S_{\mathrm{alg}}(A) = t(1-t)A[t] \subset C_{\mathrm{alg}}(A) = tA[t]$ as ideals in $A[t]$ of polynomials vanishing at 0 and 1 and at 0.

7 Homotopy Invariant, Half Exact, and Stable Functors

Most functors that we consider take values in (ab) or (ab)^{op} or in the categories ($\mathbb{C}$) or ($\mathbb{C}$)^{op}, that is, abelian group or vector-valued covariant or contravariant functors.

7.1. Definition Let $\mathscr{A}$ denote one of the categories $(C^*\backslash\mathrm{alg}) \subset (\mathrm{m}\backslash\mathrm{alg}) \subset (lc\backslash\mathrm{alg})$ and $\mathscr{C}$ be an abelian category. A functor $H : \mathscr{A} \to \mathscr{C}$ is

(a) homotopy (resp. smooth homotopy) invariant provided for a homotopy (resp. smooth homotopy) between $u', u'' : A \to B$, the induced morphisms $H(u') = H(u'') : H(A) \to H(B)$ is an isomorphism.
(b) M_2-stable (resp. stable) provided for the inclusion $A \to M_2(A)$ (resp. $A \to \mathscr{K}(A)$) induces an isomorphism $H(A) \to H(M_2(A))$ (resp. $H(A) \to H(\mathscr{K}(A))$).
(c) half exact for linearly split (resp. for positively split, for algebraically split) extensions provided for such an extension $0 \to I \to E \to B \to 0$, the sequence $H(I) \to H(E) \to H(B)$ is exact in the abelian category $\mathscr{C}$.

7.2. Example The K-theory functor $K_0 : \mathscr{A} \to (\mathrm{ab})^{\mathrm{op}}$ is homotopy invariant, is stable, and is half exact for linearly split extensions.

The next project is to extend the half exactness property to a longer sequence using properties of extensions and the homotopy and stability properties. We define connecting morphisms or boundary morphisms associated with each $S(B)$ and $S^\vee(B)$, the suspension and its dual.

7.3. Proposition *Let $0 \to I \to A \xrightarrow{u} B \to 0$ be an extension of locally convex algebras as in (6.6), and consider the comparison between the kernel I and the mapping cone $C(u)$ with $\varepsilon : I \to C(u)$ given by the formula $\varepsilon(x) = (x, 0)$ with the following commutative diagram*

$$
\begin{array}{ccccc}
I & \xrightarrow{\;j\;} & A & \longrightarrow & B \\
\downarrow{\scriptstyle\varepsilon} & & \| & & \| \\
0 \longrightarrow S(B) & \xrightarrow{\;b(u)\;} & C(u) & \xrightarrow{\;a(u)\;} & A \longrightarrow 0.
\end{array}
$$

For a half-exact homotopy invariant functor H, we have $H(C(\varepsilon)) = 0$ and $H(\varepsilon) : H(I) \to H(C(u))$ is an isomorphism. There is a long exact sequence infinite to the left of the form

$$
\xrightarrow{\;\partial\;} H(S(I)) \xrightarrow{H(Sj)} H(S(A)) \xrightarrow{H(Su)} H(S(B)) \xrightarrow{\;\partial\;} H(I) \xrightarrow{H(j)} H(A) \xrightarrow{H(u)} H(B).
$$

This follows from the following commutative diagram using the first two assertions

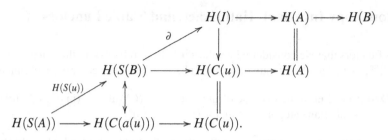

There is an argument to the effect that the composition of the maps $S(A) \xrightarrow{S(u)}$ $S(B) \longrightarrow C(a(u))$ is smoothly homotopic to the natural map $S(A) \to C(a(u))$ composed with the natural self-map of $S(A)$ reversing the orientation on $[0,1]$.

See $\langle B \rangle$, 19.4–19.6, for more explanation.

8 The Bivariant Functor $kk_*(A, B)$

We consider a functor $kk_*(A,B)$ of two variables

$$kk_* : (lc\backslash\mathrm{alg})^{\mathrm{op}} \times (lc\backslash\mathrm{alg}) \longrightarrow \mathrm{gr}(\mathbb{Z})$$

with values in the category $\mathrm{gr}(\mathbb{Z})$ of graded abelian groups which will have several properties:

(a) it will be a limit expression of the form $[A',B']$ where A' is the result of terms in an iterate universal extension of A, B' is a stabilization of B and $[A',B']$ is homotopy classes of $\mathrm{Hom}_{(lc\backslash\mathrm{alg})}(A',B')$,
(b) extensions will define classes in $kk_1(A,B)$,
(c) it will be homotopy invariant, half exact, and stable in each variable
(d) $kk_*(\mathbb{C},B) = K_*(B)$ which is already defined and $kk_*(A,\mathbb{C}) = K^*(A)$ yet to be defined as K-homology,
(e) there is a composition product with

$$kk_*(A,B) \otimes kk_*(B,C) \longrightarrow kk_*(A,C),$$

which is associative with unit in each $kk_*(A,A)$ and $kk_*(\mathbb{C},\mathbb{C}) = \mathbb{Z}$ when the split exact sequences for half-exactness are algebraically split,
(f) The pairings in Sect. 4 are examples of the composition product between $K_*(B) \otimes K^*(B) = kk_*(\mathbb{C},B) \otimes kk_*(B,\mathbb{C}) \longrightarrow kk_*(\mathbb{C},\mathbb{C}) = \mathbb{Z}$.

After establishing Bott periodicity in the next section, we have two exact hexagons associated to an extension an *kk*-theory starting with the mapping cone sequence.

8.1. Theorem *Let D be a locally convex algebra, and let*

$$(E): 0 \longrightarrow I \xrightarrow{i} E \xrightarrow{\pi} B \longrightarrow 0$$

be an extension with a continuous linear splitting. Associated to the functor $kk(D, \ )$ and $kk(\ ,D)$ there are two exact hexagons

$$
\begin{array}{ccccc}
kk_0(D,I) & \xrightarrow{kk(i)} & kk_0(D,E) & \xrightarrow{kk(\pi)} & kk_0(D,B) \\
\uparrow & & & & \downarrow \\
kk_1(D,B) & \xleftarrow{kk(\pi)} & kk_1(D,E) & \xleftarrow{kk(i)} & kk_1(D,I)
\end{array}
$$

and

$$
\begin{array}{ccccc}
kk_0(B,D) & \xrightarrow{kk(\pi)} & kk_0(E,D) & \xrightarrow{kk(i)} & kk_0(I,D) \\
\uparrow & & & & \downarrow \\
kk_1(I,D) & \xleftarrow{kk(i)} & kk_1(E,D) & \xleftarrow{kk(\pi)} & kk_1(B,D).
\end{array}
$$

Associated to the extension (E) is a classifying map $J(B) \to I$ and thus an element of $kk_1(B,I)$ denoted by $kk(E)$. The vertical arrows in the first diagram are given up to sign by multiplication on the right by $kk(E)$ and in the second diagram up to sign by multiplication on the left by $kk(E)$.

The standard theory for $KK(A,B)$ is developed in $\langle B \rangle$, 17.1–19.6.

9 Bott Map and Bott Periodicity

Some aspects of algebraic K-theory are common to operator algebra K-theory and to topological K-theory, but the most significant difference centers around Bott periodicity. The existence of the Bott map and of the inverse of the Bott map has been very nicely clarified recently in the Münster thesis of A. Thom. We present the general framework of his ideas which, as in Cuntz's proof of Bott periodicity, starts with the Töplitz extension in operator algebra K-theory and the two versions of a smooth and continuous suspension which are different from the one in the algebraic case. This is developed again in $\langle B \rangle$, 19.2, with an elementary introduction in $\langle B \rangle$, 9.1–9.3.

9.1. Remark In the suspension algebra $S = C(0,1)$ and the dual-suspension algebra $S^\vee = ker(\varepsilon : C(S^1) \to \mathbb{C})$, where $\varepsilon(f) = f(1)$, we have the smooth and polynomial versions of these algebras with values in a locally convex algebra A:

$$t(1-t)A[t] \subset S^\infty(A) \subset S(A)$$

for $S^\infty(A) = \{f \in A^\infty[0,1] \mid f^{(m)}(0) = f^{(m)}(1) = 0 \text{ for all } m\}$ and

$$(1-t)A[t,t^{-1}] \subset (S^\vee)^\infty(A) \subset S^\vee(A)$$

for $(S^\vee)^\infty(A) = \{f \in A^\infty(S^1) | f^{(m)}(1) = 0 \text{ for all } m\}$. The two continuous and the two smooth versions are isomorphic by change of variable $g \in S^\vee(A)$ is mapped to $f(t) = g(e^{2\pi it}), f \in S(A)$. On the other hand, the algebraic subalgebras do not correspond at all since there is no algebraic exponential map.

9.2. Definition Let kk_* be a kk-theory for which the dual mapping cone is zero. Then, there is an excision for the reduced Töplitz extension

$$\mathscr{K}(A) \longrightarrow C^\vee(A) \longrightarrow S^\vee(A)$$

The unstable inverse Bott morphism $\psi : SS^\vee(A) \to \mathscr{K}(A)$ is the image of the identity $id_{S^\vee(A)}$ under the composition of the boundary and the inverse boundary morphisms

$$kk_i(S^\vee(A), S^\vee(A)) \longrightarrow kk_{i-1}(S^\vee(A), \mathscr{K}(A)) \longrightarrow kk_i(S(S^\vee(A)), \mathscr{K}(A)).$$

Again $\mathscr{K}(A)$ is the compact operator stabilization of A.

9.3. Definition An inverse Bott morphism $\phi : S(S^\vee(A)) \to A$ is a morphism which composed with the natural stabilization $A \to \mathscr{K}(A)$ gives the unstable inverse Bott morphism.

9.4. Proposition *Let kk_* be a $\mathscr{K}$-stable theory with extension for the Töplitz extension. Then, there exists a unique inverse Bott morphism which is the image of the unstable inverse Bott morphism $\psi : S(S^\vee(A)) \to \mathscr{K}(A)$ under the inverse of the natural stabilization isomorphism*

$$kk_i(S^\vee(S(A)), \mathscr{K}(A)) \to kk_i(S^\vee(S(A)), A).$$

For the Bott morphism itself, we have similar considerations with the following algebra morphism induced by an inclusion

$$j : S^3 = S(\mathbb{H}) \to U(2) \subset M_2(\mathbb{C}),$$

where $j(w)$ is the left multiplication of w on the quaternions $\mathbb{H}$ and hence on $\mathbb{C}^2$ as a unitary matrix. We can view j as an element in $M_2(C_0(S^3))$.

9.5. Definition The Bott algebra morphism $C_0(S^1) \to M_2(C_0(S^3))$ is the unique morphism of C^*-algebras which carries $q = e^{2\pi i\theta}$ to the map $j \in M_2(C_0(S^3))$ and by taking the tensor product with a locally convex algebra the corresponding morphism

$$\theta : S^\vee(A) \to M_2((S^\vee)^3(A))$$

In the C^*-algebra case, the functors S and $S^\vee$ are identified.

9.6. Definition A Bott morphism $S^\vee(A) \to (S^\vee)^3(A)$ is any morphism whose composition with the stabilization morphism

$$(S^\vee)^3(A) \to M_2((S^\vee)^3(A))$$

gives the Bott algebra morphism $S^\vee(A) \to M_2((S^\vee)^3(A))$.

9.7. Proposition *Let kk_* be a M_2-stable theory. Then, there exists a Bott morphism β which is the image of the Bott algebra morphism under the inverse of the natural M_2-stable isomorphism*

$$kk_i(S^\vee(A), (S^\vee)^3(A)) \to kk_i(S^\vee(A), M_2((S^\vee)^3(A))).$$

In the case where the Bott morphism and the inverse Bott morphism exist, they are inverse morphisms under composition.

A Appendix: The Green–Julg Theorem (*by S. Echterhoff*)

In this section, we give a short KK-theoretic proof of the Green–Julg theorem, that is, we show that for any compact group G and any G-C^*-algebra B the group $KK_*^G(\mathbb{C}, B)$ is canonically isomorphic to $K_*(B \rtimes G)$.

Let G be a locally compact group and let A and B be two G-C^*-algebras. Then, the equivariant KK-groups $KK^G(A, B) =: KK_0^G(A, B)$ are defined as the set of all homotopy classes of triples $(\mathscr{E}, \Phi, T)$, where

(1) $\mathscr{E} = \mathscr{E}_0 \oplus \mathscr{E}_1$ is a $\mathbb{Z}_2$-graded Hilbert B-module endowed with a grading preserving action $\gamma : G \to \operatorname{Aut}(\mathscr{E})$;
(2) $\Phi = \begin{pmatrix} \Phi_0 & 0 \\ 0 & \Phi_1 \end{pmatrix}$ is a G-equivariant $*$-homomorphism;
(3) $T = \begin{pmatrix} 0 & P \\ Q & 0 \end{pmatrix}$ is an operator in $\mathscr{L}(\mathscr{E})$ such that

$$[\Phi(a), T], (T^* - T)\Phi(a), (T^2 - 1)\Phi(a), (\operatorname{Ad} \gamma_s(T) - T)\Phi(a) \in \mathscr{K}(\mathscr{E})$$

for all $a \in A$ and $s \in G$.

If G is compact, then it was shown by Kasparov that we may assume without loss of generality that the operator T is G-equivariant (by replacing T by $T^G = \int_G \operatorname{Ad}\gamma_s(T)\, ds$ if necessary) and that Φ is nondegenerate. In particular, if $A = \mathbb{C}$, then $KK^G(\mathbb{C}, B)$ can be described as the set of homotopy classes of pairs $(\mathscr{E}, T)$ such that T is G-invariant and

$$T^* - T, \; T^2 - 1 \in \mathscr{K}(\mathscr{E}).$$

Similarly, we can describe $K_0(B \rtimes G) = KK(\mathbb{C}, B \rtimes G)$. We start with the following easy lemma:

A.1. Lemma *Let distance G be a compact group and let B be a G-C^*-algebra. Suppose that $\mathscr{E}_B$ is a G-equivariant Hilbert B-module. Then $\mathscr{E}_B$ becomes a pre-Hilbert $B \rtimes G$-module if we define the right action of $B \rtimes G$ on $\mathscr{E}_B$ and the $B \rtimes G$-valued inner products by the formulas*

$$e \cdot f := \int_G \gamma_s \left(e \cdot f(s^{-1}) \right) ds \quad and \quad \langle e_1, e_2 \rangle_{B \rtimes G}(s) := \langle e_1, \gamma_s(e_2) \rangle_B$$

for $e, e_1, e_2 \in \mathscr{E}$ and $f \in C(G,B) \subseteq B \rtimes G$. Denote by $\mathscr{E}_{B \rtimes G}$ its completion. Moreover, if $(\mathscr{E}_B, T)$ represents an element of $KK^G(\mathbb{C},B)$ with T being G-invariant, then T extends to an operator on $\mathscr{E}_{B \rtimes G}$ such that $(\mathscr{E}_{B \rtimes G}, T)$ represents an element of $KK(\mathbb{C}, B \rtimes G)$.[1]

Proof. First note that the above defined distance right action of $C(G,B)$ on $\mathscr{E}_B$ extends to an action of $B \rtimes G$. For this, we observe that the pair (Ψ, γ), with $\Psi : B \to \mathscr{L}_{\mathscr{K}(\mathscr{E})}(\mathscr{E}_B)$ given by the formula $\Psi(b)(e) = e \cdot b^*$, is a covariant homomorphism of (B, G, β) on the left Hilbert $\mathscr{K}(\mathscr{E}_B)$-module $\mathscr{E}_B$. Then $e \cdot f = (\Psi \times \gamma(f))(e)$ for $f \in C(G,B)$, and the right-hand side clearly extends to all of $B \rtimes G$.

It is easily seen that $\langle \cdot, \cdot \rangle_{B \rtimes G}$ is a well-defined $B \rtimes G$-valued inner product which is compatible with the right action of $B \rtimes G$ on $\mathscr{E}_B$. So we can define $\mathscr{E}_{B \rtimes G}$ as the completion of $\mathscr{E}_B$ with respect to this inner product.

If $T \in \mathscr{L}_B(\mathscr{E}_B)$, then T determines an operator $T^G \in \mathscr{L}_{B \rtimes G}(\mathscr{E}_{B \rtimes G})$ by the formula

$$T^G(e) = \int_G \gamma_t \left(T(\gamma_{t^{-1}}(e)) \right) dt, \quad for\ e \in \mathscr{E}_B \subseteq \mathscr{E}_{B \rtimes G},$$

and one checks that $T \to T^G$ is a $*$-homomorphism from $\mathscr{L}_B(\mathscr{E}_B)$ to $\mathscr{L}_{B \rtimes G}(\mathscr{E}_{B \rtimes G})$. In particular, it follows that every G-invariant operator on $\mathscr{E}_B$ extends to an operator on $\mathscr{E}_{B \rtimes G}$. If $e_1, e_2 \in \mathscr{E}_B \subseteq \mathscr{E}_{B \rtimes G}$, then a short computation shows that the corresponding finite rank operator $\Theta_{e_1, e_2} \in \mathscr{K}(\mathscr{E}_{B \rtimes G})$ is given by the formula

$$\Theta_{e_1, e_2} = \tilde{\Theta}^G_{e_1, e_2}$$

if $\tilde{\Theta}_{e_1, e_2} \in \mathscr{K}(\mathscr{E}_B)$ denotes the corresponding finite rank operator on $\mathscr{E}_B$. This easily implies that the remaining part of the lemma.

A.2. Theorem *(Green–Julg Theorem) Let G be a compact group and let B be a G-C^*-algebra. Then the map*

$$\mu : KK^G(\mathbb{C},B) \to KK(\mathbb{C}, B \rtimes G); \mu \left([(\mathscr{E}_B, T)] \right) = [(\mathscr{E}_{B \rtimes G}, T)]$$

is an isomorphism.

Proof. Note first that we can apply the same formula to a homotopy, so the map is well defined. We now define a map $\nu : KK(\mathbb{C}, B \rtimes G) \to KK^G(\mathbb{C},B)$ and show that it is inverse to μ.

For this, let $L^2(G,B)$ denote the Hilbert B-module defined as the completion of $C(G,B)$ with respect to the B-valued inner product

$$\langle f, g \rangle_B = \int_G \beta_s \left(f(s^{-1})^* g(s^{-1}) \right) ds$$

---

[1] We are grateful to Walther Paravicini for pointing out a mistake in a previous version of this lemma!

and the right action of B on $L^2(G,B)$ given by $(f \cdot b)(t) = f(t)\beta_t(b)$ for $f \in C(G,B), b \in B$. There is a well-defined left action of $B \rtimes G$ on $L^2(G,B)$ given by convolution when restricted to $C(G,B) \subseteq B \rtimes G$ (and $C(G,B) \subseteq L^2(G,B)$). We even have $B \rtimes G \subseteq \mathcal{K}(L^2(G,B))$. To see this, we simply note that $\mathcal{K}(L^2(G,B)) = C(G,B) \rtimes G$ by Green's imprimitivity theorem (where G acts on $C(G)$ by left translation), and $B \rtimes G$ can be viewed as a subalgebra of $C(G,B) \rtimes G$ in a canonical way.

Let $\sigma : G \to \mathrm{Aut}(L^2(G,B))$ be defined by

$$\sigma_s(f)(t) = f(ts); \quad f \in C(G,B).$$

Then, σ is compatible with the action β of G on B. Moreover, a short computation shows that the homomorphism of $B \rtimes G$ into $\mathcal{L}(L^2(G,B))$ given by convolution is equivariant with respect to the trivial G-action on $B \rtimes G$ and the action $\mathrm{Ad}\sigma$ on $L^2(G,B)$. Now assume that $(\tilde{\mathcal{E}},T)$ represents an element of $KK(\mathbb{C},B \rtimes G)$. Then, $\tilde{\mathcal{E}} \otimes_{B \rtimes G} L^2(G,B)$ equipped with the action $\mathrm{id} \otimes \sigma$ is a G-equivariant Hilbert B-module and $(\tilde{\mathcal{E}} \otimes L^2(G,B), T \otimes 1)$ represents an element of $KK^G(\mathbb{C},B)$ (here, we use the fact that $B \rtimes G \subseteq \mathcal{K}(L^2(G,B))$). Thus, we define

$$\nu : KK(\mathbb{C},B \rtimes G) \to KK^G(\mathbb{C},B); \quad \nu([(\tilde{\mathcal{E}},T)]) = [(\tilde{\mathcal{E}} \otimes_{B \rtimes G} L^2(G,B), T \otimes 1)].$$

Again, applying the same formula to homotopies implies that ν is well defined.

To see that ν is an inverse to μ one check the following:

(a) Let $\mathcal{E}_B$ be a Hilbert B-module and let $\mathcal{E}_{B \rtimes G}$ be the corresponding Hilbert $B \rtimes G$-module as described in Lemma A.1 Then,

$$\mathcal{E}_B \odot C(G,B) \to \mathcal{E}_B; e \otimes f \mapsto e \cdot f = \int_G \gamma_s(e \cdot f(s^{-1}))\, ds$$

extends to a G-equivariant isometric isomorphism between $\mathcal{E}_{B \rtimes G} \otimes_{B \rtimes G} L^2(G,B)$ and $\mathcal{E}_B$.

(b) Let $\tilde{\mathcal{E}}$ be a Hilbert $B \rtimes G$-module. Then,

$$\tilde{\mathcal{E}} \odot C(G,B) \to \tilde{\mathcal{E}}; e \otimes f \mapsto e \cdot f$$

determines an isometric isomorphism

$$(\tilde{\mathcal{E}} \otimes_{B \rtimes G} L^2(G,B))_{B \rtimes G} \cong \tilde{\mathcal{E}}$$

as Hilbert $B \rtimes G$-modules.

Both results follow from some straightforward computations. Note that for the proof of (b) one should use the fact that for all $x \in B \rtimes G$ and $f,g \in C(G,B)$, the element of $B \rtimes G$ given by the continuous function $s \mapsto \langle x^* \cdot f, \sigma_s(g)\rangle_B$ coincides with $f^* * x * g$, where we view f,g as elements of $B \rtimes G$. This follows by direct computations for $x \in C(G,B)$, and since both expressions are continuous in x, it follows for all $x \in B \rtimes G$. Finally, it is trivial to see that the operators match up in both directions.

A.3. Remark We should remark, that the above Theorem A.2 is a special case of a more general result for crossed products by proper actions due to Kasparov and Skandalis (Theorem 5.4). There also exist important generalizations to proper groupoids by Tu (2005) (Proposition 6.25) and Paravicini (2007), where the latter provides a version within Lafforgue's Banach *KK*-theory. For the original proof of the Green–Julg theorem for compact groups we refer to Julg (1981).

References

Blackadar, B.: K-Theory for Operator Algebras, 2nd ed. Cambridge University Press, Cambridge (1998)

Cuntz, J., Meyer, R., Rosenberg, J.: Topological and Bivariant K-Theory. Birkhäuser, Basel (2007)

Valqui, C.: Universal extension and excision for topological algebras. K-Theory **22**: 145–160 (2001)

Julg, P.: *K*-thorie équivariante et produits croisés. C. R. Acad. Sci. Paris Sér. I Math. **292(13)**: 629–632 (1981)

Kasparov, G., Skandalis, G.: Groups acting properly on "bolic" spaces and the Novikov conjecture. Annals of Math. **158**: 165–206 (2003)

Tu, J.-L.: La conjecture de Novikov pour les fouilletages hyperboliques. *K*-Theory **16(2)** 129–184 (1999)

Paravicini, W.: Ph.D. Dissertation. Münster, "KK-Theory for Banach Algebras and Proper Groupoids" (2006)

Part IV
Algebra Bundles: Twisted K-Theory

In Part I, we introduced the generalities on bundles over a space. In Chaps. 1, 2, and 5, we described them as a family of spaces fibred over a base space. In Chap. 3, we saw that bundles could be described by other data using modules over a function algebra, and then in Chap. 4, the stability phenomenon of K-theory was introduced. In Part II, topological twisting coming in principal bundles and vector bundles is studied from the point of view of homotopy theory and of cohomology where classifying spaces and characteristic classes play a role, respectively.

In Part IV, we consider higher order twisting on vector bundles in terms of algebra bundles. The first such example is matrix algebra bundles clearly related to vector bundles and also operator algebra bundles with fibre $\mathscr{B}(H)$, the bounded operators on separable Hilbert space H. Matrix algebra is just the case when H is finite dimension and so $\mathscr{B}(H) = M_n(\mathbb{C})$.

Associated to an operator bundle A over a space X is a characteristic class $\alpha(A) \in H^3(X, \mathbb{Z})$ introduced by Dixmier and Douady which completely classifies the algebra bundle in the same way that the first Chern class $c_1(L)$ completely classifies line bundles up to all isomorphisms.

The next step is to consider the Fredholm elements in operator algebra bundles when the algebra bundle is trivial, this leads to ordinary K-theory, and in general, this is our first approach to twisted K-theory. Ordinary K-theory can be defined as homotopy classes of maps into Fredholm operators, and twisted K-theory can be defined as homotopy classes of cross sections of fibre bundles associated to a given algebra bundle with fibre, the Fredholm operators. In particular, the twisted K-theory groups over X depend on an extra object, an operator (algebra) bundle A and its extra parameter the characteristic class $\alpha = \alpha(A) \in H^3(X, \mathbb{Z})$. The realization as twisted K-theory and the term itself is due to Rosenberg (1989).

Chapter 18
Isomorphism Classification of Operator Algebra Bundles

We consider algebra bundles where the fibre is an algebra of bounded operators in a separable Hilbert space H over the complex numbers. If the Hilbert space H is infinite dimensional, the algebra is either $\mathscr{B}$ the algebra of all bounded operators on H or $\mathscr{K}$ the algebra of compact operators in $\mathscr{B}$, and we refer to these bundles as operator algebra bundles. If the Hilbert space H is finite dimensional, the algebra is just the n^2-dimensional matrix algebra $M_n(\mathbb{C})$, and we refer to these bundles as matrix algebra bundles.

As with vector bundles, these bundles have a fibre bundle interpretation so that the structure is governed by a related principal bundle over the automorphism group of the algebra. The classification of these algebra bundles is just a homotopy analysis of this automorphism group of the algebra.

For an algebra bundle A over X, we associate a characteristic class $\alpha(A) \in H^3(X,\mathbb{Z})$ using the homotopy classification of principal bundles. With this characteristic class, we have an isomorphism classification of infinite dimensional operator algebra bundles over X which is contained in the next two statements:

(1) Two operator algebra bundles A' and A'' over X are isomorphic if and only if $\alpha(A') = \alpha(A'') \in H^3(X,\mathbb{Z})$.
(2) Every element of $H^3(X,\mathbb{Z})$ is of the form $\alpha(A)$ for some operator algebra bundle over X.

There are two corresponding statements for the first Chern class $c_1(L)$ of complex line bundles over X.

The three-dimensionalcharacteristic class was discovered as a Čech cohomology class by Grothendieck in the finite-dimensional case and by Dixmier and Douady in the general case. The homotopy version was pointed out by Rosenberg. Operator algebra bundles were treated by Fell (1959) and by Dixmier and Douady (1963), where the Čech H^3 classification is given.

The classification of operator algebra bundles is very simple for the same reason that the only infinite-dimensional Hilbert bundles are the trivial ones. This is the Kuiper theorem, see 22(2.1), which says that the automorphism group of an

D. Husemöller et al.: *Isomorphism Classification of Operator Algebra Bundles*, Lect. Notes Phys. **726**, 229–239 (2008)
DOI 10.1007/978-3-540-74956-1_19 © Springer-Verlag Berlin Heidelberg 2008

infinite-dimensional separable Hilbert space is contractible, and in particular, then its classifying space is contractible also.

Matrix algebra bundles were considered by Serre and Grothendieck, and the subject was treated as an introduction to Grothendieck's three Seminar Bourbaki exposés which appeared also in the collection *Dix Exposés*. The same homotopy analysis used for operator algebra bundles applies to M_n-matrix algebra bundles A over X. The result is a characteristic class $\alpha_n(A) \in {}_nH^3(X,\mathbb{Z})$ with values in the subgroup of $a \in H^3(X,\mathbb{Z})$ with $n \cdot a = 0$, that is, n-torsion elements. Grothendieck gave a description of this class using sheaf theory. In the finite-dimensional case, we do not obtain an isomorphism classification theory as with operator algebras, for all matrix algebra bundles of the form $\mathrm{End}(E)$ where E is a vector bundle have the property that $\alpha_n(\mathrm{End}(E)) = 0$. To each matrix bundle A', there is an operator algebra bundle A with $\alpha(A) = \alpha(A')$, and A is unique up to isomorphism. This bundle A associated to A' can be constructed by tensoring the trivial operator bundle A with A'.

This leads to the following refinement of the classification theorem

(3) An operator algebra bundle A is of the form of a M_n-matrix algebra bundle tensor with the trivial operator algebra bundle if and only if $\alpha(A)$ is an n-torsion element in $H^3(X,\mathbb{Z})$.

Returning to line bundles, we recall the related result which says that a line bundle L over X is flat if and only if $c_1(L)$ is a torsion element in $H^2(X,\mathbb{Z})$.

For the convenience of the reader, certain features of vector bundle theory which are treated in Chaps. 2, 5 and 7 are repeated in a parallel manner to algebra bundle theory.

1 Vector Bundles and Algebra Bundles

1.1. Definition A vector bundle E over a space X is a bundle $p : E \to X$ together with bundle morphisms $E \times_X E \to E$ and $\mathbb{C} \times E \to E$ over X such that these operations make each fibre into a vector space over $\mathbb{C}$. Moreover, each point has an open neighborhood U such that $E|U$ is isomorphic to the product vector bundle $U \times \mathbb{C}^n$. If all the fibres of E are n-dimensional, then E is called an n-dimensional, vector bundle.

1.2. Notation Let $\mathrm{Vect}^n(X)$ denote the isomorphism classes of n-dimensional vector bundles over X. For a map $f : X \to Y$ and a vector bundle E over Y, the induced bundle $f^*(E)$ has a natural vector bundle structure with morphism $f^*(E) \to E$ over f such that on the fibres it restricts to an isomorphism of vector spaces. Hence, $\mathrm{Vect}^n : (\mathrm{top})^{\mathrm{op}} \to (\mathrm{set})$ is a functor from the opposite category of spaces (top) to the category of sets (set).

1.3. Definition A matrix algebra bundle A over a space X is a bundle $p : A \to X$ together with bundle morphisms $A \times_X A \to A$ and $\mathbb{C} \times A \to A$ such that these operations make each fibre into a matrix algebra over $\mathbb{C}$. Moreover, each point has an open neighborhood U such that $A|U$ is isomorphic to the product matrix algebra bundle $U \times M_n(\mathbb{C})$. If all the fibres of A are n by n matrix algebras, then A is called an n by n matrix algebra bundle.

1.4. Notation Let $\mathrm{Alg}^n(X)$ denote the isomorphism classes of n by n matrix algebra bundles over X. For a map $f : X \to Y$ and a matrix algebra bundle A over Y, the induced bundle $f^*(A)$ has a natural matrix algebra bundle structure with morphism $f^*(A) \to A$ over f such that the restrictions to the fibres are isomorphisms of algebras. Hence, $\mathrm{Alg}^n : (\mathrm{top})^{\mathrm{op}} \to (\mathrm{set})$ is a functor.

1.5. Definition There is a natural morphism of functors

$$\mathrm{End} : \mathrm{Vect}^n \longrightarrow \mathrm{Alg}^n$$

which assigns to each vector bundle E the endomorphism matrix algebra bundle $\mathrm{End}(E) = E \otimes E^\vee$ as a vector bundle with multiplication given by using the evaluation morphism $E^\vee \otimes E \to X \times \mathbb{C}$.

This operation commutes with the induced bundle construction so that it defines a morphism of functors. These $\mathrm{End}(E)$ are examples of finite-dimensional algebra bundles reflecting the vector bundle structure of E.

1.6. Remark Let H be a separable infinite-dimensional Hilbert space, and let $\mathscr{B} = \mathscr{B}(H)$ be the algebra of bounded operators on H with the ideal $\mathscr{K} = \mathscr{K}(H)$ of compact operators. As in definition (1.1), we can define Hilbert space vector bundles which are locally of the form $U \times H$, and as in definition (1.3), we can define operator algebra bundles in two ways which are locally of the form $U \times \mathscr{B}$ or $U \times \mathscr{K}$.

1.7. Remark In fact, all infinite-dimensional Hilbert space bundles are trivial, and the reason stems, as we have mentioned before, from the contractibility of the automorphism group of the Hilbert space H which is used to glue locally trivial bundles together into a global bundle. This follows clearly from the fibre bundle structure of all the bundles under consideration.

We leave the description of the categories of algebra bundles to the reader.

2 Principal Bundle Description and Classifying Spaces

2.1. Notation For a topological group G and a space X, we denote the set of isomorphism classes of locally trivial principal G-bundles by $\mathrm{Prin}_G(X)$. Further, if Y is a left G-space and P is a principal G-bundle, then the associated fibre bundle $P[Y] = P \times^G Y = P \times Y/G$, the quotient of $P \times Y$ by the relation (x, y), is equivalent to $(xs, s^{-1}y)$ for all $s \in G$.

2.2. Remark In order to use principal bundles, we recall that the group of automorphisms of the fibres of vector bundles and matrix algebra bundles are given, respectively, by

$$GL(n,\mathbb{C}) = \mathrm{Aut}(\mathbb{C}^n)$$

and

$$PGL(n,\mathbb{C}) = GL(n,\mathbb{C})/GL(1,\mathbb{C}) = \mathrm{Aut}(M_n\mathbb{C}).$$

Here, $GL(n,\mathbb{C})$ acts on $M_n(\mathbb{C})$ by inner automorphisms, that is, for $A \in GL(n,\mathbb{C})$ and $X \in M_n(\mathbb{C})$, AXA^{-1} is the inner automorphism action of A in $X \in M_n(\mathbb{C})$.

2.3. Assertion For each principal $GL(n,\mathbb{C})$-bundle P over X and the group $GL(n,\mathbb{C})$ acting by matrix multiplication on $\mathbb{C}^n$ the associated fibre bundle $P[\mathbb{C}^n]$ is a vector bundle with vector operations from the associated fibre $\mathbb{C}^n$. This defines a bijection from

$$\mathrm{Prin}_{GL(n,\mathbb{C})}(X) \longrightarrow \mathrm{Vect}^n(X)$$

which is an isomorphism of functors. The inverse of the fibre bundle construction is the principal bundle $P(E)$ of all maps $\mathbb{C}^n \to E$ which map isomorphically onto some fibre and $GL(n,\mathbb{C})$ action given by right composition.

2.4. Assertion For each principal $PGL(n,\mathbb{C})$-bundle P over X and the group $PGL(n,\mathbb{C})$ acting by inner automorphisms on $M_n(\mathbb{C})$, the associated fibre bundle $P[M_n(\mathbb{C})]$ is a matrix algebra bundle with algebra operations from the associated fibre $M_n(\mathbb{C})$. This defines a bijection from

$$\mathrm{Prin}_{PGL(n,\mathbb{C})}(X) \longrightarrow \mathrm{Alg}^n(X)$$

which is an isomorphism of functors. The inverse of the fibre bundle construction is the principal bundle $P(A)$ of all maps $M_n(\mathbb{C}) \to A$ which map isomorphically onto some fibre and $PGL(n,\mathbb{C})$ action given by right composition of inner automorphisms.

2.5. Assertion By dividing out the diagonal subgroup of $GL(n,\mathbb{C})$, we obtain a quotient morphism $GL(n,\mathbb{C}) \to PGL(n,\mathbb{C})$ defining the projective linear group $PGL(n,\mathbb{C})$ as a left $GL(n,\mathbb{C})$ space. Hence, for each principal bundle P over X for $GL(n,\mathbb{C})$, the associated fibre bundle $P[PGL(n,\mathbb{C})]$ is a principal bundle over X for $PGL(n,\mathbb{C})$. This defines a morphism of functors $\theta : \mathrm{Prin}_{GL(n,\mathbb{C})} \to \mathrm{Prin}_{PGL(n,\mathbb{C})}$ between principal bundle sets. In this way, we obtain a commutative diagram of morphisms of functors $(\mathrm{top})^{\mathrm{op}} \to (\mathrm{set})$ using the End functor in (1.5)

$$\begin{array}{ccc} \mathrm{Prin}_{GL(n,\mathbb{C})}(X) & \longrightarrow & \mathrm{Vect}^n(X) \\ \theta \downarrow & & \downarrow \mathrm{End} \\ \mathrm{Prin}_{PGL(n,\mathbb{C})}(X) & \longrightarrow & \mathrm{Alg}^n(X) \end{array}$$

2.6. Remark For infinite dimensional Hilbert space vector bundles and operator algebra bundles, we have the same construction using $GL(1,\mathscr{B}) = \mathrm{Aut}(H)$, the group of invertible bounded operators. Using the fibre bundle construction, we have a bijection

$$\mathrm{Prin}_{GL(1,\mathscr{B})}(X) \longrightarrow \mathrm{Alg}^H(X)$$

which is an isomorphism of functors of X. The functor $\mathrm{Alg}^H(X)$ has two interpretations either as bundles modeled on $\mathscr{B}$ or $\mathscr{K}$.

Now, the classification of vector bundles and algebra bundles is reduced to the classification of principal bundles. In this form, there is a homotopy classification of principal bundles, and this we take up in the next section.

3 Homotopy Classification of Principal Bundles

3.1. Remark Associated with the two groups and the group morphism $GL(n,\mathbb{C}) \to PGL(n,\mathbb{C}) = GL(n,\mathbb{C})/GL(1,\mathbb{C})$ are the classifying spaces $BGL(n,\mathbb{C})$ and $BPGL(n,\mathbb{C})$ with universal principal bundles $EGL(n,\mathbb{C})$ and $EPGL(n,\mathbb{C})$ for the groups $GL(n,\mathbb{C})$ and $PGL(n,\mathbb{C})$, respectively. The induced morphism $BGL(n,\mathbb{C}) \to BPGL(n,\mathbb{C})$ on the classifying spaces is just the functor B applied to the quotient morphism, or it is the classifying map for the principal $PGL(n,\mathbb{C})$ bundle $EGL(n,\mathbb{C})[PGL(n,\mathbb{C})]$ over $BGL(n,\mathbb{C})$. The induced universal bundle over a space X leads to the two new horizontal arrows in the following extension of (2.5).

3.2. Assertion In the following commutative diagram, the horizontal functions are all bijections

$$
\begin{array}{ccccc}
[X,BGL(n,\mathbb{C})] & \longrightarrow & \mathrm{Prin}_{GL(n,\mathbb{C})}(X) & \longrightarrow & \mathrm{Vect}^n(X) \\
\downarrow & & \theta\downarrow & & \downarrow{\scriptstyle\mathrm{End}} \\
[X,BPGL(n,\mathbb{C})] & \longrightarrow & \mathrm{Prin}_{PGL(n,\mathbb{C})}(X) & \longrightarrow & \mathrm{Alg}^n(X).
\end{array}
$$

The first vertical arrow is induced by the induced quotient map $BGL(n,\mathbb{C}) \to BPGL(n,\mathbb{C})$.

3.3. Remark There are compact versions of these groups and related classifying spaces. The unitary group $U(n)$ is a subgroup of $GL(n,\mathbb{C})$ and the projective unitary group $PU(n) = U(n)/U(1)$ is a subgroup of $PGL(n,\mathbb{C})$. The natural inclusions are seen to be homotopy equivalences with the Gram–Schmidt process, and hence, the inclusions of classifying spaces are homotopy equivalences

$$BU(n) \longrightarrow BGL(n,\mathbb{C}) \quad \text{and} \quad BPU(n) \longrightarrow BPGL(n,\mathbb{C}).$$

Again, there are universal bundles and the remarks in (2.1) carry over to the compact groups. The induced universal bundle over a space X leads to the two new horizontal arrows in the following extension of (3.2)

3.4. Assertion In the following commutative diagram, the horizontal functions are all bijections

$$
\begin{array}{ccccccc}
[X,BU(n)] & \longrightarrow & [X,BGL(n,\mathbb{C})] & \longrightarrow & \mathrm{Prin}_{GL(n,\mathbb{C})}(X) & \longrightarrow & \mathrm{Vect}^n(X) \\
\downarrow & & \downarrow & & \theta \downarrow & & \downarrow \text{End} \\
[X,BPU(n)] & \longrightarrow & [X,BPGL(n,\mathbb{C})] & \longrightarrow & \mathrm{Prin}_{PGL(n,\mathbb{C})}(X) & \longrightarrow & \mathrm{Alg}^n(X).
\end{array}
$$

The first vertical arrow is induced by the induced quotient map $BU(n) \to BPU(n)$.

3.5. Remark There is another representation of $PGL(n,\mathbb{C})$ using the special linear group $SL(n,\mathbb{C}) = \ker(\det : GL(n,\mathbb{C}) \to GL(1,\mathbb{C}))$. Since up to a scalar matrix every matrix has determinant 1, the restriction of $GL(n,\mathbb{C}) \to PGL(n,\mathbb{C})$ to $SL(n,\mathbb{C}) \to PGL(n,\mathbb{C})$ is surjective with kernel $\mu(n,\mathbb{C})$, the nth roots of unity in $\mathbb{C}$. Hence, we have

$$
PGL(n,\mathbb{C}) = SL(n,\mathbb{C})/\mu(n,\mathbb{C}).
$$

There is also the compact version with $SU(n) \subset SL(n,\mathbb{C})$, the special unitary subgroup, where we have two group extensions one for $SU(n)$ and one for $U(n)$

$$
\begin{array}{ccccccccc}
1 & \longrightarrow & \mathbb{Z}/n\mathbb{Z} & \longrightarrow & SU(n) & \longrightarrow & PSU(n) & \longrightarrow & 1 \\
& & \cap & & \cap & & \| & & \\
1 & \longrightarrow & U(1) & \longrightarrow & U(n) & \longrightarrow & PU(n) & \longrightarrow & 1.
\end{array}
$$

These group extensions are contained in two group extensions one for $SL(n,\mathbb{C})$ and one for $GL(n,\mathbb{C})$

$$
\begin{array}{ccccccccc}
1 & \longrightarrow & \mu(n,\mathbb{C}) & \longrightarrow & SL(n,\mathbb{C}) & \longrightarrow & PSL(n,\mathbb{C}) & \longrightarrow & 1 \\
& & \cap & & \cap & & \| & & \\
1 & \longrightarrow & GL(1,\mathbb{C}) & \longrightarrow & GL(n,\mathbb{C}) & \longrightarrow & PGL(n,\mathbb{C}) & \longrightarrow & 1.
\end{array}
$$

3.6. Fibre Sequences for Classifying Spaces We have the following diagram for the classifying spaces

$$
\begin{array}{ccccccc}
K(\mathbb{Z}/n,1)=B(\mathbb{Z}/n) & \longrightarrow & BSU(n) & \longrightarrow & BPU(n) & \overset{\gamma}{\longrightarrow} & B(B(\mathbb{Z}/n))=K(\mathbb{Z}/n,2) \\
\delta \downarrow & & \text{inc} \downarrow & & \text{id} \downarrow & & \downarrow \delta \\
K(\mathbb{Z},2)=BU(1) & \longrightarrow & BU(n) & \longrightarrow & PBU(n) & \longrightarrow & B(BU(1))=K(\mathbb{Z},3),
\end{array}
$$

where the morphisms δ are Bockstein maps coming from the short exact sequence of groups $0 \to \mathbb{Z} \overset{n}{\to} \mathbb{Z} \to \mathbb{Z}/n\mathbb{Z} \to 0$ as the connecting morphism $\delta : K(\mathbb{Z}/n\mathbb{Z},q) \to K(\mathbb{Z},q+1)$.

3.7. Matrix Algebra Characteristic Classes We define two related characteristic classes for matrix algebra bundles using the fibre sequences in (3.6)

$$\beta_n : \text{Alg}^n(X) \longrightarrow H^2(X, \mathbb{Z}/n\mathbb{Z}) \quad \text{and} \quad \alpha_n : \text{Alg}^n(X) \longrightarrow H^3(X, \mathbb{Z}).$$

To a matrix algebra bundle A, we associate a principal $PU(n)$-bundle P so that the fibre bundle $P[M_n(\mathbb{C})] = A$ and then classify the principal $PU(n)$ bundle A with a homotopy class $[f] \in [X, BPU(n)]$. Then, we define

$$\beta_n(A) = \gamma[f] \in H^2(X, \mathbb{Z}/n\mathbb{Z}) \quad \text{and} \quad \alpha_n(A) = \delta\beta_n(A) \in H^3(X, \mathbb{Z}).$$

3.8. Remark The classes $\beta_n(A)$ and $\alpha_n(A)$ are n-torsion classes. From the exact fibre sequence, we have the exactness statement that $\beta_n(A) = 0$ if and only if A is isomorphic to a bundle of the form $\text{End}(E)$, where $\Lambda^n E$ is a trivial line bundle, and $\alpha_n(A) = 0$ if and only if A is isomorphic to a bundle of the form $\text{End}(E)$ for some vector bundle E.

Observe that the image of β_n depends on n while the image of α_n is in $H^3(X, \mathbb{Z})$ independent of n. Of course α_n has the property that it lies in the subgroup $_n H^3(X, \mathbb{Z})$ of n-torsion points. In the next section, we will see that all the elements of $H^3(X, \mathbb{Z})$ have an interpretation in terms of operator algebra bundles, and then we will relate $\alpha_n(A)$ of matrix algebra bundles to the class $\alpha(A)$ of operator algebra bundles.

4 Classification of Operator Algebra Bundles

4.1. Notation Let H be a separable Hilbert space. The group $U(H)$ of unitary automorphisms of H endowed with the norm topology is a Banach Lie group. Its central subgroup is $U(1)$ the circle group of complex numbers of absolute value 1. The group of projective unitary automorphisms of H, denoted by $PU(H)$, is the quotient $U(H)/U(1)$ with the quotient topology. We obtain a central extension of Banach Lie groups

$$1 \longrightarrow U(1) \longrightarrow U(H) \longrightarrow PU(H) \longrightarrow 1$$

and a principal $U(1)$-bundle $U(H) \to PU(H)$. When H is n-dimensional the groups are denoted by $U(n) = U(H)$ and $PU(n) = PU(H)$.

The entire classification discussion of the operator algebra bundles is based on the following theorem of Kuiper which we quote.

4.2. Kuiper's Theorem The group $U(H)$ is a contractible space for an infinite-dimensional separable Hilbert space.

The reference is Kuiper (1965).

The corollary for an infinite-dimensional Hilbert space bundles is the following triviality result.

4.3. Corollary *The classifying space $BU(H)$ is contractible for an infinite-dimensional separable Hilbert space H, and hence, any Hilbert space vector bundle is trivial, see 8(6.3).*

The corollary for operator algebra bundle is the following classification by a single-characteristic class.

4.4. Corollary *The group $PU(H)$ as a space is a classifying space for the circle. In particular, $PU(H)$ is a $K(\mathbb{Z},2)$ space and its classifying space $BPU(H)$ is a $K(\mathbb{Z},3)$ space. Thus, principal bundles over a space X for the group $PU(H)$ with the norm topology are classified by the set, in fact the group,*

$$[X, BPU(H)] = [X, K(\mathbb{Z},3)] = H^3(X,\mathbb{Z})$$

of homotopy classes of maps $X \to K(\mathbb{Z},3)$ which is ordinary third cohomology.

Again, we have an important relation between classification of principal bundles and cohomology.

4.5. Notation We denote by $\alpha(A) \in H^3(X,\mathbb{Z})$ for the operator algebra bundle $P[\mathscr{B}(H)]$ or $P[\mathscr{K}(H)]$, where P is the principal bundle associated with an algebra bundle A, and $\alpha : \mathrm{Vect}^H(X) \to H^3(X,\mathbb{Z})$ is an isomorphism of functors.

In order to compare the characteristic classes α for operator algebra bundles and α_n for matrix algebra bundles, we consider the tensor product pairing $\otimes$ of two groups into a third.

4.6. Example There are two tensor product pairings $\otimes : \mathbb{Z} \times \mathbb{Z} \to \mathbb{Z}$ and $\otimes : (\mathbb{Z}/n'\mathbb{Z}) \times (\mathbb{Z}/n''\mathbb{Z}) \to \mathbb{Z}/n'n''\mathbb{Z}$ for cyclic groups, where the second pairing is the quotient of the first. For the unitary groups, we have $\otimes : U(H') \otimes U(H'') \to U(H' \otimes H'')$, which is a group morphism from the formula $(A' \otimes B')(A'' \otimes B'') = (A'A'') \otimes (B'B'')$. This defines a quotient pairing $\otimes : PU(H') \otimes PU(H'') \to PU(H' \otimes H'')$, and in the case where H' and H'' are finite-dimensional the pairing on the unitary groups induces a pairing $\otimes : SU(H') \times SU(H'') \to SU(H' \otimes H'')$ on the special unitary groups. All of the morphisms $\otimes$ are group morphisms in each variable.

4.7. Remark If we apply the classifying space construction to the tensor pairing $\mathbb{Z} \times \mathbb{Z} \to \mathbb{Z}$ and compose with a product map, then the composite

$$B\mathbb{Z} \times B\mathbb{Z} \to B(\mathbb{Z} \times \mathbb{Z}) \stackrel{B(\otimes)}{\to} B\mathbb{Z}$$

is just the pairing $\otimes : U(1) \times U(1) \to U(1)$.

Now, we apply the pairing maps to the fibre space sequence

$$U(1) \longrightarrow U(H) \longrightarrow PU(H) \longrightarrow BU(1).$$

4.8. Remark We have the following diagram of tensor product pairings for two Hilbert spaces H' and H''

$$
\begin{array}{ccccc}
BU(1) \times BU(1) & \longrightarrow & B(U(1) \times U(1)) & \xrightarrow{B(\otimes)} & BU(1) = K(\mathbb{Z},2) \\
\downarrow & & \downarrow & & \downarrow \\
BU(H') \times BU(H'') & \longrightarrow & B(U(H') \times U(H'')) & \xrightarrow{B(\otimes)} & BU(H' \otimes H'') \\
\downarrow & & \downarrow & & \downarrow \\
BPU(H') \times BPU(H'') & \longrightarrow & B(PU(H') \times PU(H'')) & \xrightarrow{B(\otimes)} & BPU(H' \otimes H'') \\
\downarrow & & \downarrow & & \downarrow \\
B^2U(1) \times B^2U(1) & \longrightarrow & B^2(U(1) \times U(1)) & \xrightarrow{B(\otimes)} & B^2U(1) = K(\mathbb{Z},3).
\end{array}
$$

4.9. Remark If P' is a principal G'-bundle over X and if P'' is a principal G''-bundle over X, then the fibre product $P' \times_X P''$ over X is a principal $(G' \times G'')$-bundle over X. If $G' \times G'' \to G$ is a morphism of topological groups, then $(P' \times_X P'')[G]$ is a principal G-bundle over X.

If $A' = P'[\mathscr{B}(H')]$ and $A'' = P''[\mathscr{B}(H'')]$ are two operator algebra bundles over X with groups $PU(H')$ and $PU(H'')$, respectively, then we can construct a tensor product bundle $A' \otimes A''$ defined as $P[\mathscr{B}(H' \otimes H'')]$ and using the fact that the above pairings are just the unique up to homotopy H-space structure on $K(\mathbb{Z},n)$ for $n = 1, 2$, and 3. Note the algebraic tensor product $A'_x \otimes A''_x$ is in general only contained in $(A' \otimes A'')_x$. We obtain the following theorem.

4.10. Theorem *For two operator algebra bundles A' and A'' on X, we have $\alpha(A' \otimes A'') = \alpha(A') + \alpha(A'')$.*

4.11. Remark The matrix algebra bundles and the operator algebra bundles have principal bundles with structure groups linked by the following diagram

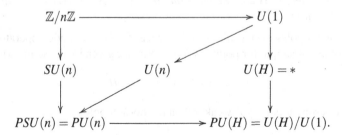

For classifying spaces, we have the following diagram of fibrations given in terms of vertical sequences

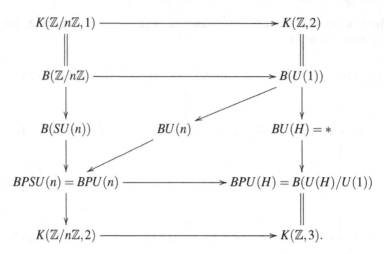

4.12. Theorem *Let H be a separable infinite-dimensional Hilbert space. The characteristic class isomorphism $\alpha : Alg^H(X) \to H^3(X, \mathbb{Z})$ of functors has the additional property that $\alpha(A)$ is a torsion class of order n if and only if there exists an M_n-matrix algebra bundle A' with A isomorphic to A' tensored with the trivial operator algebra bundle over X.*

4.13. Remark As a general assumption, in this section, the unitary group $U(H)$ is endowed with the norm topology which can also be described as the topology of uniform convergence on the unit ball of H. Under this assumption, $U(H)$ is a Banach Lie group in a natural way, and it has been shown in (4.4) that each cohomology class $\alpha \in H^3(X, \mathbb{Z})$ can be realized by a projective bundle on X with structure group $PU(H) = U(H)/U(1)$ in the norm topology. The norm topology is, however, not the only natural choice of a topology on $U(H)$ and $PH(U) = U(H)/U(1)$ as is indicated by the following observation.

4.14. Lemma *For the trivial vector bundle $X \times H$, let $\Phi : X \times H \to X \times H$ be a candidate of an automorphism given by $(x, h) \longmapsto (x, \phi(x)(h))$ with $\phi(x) \in U(H)$ for all $x \in X$. Then, Φ is continuous if and only if $\phi : X \to U(H)$ is continuous in the strong topology.*

The strong topology on a subset of the space $\mathscr{B}(H)$ of bounded operators on H is the topology of simple (or pointwise) convergence given by the seminorms

$$p_h(A) = \|A(h)\| \, , A \in \mathscr{B}(H),$$

$h \in H$. The strong topology is weaker than the norm topology.

By (4.14), the natural structure group of Hilbert space vector bundles is $U(H)$ with the strong topology. What do we know about $U(H)$ with this topology?

4.15. Proposition *The unitary group $U(H)$ with the strong topology is a topological group. Moreover, the strong topology agrees with the compact open topology and is metrizable (if H is separable).*

The first statement is in sharp contrast to several claims in the corresponding literature on quantization of symmetries and in other publications. Even in recent publications as, for example, in the article *Twisted K-theory* of Atiyah and Segal (2005), it is stated that $U(H)$ is not a topological group with respect to the strong topology. But the proof is very simple. For example, $U \mapsto U^{-1}$ is continuous in $U_0 \in \mathscr{U}(H)$, since for $h \in H$ and $\varepsilon > 0$ one defines $g = U_0^{-1}(h)$ and obtains for all $U \in U(H)$:

$$\left\| U_0^{-1}(h) - U^{-1}(h) \right\| = \left\| g - U^{-1}U_0(g) \right\| = \left\| U(g) - U_0(g) \right\| .$$

Hence, $\left\| U_0^{-1}(h) - U^{-1}(h) \right\| < \varepsilon$ for all $U \in U(H)$ with $\|U_0(g) - V(g)\| < \varepsilon$. That $U(H)$ is a topological group in the strong topology in contrast to $GL(H)$ can be attributed to the fact that $U(H)$ is equicontinuous. The equicontinuity of $U(H)$ as a set of operators also accounts for the coincidence of the strong and the compact open topology and the metrizability.

A much deeper result is the contractibility of $U(H)$ in the strong topology which is shown in the above mentioned article. As a consequence much of classification results are parallel to the norm topology case, in particular the results in (4.3) and (4.4).

4.16. Corollary *Let $U(H)$ be the topological group with the strong topology and let $PU(H) = U(H)/U(1)$ with the quotient topology. Then, any Hilbert space vector bundle over X with structure group $U(H)$ is trivial. And the principal bundles over X for the group $PU(H)$ are classified by $H^3(X,\mathbb{Z})$.*

References

Atiyah, M.F., Segal, G.B.: Twisted K-Theory and cohomology (2005) (arXiv:math.KT/0510674)
Dixmier, J. Douady, A.: Champs continus d'espaces Hilbertiens et de C*-algèbres. Bull. Soc. Math. France **91**: 2227–284 (1963)
Kuiper, N.H.: The homotopy type of the unitary group of Hilbert space. Topology 3:19–30 (1965)

Chapter 19
Brauer Group of Matrix Algebra Bundles and K-Groups

Before using the characteristic class isomorphism

$$\alpha : \mathrm{Alg}^H(X) \longrightarrow H^3(X, \mathbb{Z})$$

to study its role in extending vector bundle K-theory, we analyze further the matrix bundle theory using the restriction $\alpha_n : \mathrm{Alg}^n(X) \to_n H^3(X, \mathbb{Z})$ of α to $\mathrm{Alg}^n(X)$ which is a group morphism in the corollary (1.4). In the context of sheaf theory, we have already derived a multiplicative property of the morphisms α_n and β_n. Now, we carry this out in the context of classifying space theory.

This leads naturally into the question of the Brauer group of all matrix algebra bundles with the relation A' and A'' are Brauer equivalent provided that $A' \otimes \mathrm{End}(E')$ and $A'' \otimes \mathrm{End}(E'')$ are isomorphic over X for two vector bundles E' and E''. The result is the Brauer group $B(X)$ of a space X which is a form of K-group or Klassengruppe in the sense of Grothendieck. We study the stability properties of this group.

The reader with a special interest in twisted K-theory can go directly to the next chapter. The last section discusses a sheaf theory approach to matrix algebra bundles for completeness of the discussion. It is not used in the later chapters.

1 Properties of the Morphism α_n

The group morphism property of α_n follows the line of argument as we established the same property of α in 20(4.8) using 20(4.9) to obtain 20(4.10).

1.1. Remark Now, we apply the pairing maps of 20(4.6) to the fibre space sequence associated with matrix algebra bundles

$$\mathbb{Z}/m\mathbb{Z} \longrightarrow SU(H) \longrightarrow PU(H) \longrightarrow B(\mathbb{Z}/m\mathbb{Z}),$$

$\dim H = m$ in the case of $\otimes : (\mathbb{Z}/n'\mathbb{Z}) \times (\mathbb{Z}/n''\mathbb{Z}) \to (\mathbb{Z}/n'n''\mathbb{Z})$.

D. Husemöller et al.: *Brauer Group of Matrix Algebra Bundles and K-Groups*, Lect. Notes Phys. **726**, 241–249 (2008)
DOI 10.1007/978-3-540-74956-1_20   © Springer-Verlag Berlin Heidelberg 2008

1.2. Remark We have the following diagram of tensor product pairings

$$
\begin{array}{ccccc}
B(\mathbb{Z}/n') \times B(\mathbb{Z}/n'') & \longrightarrow & B((\mathbb{Z}/n') \times (\mathbb{Z}/n'')) & \xrightarrow{B(\otimes)} & B(\mathbb{Z}/n'n'') = K(\mathbb{Z}/n'n'', 1) \\
\downarrow & & \downarrow & & \downarrow \\
BSU(n') \times BSU(n'') & \longrightarrow & BSU(n') \times SU(n'') & \xrightarrow{B(\otimes)} & BSU(n'n'') \\
\downarrow & & \downarrow & & \downarrow \\
BPU(n') \times BPU(n'') & \longrightarrow & BPU(n') \times PU(n'') & \xrightarrow{B(\otimes)} & BPU(n'n'') \\
\downarrow & & \downarrow & & \downarrow \\
B^2(\mathbb{Z}/n') \times B^2(\mathbb{Z}/n'') & \longrightarrow & B^2((\mathbb{Z}/n') \times (\mathbb{Z}/n'')) & \xrightarrow{B(\otimes)} & B^2(\mathbb{Z}/n'n'') \to K(\mathbb{Z}, 3).
\end{array}
$$

Recall from 1(4.9) that $(P' \times_X P'')[G]$ is a principal G-bundle over X if $G' \times G'' \to G$ is a morphism of topological groups, if P' is a principal G'-bundle over X, and if P'' is a principal G''-bundle over X. The fibre product $P' \times_X P''$ is a principal $(G' \times G'')$-bundle.

If $A' = P'[M_{n'}(\mathbb{C})]$ and $A'' = P''[M_n(\mathbb{C})]$ are two matrix algebra bundles over X with groups $PU(n')$ and $PU(n'')$, respectively, then we can construct a tensor product bundle using $P = P' \times_X P''$ by

$$
A' \otimes A'' = P[M_{n'}(\mathbb{C}) \otimes M_{n''}(\mathbb{C})] = P[M_{n'n''}(\mathbb{C})].
$$

Using the fact that the above pairings are just the unique maps up to homotopy which are given by the H-space structure on $K(\mathbb{Z}/m\mathbb{Z}, n)$ for $n = 1, 2$, and 3, we obtain the following theorem.

1.3. Theorem *For an $M_{n'}$-matrix algebra bundle A' and an $M_{n''}$-matrix algebra bundle A'' on X, we have*

$$
\alpha_{n'n''}(A' \otimes A'') = \alpha_{n'}(A') + \alpha_{n''}(A'')
$$

in $H^3(X, \mathbb{Z})$.

The exactness properties of the morphism α_n can be derived from the fibre sequences introduced in 20(3.6).

1.4. Theorem *There is an exact sequence of functors $(\mathrm{top})^{\mathrm{op}} \to (\mathrm{set})_*$, the category of pointed sets*

$$
\mathrm{Vect}^n \xrightarrow{\varepsilon} \mathrm{Alg}^n \xrightarrow{\alpha_n} H^3(\ , \mathbb{Z}) \xrightarrow{n} H^3(\ , \mathbb{Z}),
$$

where $\varepsilon = \mathrm{End}$ and n is multiplication by n.

Proof. The exactness of the first three terms can be seen from the classifying space point of view in 20(3.6). The exactness of the last three terms is the result of bringing together two exact sequences

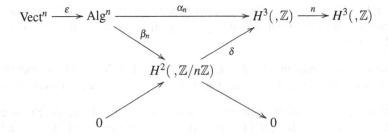

in 20(3.6) for classifying spaces. This proves the theorem.

In terms of the characteristic class $\alpha_n(A)$ of an M_n-matrix algebra bundle isomorphism class A on X, we can interpret part of (1.5) in the following corollary.

1.5. Corollary *The characteristic class $\alpha_n(A)$ of an M_n-matrix algebra bundle A is an n-torsion class, that is, $n\alpha_n(A) = 0$. The algebra bundle A is isomorphic to an endomorphism class $End(E)$ if and only if $\alpha_n(A) = 0$.*

1.6. Remark Observe that everything refers to a dimension n of the M_n-matrix algebra bundles, and the resulting characteristic class is an n-torsion class. In 20(4.12), we saw how to get around this with a tensor product type of equivalence which we consider now at length starting in the next section.

2 From Brauer Groups to Grothendieck Groups

2.1. Generalities on Class Groups K-theory in its broadest setting can be conceived as starting with a group-valued invariant which is not strong enough for an isomorphism classification of objects in a category $\mathscr{C}$. Such an invariant I : $\mathrm{Ob}(\mathscr{C}) \to A$ would lead naturally to a class group $K(\mathscr{C}) = \mathrm{Ob}(\mathscr{C})/(\text{relation})$ with a universal type invariant $\mathrm{Ob}(\mathscr{C}) \to K(\mathscr{C})$ factoring I with a group morphism $I' : K(\mathscr{C}) \to A$. The algebraic structure of invariants of type I would be encoded in the class group $K(\mathscr{C})$, and the study of the invariant I and the study of $K(\mathscr{C})$ would be thought of as being equivalent.

2.2. Example An example of this phenomenon outside the subject of these notes is the divisor class group related to a smooth complete algebraic curve C. The divisor group $\mathrm{Div}(C)$ is the abelian group generated by points of C. Each point of the curve has a degree over the field, and the degree function extends to a group morphism $\mathrm{Div}(C) \to \mathbb{Z}$. The divisor of a function which is the weighted sum of zeros and poles of the function has degree zero with this extension, and it is natural to consider the quotient of the group of divisors modulo the subgroup of divisors of functions. It is this divisor class group which plays an important role in the theory of curves.

Now, we come to two types of class groups of algebra bundles.

2.3. Definition Two matrix algebra bundles A and A' over X are Brauer equivalent provided there exists two vector bundles E and E' with $A \otimes End(E)$ and $A' \otimes End(E')$

isomorphic. The set $\mathrm{Br}(A)$ of Brauer equivalence class $\mathrm{cl}(A)$ of matrix algebra bundles together with the group structure given by $\mathrm{cl}(A) \cdot \mathrm{cl}(A') = \mathrm{cl}(A \otimes A')$ is called the Brauer group.

From the isomorphism $A^{\mathrm{op}} \otimes A \to \mathrm{End}(\mathrm{Vect}(A))$, where $\mathrm{Vect}(A)$ is the underlying vector bundle of the algebra, we see that $\mathrm{cl}(A^{\mathrm{op}}) = \mathrm{cl}(A)^{-1}$ which gives the inverse.

2.4. Remark In line with the generalization of (2.1), we can define the characteristic class $\alpha : \mathrm{Br}(A) \to \mathrm{Tors}(H^3(X, \mathbb{Z}))$ on the Brauer classes by $\alpha(\mathrm{cl}(A)) = \alpha_n(A)$, where A is an M_n-matrix algebra representative of the class $\mathrm{cl}(A) \in \mathrm{Br}(A)$.

We have the following theorem of Serre which we just quote now.

2.5. Theorem *If X is a finite complex, then the characteristic class $\alpha : Br(A) \to Tors(H^3(X, \mathbb{Z}))$ is an isomorphism.*

Recall that reindexing of $M_n M_q$ gives $M_n(M_q(\mathbb{C})) = M_{nq}(\mathbb{C})$.

2.6. Definition Two matrix algebras A and A' over X are stably equivalent provided there exists n and n' with $M_n(A)$ and $M_{n'}(A')$ isomorphic. The set $KP(X)$ of stable equivalence classes $\{A\}$ of matrix algebra bundles A together with the semigroup structure given by $\{A'\} \cdot \{A''\} = \{A' \otimes A''\}$ is called the projective K-group.

There is a natural morphism $KP(X) \to \mathrm{Br}(X)$, where $\{A\} \mapsto \mathrm{cl}(A)$. We consider $KP(X)$ and $\mathrm{Br}(X)$ as K-groups in the sense of Grothendieck further in the next three sections.

3 Stability I: Vector Bundles

Now, we return to Grothendieck who pointed out the importance of the class theory for coherent sheaves and vector bundles.

3.1. Definition Two vector bundles E and E' over X are K-equivalent provided that there exists a trivial bundle T with $E \oplus T$ and $E' \oplus T$ isomorphic. Let $[E]$ denote the K-equivalence class of E on X.

If $[E]$ is a K-equivalence class on Y and if $f : X \to Y$ is a continuous function, then $f^*[E] = [f^* E]$ is well-defined K-equivalence class on X which defines a functor $[\ ]$ on $(\mathrm{top})^{\mathrm{op}}$ with values in the category of semirings.

3.2. Notation Let $[X]$ denote the set of K-equivalence classes $[E]$ of bundles on X with the semiring structure

$$[E'] + [E''] = [E' \oplus E''] \quad \text{and} \quad [E'] \cdot [E''] = [E' \otimes E''].$$

Let $K(X)$ denote the universal abelian group. Then, $K(X)$ is also the universal commutative ring on the semigroup $[X]$. Further, the universal constructions lead to a functor $K : (\mathrm{top})^{\mathrm{op}} \to (\mathrm{c \backslash rg})$ with values in the category of commutative rings.

This is the fundamental class group or ring, *Klassengruppe*, of Grothendieck.

3.3. Definition Two vector bundles E and E' over X are stably equivalent provided there exist two trivial bundles T and T' with $E \oplus T$ and $E' \oplus T'$ isomorphic.

If E and E' are K-equivalent, they have the same fibre dimension at each point.

3.4. Remark Two K-equivalent vector bundles E and E' on X give the same element in $K(X)$. Let X be a connected space, let $* \in X$ be a base point of X with inclusion map $* \to X$ inducing rk $: K(X) \to K(*) = \mathbb{Z}$ given by dimension at $*$, and let $\tilde{K}(X) = \ker(\mathrm{rk})$. Then E and E' are stably equivalent if and only if $[E] - [E'] = 0 \in \tilde{K}(X)$.

Now, we interpret the sequence of isomorphisms for dimension n

$$[X, BU(n)] \longrightarrow [X, BGL(n, \mathbb{C})] \longrightarrow \mathrm{Prin}_{GL(n, \mathbb{C})}(X) \longrightarrow \mathrm{Vect}^n(X)$$

of 2(2.3) in terms of $K(X)$. It is carried out with the inclusions of groups $U(n) \to U(m+n)$ and $j : BU(n) \to BU(n+m)$ of classifying spaces. If a bundle E over X corresponds to a map $f : X \to BU(n)$ and if T is a trivial bundle of dimension m, then $E \oplus T$ over X corresponds to $jf : X \to BU(m+n)$.

3.5. Assertion The isomorphisms of 4(3.4) fit into an inductive system corresponding to Whitney sum with a trivial bundle with inductive limit $K(X)$, that is,

$$\varinjlim_n [X, BU(n)] \longrightarrow \varinjlim_n [X, BGL(n, \mathbb{C})] \longrightarrow \varinjlim_n \mathrm{Vect}^n(X)\hat{} = K(X).$$

For a path-connected space X with base point we the base-point-preserving homotopy classes $\tilde{K}(X)$, that is,

$$\varinjlim_n [X, BU(n)]_* \longrightarrow \varinjlim_n [X, BGL(n, \mathbb{C})]_* \longrightarrow \tilde{K}(X).$$

There is a sheaf cohomology description discussed later in Sect. 6.

4 Stability II: Characteristic Classes of Algebra Bundles and Projective K-Group

4.1. Definition For two natural numbers m and n, the matrix morphism of functors $\mathrm{Alg}^n(X) \to \mathrm{Alg}^{mn}(X)$ is defined by associating to an n by n matrix algebra bundle A the mn by mn matrix algebra bundle $M_m(A) = A \otimes (X \times M_m(\mathbb{C})) = A \otimes M_m(X \times \mathbb{C})$, which is the tensor product with the m by m trivial algebra bundle.

4.2. Remark Recall that there are natural embeddings of cyclic groups $(\frac{1}{n}\mathbb{Z})/\mathbb{Z} \subset (\frac{1}{nm}\mathbb{Z})/\mathbb{Z} \subset \mathbb{Q}/\mathbb{Z}$. Using the multiplicative property of the characteristic class $\alpha_n : \mathrm{Alg}^n() \to H^3(, \mathbb{Z})$, we see that $\beta_n : \mathrm{Alg}^n() \to H^2(, (\frac{1}{n}\mathbb{Z})/\mathbb{Z})$ fits into the following commutative diagram by (1.4)

$$\begin{CD} \mathrm{Alg}^n() @>>> \mathrm{Alg}^{mn}() \\ @AAA @AAA \\ BPU(n) @>>> BPU(mn) \\ @V{\beta_n}VV @V{\beta_{mn}}VV \\ K((\tfrac{1}{n}\mathbb{Z})/\mathbb{Z},2) @>>> K((\tfrac{1}{mn}\mathbb{Z})/\mathbb{Z},2) \end{CD}$$

Now, we take the direct limit of these diagrams to obtain a basic first characteristic class on $KP(X)$.

4.3. Remark Using the inductive limit definition where

$$KP(X) = \varinjlim_{m,n} (\mathrm{Alg}^n(X) \to \mathrm{Alg}^{mn}(X)),$$

we define the inductive limit characteristic class $\gamma : KP(X) \to H^2(X,\mathbb{Q}/\mathbb{Z})$. Here, we use the formulas

$$K(\mathbb{Q}/\mathbb{Z},2) = \varinjlim_{m,n} \left(K\left(\left(\frac{1}{n}\mathbb{Z}\right)/\mathbb{Z},2 \right) \longrightarrow K\left(\left(\frac{1}{mn}\mathbb{Z}\right)/\mathbb{Z},2 \right) \right)$$

and

$$H^2(X,\mathbb{Q}/\mathbb{Z}) = \varinjlim_{m,n} \left(H^2\left(X, \left(\frac{1}{n}\mathbb{Z}\right)/\mathbb{Z} \right) \longrightarrow H^2\left(X, \left(\frac{1}{mn}\mathbb{Z}\right)/\mathbb{Z} \right) \right).$$

5 Rational Class Groups

5.1. Remark Since the first paper of Atiyah–Hirzebruch on topological K-theory, we know that the rational Chern character

$$\mathrm{ch}_\mathbb{Q} : K(X)_\mathbb{Q} \to H^{**}(X)_\mathbb{Q}$$

is an isomorphism of ring-valued functors. Another way to express this is that the localization $BU_\mathbb{Q}$ of BU at the rational numbers decomposes as a product of $K(\pi,n)$-spaces, that is, we have a decomposition $BU_\mathbb{Q} = \prod_{i>0} K(\mathbb{Q},2i)$.

5.2. Assertion Above, we have the map $BPU \to K(\mathbb{Q}/\mathbb{Z},2)$, and there are other factors, as in $BU_\mathbb{Q}$ but without tensoring over the rational numbers, leading to the homotopy equivalence

$$BPU \longrightarrow K(\mathbb{Q}/\mathbb{Z},2) \times \prod_{i>1} K(\mathbb{Q},2i).$$

For this, see Grothendieck (1995).

5.3. Remark For $x \in \text{Tors}(H^3(X, \mathbb{Z}))$, we have $x = \delta(y)$, where $y \in H^2(X, \mathbb{Q}/\mathbb{Z})$ corresponding to a map $X \to K(\mathbb{Z}/n\mathbb{Z}, 2) \to K(\mathbb{Q}/\mathbb{Z}, 2)$ for a large enough n. This gives the principal $PU(n)$-bundle corresponding to $x \in H^3(X, \mathbb{Z})$.

6 Sheaf Theory Interpretation

Let X be a space, and let $\mathscr{C}_X$ denote the sheaf of germs of complex-valued functions on X. This section is for the reader with background in sheaf theory.

6.1. Remark The set $\text{Vect}^n(X)$ is isomorphic to the set of isomorphism classes of $\mathscr{C}_X$-modules of sheaves locally isomorphic to $\mathscr{C}_X^n$ and the set $\text{Alg}^n(X)$ is isomorphic to the set of isomorphism classes of $\mathscr{C}_X$ algebras locally isomorphic to $M_n(\mathscr{C}_X)$.

6.2. Assertion Using sheaf cohomology by which local charts are glued together with 1-cocycles, we obtain a sheaf cohomology version of the commutative diagram 20(2.5) of the following form, where the horizontal arrows are bijections

$$
\begin{array}{ccc}
H^1(X, GL(n, \mathscr{C}_X)) & \longrightarrow & \text{Vect}^n(X) \ . \\
\theta \downarrow & & \downarrow \text{End} \\
H^1(X, PGL(n, \mathscr{C}_X)) & \longrightarrow & \text{Alg}^n(X).
\end{array}
$$

In this case, the vertical morphism θ is one of the morphisms in the sheaf cohomology exact sequence associated to the short exact sequence of sheaves

$$ 1 \longrightarrow GL(1, \mathscr{C}_X) \longrightarrow GL(n, \mathscr{C}_X) \longrightarrow PLG(n, \mathscr{C}_X) \longrightarrow 1 \ . $$

It is useful to have the following terms of the cohomology exact sequence including θ, namely

$$ \cdots \to H^0(X, PGL(n, \mathscr{C}_X)) \to H^1(X, GL(1, \mathscr{C}_X)) \to H^1(X, GL(n, \mathscr{C}_X)) \xrightarrow{\theta} $$

$$ H^1(X, PGL(n, \mathscr{C}_X)) \to H^2(X, GL(1, \mathscr{C}_X)) \to \cdots . $$

6.3. Remark Returning to 20(3.5), we introduce another exact sequence of sheaves using the special linear group

$$ 1 \longrightarrow \mu(n, \mathscr{C}_X) \longrightarrow SL(n, \mathscr{C}_X) \longrightarrow PLG(n, \mathscr{C}_X) \longrightarrow 1 \ . $$

Again, the following terms of the cohomology exact sequence with θ are

$$ \cdots \to H^0(X, PGL(n, \mathscr{C}_X)) \to H^1(X, \mu(n, \mathscr{C}_X)) \to H^1(X, SL(n, \mathscr{C}_X)) \xrightarrow{\theta} $$

$$ H^1(X, PGL(n, \mathscr{C}_X)) \to H^2(X, \mu(n, \mathscr{C}_X)) \to \cdots . $$

6.4. Remark In 20(3.6), the fibre sequence of classifying spaces led naturally to $K(\pi,n)$ spaces which classify ordinary cohomology, and the morphism θ on a space is studied with cohomology with coefficients in $\mathbb{Z}$ or a finite group. The exact sequence with sheaf cohomology gives the same analysis of the morphism θ defined by assigning to a vector bundle the matrix algebra bundle $\text{End}(E)$. For this, we have to analyze the cohomology group

$$H^i(X,GL(1,\mathscr{C}_X)) = H^i(X,\mathscr{C}_X^*)$$

by the exponential sequence $\langle\exp\rangle$ with kernel $\mathbb{Z}$

$$\langle\exp\rangle \qquad\qquad 0 \longrightarrow \mathbb{Z}_X \longrightarrow \mathscr{C}_X \xrightarrow{\exp} \mathscr{C}_X^* \longrightarrow 1$$

and by the *n*th power sequence $\langle n\rangle$ with kernel the roots of unity

$$\langle n\rangle \qquad\qquad 0 \longrightarrow \mu(n,\mathscr{C}_X) \longrightarrow \mathscr{C}_X^* \xrightarrow{n} \mathscr{C}_X^* \longrightarrow 1 \;.$$

6.5. Remark The related cohomology sequences take the following form: for $\langle\exp\rangle$

$$\ldots \longrightarrow H^0(X,\mathscr{C}_X^*) \longrightarrow H^1(X,\mathbb{Z}_X) \longrightarrow H^1(X,\mathscr{C}_X)$$

$$\xrightarrow{\exp} H^1(X,\mathscr{C}_X^*) \longrightarrow H^2(X,\mathbb{Z}_X) \longrightarrow \ldots$$

and for $\langle n\rangle$

$$\ldots \longrightarrow H^0(X,\mathscr{C}_X^*) \longrightarrow H^1(X,\mu(n,\mathscr{C}_X)) \longrightarrow H^1(X,\mathscr{C}_X^*)$$

$$\xrightarrow{n} H^1(X,\mathscr{C}_X^*) \longrightarrow H^2(X,\mu(n,\mathscr{C}_X)) \longrightarrow \ldots$$

For paracompact spaces, the sheaf $\mathscr{C}_X$ has no higher cohomology, and we have the natural isomorphism

$$H^q(X,\mathscr{C}_X^*) \longrightarrow H^{q+1}(X,\mathbb{Z}_X) = H^{q+1}(X,\mathbb{Z}) \quad \text{for } q > 0.$$

With this isomorphism to standard singular integral cohomology, we see the relation with the classifying space sequence of 20(3.6)

$$K(\mathbb{Z},2) = BU(1) \longrightarrow BU(n) \longrightarrow BPU(n) \longrightarrow B(BU(1)) = K(\mathbb{Z},3)$$

and the sheaf cohomology sequence

$$H^2(X,\mathbb{Z}) \hookleftarrow H^1(X,GL(1,\mathscr{C}_X)) \to H^1(X,GL(n,\mathscr{C}_X)) \xrightarrow{\theta}$$

$$H^1(X,PGL(n,\mathscr{C}_X)) \xrightarrow{\delta} H^2(X,GL(1,\mathscr{C}_X)) \hookleftarrow H^3(X,\mathbb{Z}).$$

6.6. Remark Using the relation $PGL_n = SL_n/\mu_n$, we have the cohomology sequence with mod *n* reduction *r*

$$H^1(X, \mu(n, \mathscr{C}_X)) \to H^1(X, SL(n, \mathscr{C}_X)) \xrightarrow{\theta} H^1(X, PGL(n, \mathscr{C}_X)) \xrightarrow{\delta'} H^2(X, \mu(n, \mathscr{C}_X)),$$

together with the maps

$$H^1(X, \mu(n, \mathscr{C}_X)) \xrightarrow{r\delta} H^2(X, \mathbb{Z}/n\mathbb{Z}) \text{ and } H^2(X, \mu(n, \mathscr{C}_X)) \to H^3(X, \mathbb{Z}).$$

Reference

Grothendieck, A.: Le groupe de Brauer. Part I. Séminaire Bourbaki, Vol. 9, Exp. 290: 199–219 (1995)

Chapter 20
Analytic Definition of Twisted K-Theory

As pointed out in the previous chapter, there are several approaches to ordinary complex K-theory. Each of these approaches leads to a version of twisted K-theory. The basic data for twisted K-theory is an operator algebra bundle A over X. The twisted K-group is denoted by $K_\alpha(X)$, where the characteristic class α of A is part of the data $\alpha(A) = \alpha \in H^3(X, \mathbb{Z})$.

First, we consider twisted K-theory in terms of the Fredholm subbundle $\mathrm{Fred}(A) \subset A$ over X. We relate mappings from X into the Fredholm operators $\mathrm{Fred}(H)$ and cross sections of the Fredholm subbundle over X. The twisted K-group of X relative to the algebra bundle A is given by fibre homotopy classes of cross sections of $\mathrm{Fred}(A)$.

Ordinary K-theory is also described by mappings into a space $\mathrm{Fred}(H)$. Mappings into $\mathrm{Fred}(H)$ can be thought as cross sections of the trivial bundle with fibre $\mathrm{Fred}(H)$. By replacing the trivial bundle with fibre $\mathrm{Fred}(H)$ with a nontrivial bundle with fibre $\mathrm{Fred}(H)$, we obtain a twisted K-theory from classes of cross sections of this nontrivial bundle.

Secondly, we introduce the C^*-algebra of sections $\Gamma(X, \mathscr{K}(A))$ of the algebra bundle. The K-theory of a C^*-algebra is the twisted K-group relative to the algebra bundle A over X.

Operator bundles with fibre $\mathscr{B} = \mathscr{B}(H)$ or with fibre $\mathscr{K} = \mathscr{K}(H)$ are equivalent since they both have an automorphism group $PU(H)$ up to homotopy. The bundles with fibre $\mathscr{B}$ are more useful for a Fredholm operator approach to twisted K-theory while the bundles with a fibre $\mathscr{K}$ appear more naturally in the operator K-theory approach to twisted K-theory.

1 Cross Sections and Fibre Homotopy Classes of Cross Sections

1.1. Definition A cross section s of a bundle $p : E \to B$ is map $s : B \to E$ with $ps = id_B$. Let $\Gamma(B, E)$ denote the set of all cross sections of E over B.

The condition $ps = id_B$ is equivalent to $s(b) \in E_b$ for all $b \in B$.

D. Husemöller et al.: *Analytic Definition of Twisted K-Theory*, Lect. Notes Phys. **726**, 251–253 (2008)
DOI 10.1007/978-3-540-74956-1_21

1.2. Example For the product bundle $p : B \times Y \to B$, the set of cross sections $\Gamma(B, B \times X)$ can be identified with the set $\mathrm{Map}(B, Y)$ of mappings form B to Y. A cross section of the product bundle has the form $s(b) = (b, f(b))$, where $f : B \to Y$ is a continuous function, and conversely, f determines s_f uniquely by the formula $s_f(b) = (b, f(b))$.

For many invariants of continuous functions $f : X \to Y$ between two spaces X and Y, the function itself is less important than the deformation class or homotopy class of the mapping. This leads to the following two basic definitions.

1.3. Definition A homotopy between two maps $f', f'' : X \to Y$ is a map $h : X \times [0, 1] \to Y$ such that $f'(x) = h(x, 0)$ and $f''(x) = h(x, 1)$. Two maps $f', f'' : X \to Y$ are homotopic provided there exists a homotopy from f' to f''.

Homotopy of maps is an equivalence relation on the set $\mathrm{Map}(X, Y)$. We denote the set of homotopy classes $[f] : X \to Y$ by $[X, Y]$. The set $[X, Y]$ is a quotient of $\mathrm{Map}(X, Y)$.

1.4. Definition Let $s' s'' : B \to E$ be two cross sections of $p : E \to B$. A fibre homotopy between s' and s'' is a homotopy h such that $ph(b, t) = b$ for all $b \in B$ and $t \in [0, 1]$. Two cross sections $s' s'' : B \to E$ are fibre homotopic (or fibre homotopically equivalent) provided there exists a fibre homotopy map from s' to s''.

Fibre homotopy of sections is an equivalence relation on the set $\Gamma(B, E)$ of cross sections of $p : E \to B$. Let $[\Gamma](B, E)$ denote the set of fibre homotopy classes of cross sections of $p : E \to B$. Then $[\Gamma](B, E)$ is a quotient set of $\Gamma(B, E)$.

1.5. Remark For the trivial bundle $p : B \times Y \to B$, the function which assigns to a map $f : B \to Y$ in $\mathrm{Map}(B, Y)$ the cross section $s_f(b) = (b, f(b))$ in $\Gamma(B, B \times Y)$ carries homotopy equivalent f to fibre homotopy equivalent cross sections. This defines a natural bijection $[B, Y] \to [\Gamma](B, B \times Y)$ of the bijection $\mathrm{Map}(B, Y) \to \Gamma(B, B \times Y)$.

2 Two Basic Analytic Results in Bundle Theory and K-Theory

2.1. Theorem (Kuiper) *Let H be a separable infinite-dimensional Hilbert space. The unitary group $U(H)$ with the norm topology is a contractible group.*

2.2. Theorem *(Atiyah–Jänich). Let H be a separable infinite-dimensional Hilbert space. There is a natural morphism of functors on compact spaces $[X, \mathrm{Fred}(H)] \to K(X)$ with the following property: If $f : X \to \mathrm{Fred}(H)$ is a map with $\ker(f(x))$ and $\mathrm{coker}(f(x))$ locally of constant dimension, then the image of $[f]$ is $E' - E''$, where E' and E'' are the subbundles of the trivial $X \times H \to X$ with fibres*

$$E'_x = x \times \ker(\mathrm{f}(x)) \quad \text{and} \quad \mathrm{E}'_x = x \times \ker(\mathrm{f}(x)^*).$$

This map $[X, \mathrm{Fred}(H)] \to \mathrm{K}(X)$ is an isomorphism of functors on the opposite category of compact spaces and continuous map.

Recall that bundles of C^*-algebras with fibres isomorphic to $\mathscr{B}(H)$ up to isomorphism are classified by elements in $[X, B(PU(H))]$. Moreover, for a separable infinite-dimensional H, the space $BPU(H)$ is a $K(\mathbb{Z}, 3)$.

2.3. Corollary *For a separable infinite-dimensional H, the bundles of C^*-algebras with fibres isomorphic to $\mathscr{B}(H)$ are classified by an element $t \in H^3(X,\mathbb{Z})$.*

3 Twisted *K*-Theory in Terms of Fredholm Operators

For each $t \in H^3(X,\mathbb{Z})$, we define $K_t(X)$ the twisted K-theory over X using Fredholm operators in the related $\mathscr{B}(H)$-bundle $A(t)$ or related principal $PU(H)$ bundle $P(t)$, where $P(t)[\mathscr{B}(H)]$ and $A(t)$ are isomorphic as $\mathscr{B}(H)$ bundles. We use the following remark in order to bring in Fredholm operators.

3.1. Remark The automorphisms in $PU(H)$ or more broadly $PGL(H)$ preserve the open subspace $\mathrm{Fred} \subset \mathscr{B}(H)$. Hence, we have a subbundle $P(t)[\mathrm{Fred}]$ or $\mathrm{Fred}(A(t))$ where the fibre $\mathrm{Fred}(A(t))_x = \mathrm{Fred}(A(t)_x)$ which is isomorphic to $\mathrm{Fred}(H)$.

Thus, the Fredholm elements in the fibre of a bundle of $\mathscr{B}(H)$ algebras have a meaning independent of the local chart of the bundle. Now building on the Atiyah–Jänich theorem, we can extend the notion of K-theory to a twisted version using $\mathscr{B}(H)$ algebras.

3.2. Definition For $t \in H^3(X,\mathbb{Z})$, the twisted K-group $K_t(X)$ of X relative to the twisting t is $[\Gamma](X,\mathrm{Fred}(A(t))$.

In the previous definition of the group $K_t(X)$, there is a choice made for the algebra bundle $A(t)$ which is unique up to isomorphism. In fact, we could define the twisted K-theory relative to operator algebra bundles A by $K_A(X) = [\Gamma](X,\mathrm{Fred}(A))$, and with this form of the definition, we have the following functoriality.

3.3. Functoriality For untwisted K-theory, any map $f : X \to Y$ induces $f^* : K(Y) = [Y,BU] \to [X,BU] = K(X)$ by composition from the right. For cross sections $f^* = \Gamma(f) : \Gamma(Y,E) \to \Gamma(X,f^*(E))$ or fibre homotopy classes of cross sections $[\Gamma](f) : [\Gamma](Y,E) \to [\Gamma](X,f^*(E))$ are defined. This means that functoriality is for maps or homotopy classes of maps where the twisting by the bundles are related by inverse images under mappings in the category with twistings.

Chapter 21
The Atiyah–Hirzebruch Spectral Sequence in K-Theory

Now that we have K-theory and its generalization, we discuss the Atiyah–Hirzebruch spectral sequence relating the ordinary cohomology to K-theory. Since cohomology is known for many spaces, with this spectral sequence, it is possible in some case to calculate the K-theory or more generally the twisted K-theory of the space by this spectral sequence.

The spectral sequence is a tool introduced into algebraic topology at the end of 1940 originating with the ideas of Jean Leray. He was studying the effect of mapping on homology in order to extend the existing fixed point theorems used in the theory of nonlinear partial differential equations. A spectral sequence is a family of chain complexes linked together via their homology, and we introduce the concept via the exact couple originating with William Massey.

Any filtered space X with filtration $X_0 \subset X_1 \subset \ldots \subset X_n \subset \ldots \subset X = \bigcup_n X_n$ has a spectral sequence for any type of K-theory or cohomology theory $E^*(X)$ which analyzes the group E^* in terms of the value on the relative layers $E^*(X_n, X_{n-1})$ from the filtration X. We begin by explaining what is an exact couple and its derived couple and then consider the related spectral sequence. Then this is applied to the K-theory of a filtered space. With a spectral sequence, it is usually the E_2 term which lends itself to calculation. This is the case with the K-theory spectral sequence where the E_2 term is calculated in terms of ordinary cohomology, and the spectral sequence itself "converges" to the K-theory.

Finally, we illustrate how the spectral sequence leads to interesting special calculations in K-theory and string theory. The twisted K-theory spectral sequence was first studied in Rosenberg (1989).

1 Exact Couples: Their Derivation and Spectral Sequences

1.1. Definition An exact couple $(A, E, \alpha, \beta, \gamma)$ is a pair of objects A and E with morphisms giving a diagram

D. Husemöller et al.: *The Atiyah–Hirzebruch Spectral Sequence in K-Theory*, Lect. Notes Phys. **726**, 255–264 (2008)
DOI 10.1007/978-3-540-74956-1_22   © Springer-Verlag Berlin Heidelberg 2008

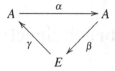

such that $\text{im}(\alpha) = \text{ker}(\beta)$, $\text{im}(\beta) = \text{ker}(\gamma)$, and $\text{im}(\gamma) = \text{ker}(\alpha)$.

This must take place in a category where images and kernel are defined, for example, the category of abelian groups or of graded k-modules. In fact, any abelian category is suitable.

Usually, there is a question of degree or bidegree as with complexes because a boundary operator in algebra usually changes the degree down as in homology or up as in cohomology and K-theory.

1.2. Notation In homology, a complex C is graded and has a boundary operator $d : C \to C$ with $d : C_i \to C_{i-1}$ of degree -1, and cohomology of complex C is graded and has a boundary operator $d : C \to C$ with $d : C^i \to C^{i+1}$ of degree $+1$. Always up and lower notations for degrees are related by the rule $C^i = C_{-i}$ and $C_i = C^{-i}$ giving the translation between upper and lower indices. In the case of exact couples, we have bidegrees and exact couples with homological and cohomological gradings. Since we are just interested in the cohomological case, we will only introduce that case in the next definition.

1.3. Definition A graded exact couple of type r is an exact couple $(A_r, E_r, \alpha, \beta, \gamma)$, where A_r and E_r are bigraded k-modules with α of bidegree $(r, -r)$, β of bidegree $(0,0)$, and γ of bidegree $(0, +1)$, that is, we have

$$\alpha : A^{i,j} \longrightarrow A^{i+r,j-r}, \quad \beta : A^{i,j} \longrightarrow E^{i,j}, \quad \text{and} \quad \gamma : E^{i,j} \longrightarrow A^{i,j+1}.$$

2 Homological Spectral Sequence for a Filtered Object

2.1. Notation Let $* = X_1 \subset X_0 \subset X_1 \subset \ldots \subset X_p \subset \ldots \subset X$ denote either

(a) a filtered chain complex with the algebraic object X_p redenoted by $F_p X$ so as not to mix filtration and degree indices or

(b) a filtered space X, where $X_n = X$ for some n and the homology theory has excision, that is, $H_*(X_p, X_q) \to H_*(X_p/X_q)$ is an isomorphism for $p > q$.

2.2. Definition In both cases for notation (2.1), the filtration on the homology is defined as the image

$$F_p H_n(X) = \text{im}(H_n(X_p) \longrightarrow H_n(X)).$$

Note that for the algebraic object the pth term in the filtration is denoted by $F_p H(X)$.

2.3. Remark We have a level exact sequence

$$0 \longrightarrow X_{p-1} \longrightarrow X_p \longrightarrow X_p/X_{p-1} = E_p^0(X) \longrightarrow 0$$

and a related exact triangle for graded objects

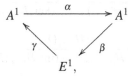

Observe that there is a shift in homological degrees to lead to an exact triangle.

Now, we assemble further allowing a shift in the filtration degree to obtain the homology exact couple.

2.4. Definition The homology exact couple associated with the data (2.1) is given by the diagram

$$
\begin{array}{ccc}
A^1 & \xrightarrow{\ \ \alpha\ \ } & A^1 \\
& \diagdown{\gamma} \quad \diagup{\beta} & \\
& E^1, &
\end{array}
$$

where $A_{p,q}^1 = H_{p+q}(X_p)$ and $E_{p,q}^1 = H_{p+q}(X_p/X_{p-1})$ and the bidegrees of the three morphisms in the exact couple are

$$\deg(\alpha) = (+1,-1), \quad \deg(\beta) = (0,0), \quad \text{and} \quad \deg(\gamma) = (-1,0).$$

As usual, the fact that we have an exact couple means that

$$\mathrm{im}(\alpha) = \ker(\beta), \quad \mathrm{im}(\beta) = \ker(\gamma), \quad \text{and} \quad \mathrm{im}(\gamma) = \ker(\alpha).$$

We know that we can derive the above exact couple from the discussion in the first section, but we will only need to consider the $E^2 = H(E^1, d^1)$ to see the other terms of the spectral sequence.

2.5. Remark The differential $d^1 : E^1 \to E^1$ is given by $d^1 = \beta\gamma$ and the bidegree $(-1,0)$, that is, $d^1 : E_{p,q}^1 \to E_{p-1,q}^1$. This means that the E^2 term is given by

$$E^2 = H(E^1, \beta\gamma) = \frac{\ker(\beta\gamma)}{\mathrm{im}(\beta\gamma)} = \frac{\gamma^{-1}(\ker(\beta))}{\beta(\mathrm{im}(\gamma))} = \frac{\gamma^{-1}(\mathrm{im}(\alpha))}{\beta(\ker(\alpha))}.$$

From general considerations in the first section, we can calculate the E^r term of the spectral sequence by the formula

$$E^r = H(E^{r-1}, d^{r-1}) = \frac{\gamma^{-1}(\mathrm{im}(\alpha^{r-1}))}{\beta(\ker(\alpha^{r-1}))}.$$

Now, we are interested in what might be called the elementary convergence properties of the above spectral sequence arising in homology. This means that $E_{p,q}^r = E_{p,q}^{r+1} = E_{p,q}^{r+2} = \ldots$ for some r.

2.6. Finiteness Assumption Returning to the notation of (2.1), we assume that for (a) $F_p H_n(X) = X_n$ for $p > n$ or (b) $X_m = X$ for some m which is interpreted as a dimension. There we have $E_{p,q}^r = E_{p,q}^{r+1} = E_{p,q}^{r+2} = \ldots = E_{p,q}^\infty$ by definition.

The final step in this discussion is to calculate this limit or stable E^∞ in terms of $F_p H(X)$ of (2.2).

2.7. Theorem *Let X be a filtered object satisfying the finiteness assumption of (2.6). Then*

$$E_{p,q}^\infty = \frac{F_p H_{p+q}(X)}{F_{p-1} H_{p+q}(X)} = E_{p,q}^0(H(X)).$$

Proof. We calculate for large r

$$E_{p,q}^\infty = E_{p,q}^r = \frac{\gamma^{-1}(\mathrm{im}(\alpha^{r-1}))}{\beta(\ker(\alpha^{r-1}))} = \frac{\mathrm{im}(\beta)}{\beta(\ker(H_{p+q}(X_p) \to H_{p+q}(X)))}$$

$$\xrightarrow{\beta'} \frac{H_{p+q}(X_p)}{(\ker(H_{p+q}(X_p) \to H_{p+q}(X)) + (\mathrm{im}(H_{p+q}(X_{p-1}) \to H_{p+q}(X_p))},$$

where β' is induced by β and is an isomorphism. This quotient is mapped isomorphically by the induced map $H(X_p) \to H(X)$ giving

$$E_{p,q}^\infty = \frac{\mathrm{im}(H_{p+q}(X_p) \to H_{p+q}(X))}{\mathrm{im}(H_{p+q}(X_{p-1}) \to H_{p+q}(X))} = \frac{F_p H_{p+q}(X)}{F_{p-1} H_{p+q}(X)} = E_{p,q}^0(H(X)).$$

3 K-Theory Exact Couples for a Filtered Space

Continuing with the notation (2.1)(b) for a filtered space, we apply the K-theory functor to the triple of spaces $X_{p-1} \subset X_p \subset X$ and related reduced spaces $X_p/X_{p-1} \to X/X_{p-1} \to X/X_p$. What we define for the K functor K^* can be carried out for any cohomology theory E^*.

3.1. Definition The filtration $F^p K^*(X)$ on the K-theory of X is defined by either the image or kernel

$$F^p(K^m X) = \mathrm{im}(K^m(X/X_{p-1}) \to K^m(X)) = \ker(K^m(X) \to K^m(X_{p-1})).$$

3.2. Remark Observe that the factorization

$$X \longrightarrow X/X_{p-1} \longrightarrow X/X_p$$

shows that $F^p K^m(X) \supset F^{p+1} K^m(X)$ so that the filtration is decreasing.

3.3. Remark We have an exact sequence

$$* \longrightarrow X_p/X_{p-1} \longrightarrow X/X_{p-1} \longrightarrow X/X_p \longrightarrow *$$

and a related exact triangle for the *K*-groups

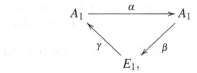

$$A_1^{p+1,q-1} = K^{p+q}(X/X_p) \longrightarrow K^{p+q}(X/X_{p-1}) = A_1^{p,q} \; .$$

$$E_1^{p,q-1} = K^{p+q+1}(X_p/X_{p-1}) \qquad K^{p+q}(X_p/X_{p-1}) = E_1^{p,q}$$

Observe that there is a shift in homological degrees to lead to an exact triangle, and by assembling the filtration degree, we obtain a cohomology exact couple.

3.4. Definition The *K*-theory exact couple is given by

$$
\begin{array}{ccc}
A_1 & \xrightarrow{\;\;\alpha\;\;} & A_1 \\
& \gamma \searrow \quad \swarrow \beta & \\
& E_1, &
\end{array}
$$

where $A_1^{p,q} = K^{p+q}(X/X_{p-1})$ and $E_1^{p,q} = K^{p+q}(X_p/X_{p-1})$ with $\deg(\alpha) = (-1,+1)$, $\deg(\beta) = (0,0)$, and $\deg(\gamma) = (-1,0)$.

From the discussion in the first section, we know that we can derive this couple any number of times yielding a spectral sequence with $E_{r+1} = H(E_r, d_r)$. We consider the exact formulas for $r = 1$ and the E_2 term.

Now, we are interested in what might be called the elementary convergence properties of the above spectral sequence arising in homology. This means that $E_r^{p,q} = E_{r+1}^{p,q} = E_{r+2}^{p,q} = \ldots$ for some r, and in this case, we define $E_r^{p,q}{}_\infty = E^{p,q}$.

Returning to the finiteness assumption (2.6)(b), we complete the discussion by calculating the E_∞ term in terms of $F^p K(X)$ introduced in (3.2).

3.5. Theorem *Let X be a filtered object satisfying the finiteness assumption of (2.6). Then*

$$E_\infty^{p,q} = \frac{F^p K^{p+q}(X)}{F^{p+1} K^{p+q}(X)} = E_0^{p,q}(H(X)).$$

Proof. We calculate for large *r*

$$E_\infty^{p,q} = E_r^{p,q} = \frac{\gamma^{-1}(\operatorname{im}(\alpha^{r-1}))}{\beta(\ker(\alpha^{r-1}))} = \frac{\operatorname{im}(\beta)}{\beta(\ker(K^{p+q}(X) \to K^{p+q}(X_{p-1})))}$$

$$\xrightarrow{\;\beta'\;} \frac{K^{p+q}(X/X_{p-1})}{(\ker(K^{p+q}(X/X_{p-1}) \to K^{p+q}(X)) + (\operatorname{im}(K^{p+q}(X/X_p) \to K^{p+q}(X/X_{p-1})))},$$

where β' is induced by β and is an isomorphism. This quotient is mapped isomorphically by the induced map $K(X/X_{p-1}) \to K(X)$ giving

$$E_\infty^{p,q} = \frac{\mathrm{im}(K^{p+q}(X/X_{p-1}) \to K^{p+q}(X))}{\mathrm{im}(K^{p+q}(X/X_p) \to K^{p+q}(X))} = \frac{F^p K^{p+q}(X)}{F^{p+1} K^{p+q}(X)} = E_0^{p,q}(K(X)).$$

This establishes the theorem.

3.6. Remark This result depends just on exactness and the excision isomorphism $E^*(X/Y) \to E^*(X,Y)$ for any cohomology theory E^*.

4 Atiyah–Hirzebruch Spectral Sequence for K-Theory

Now we return to the ideas of Chap. 9 and CW-complexes which are filtered spaces X with X_p/X_{p-1} equal to a wedge of spheres all of dimension p. In 9(4.4), we introduced cellular chains where the spheres in X_p/X_{p-1} appear as basis elements.

4.1. Remark Let X be a CW-complex. The K-theory spectral sequence defined by filtration by cells has the E_1 term

$$E_1^{p,q} = K^{p+q}(X_p/X_{p-1}).$$

If it has only $m(p)$ p-cells, that is if $X_p/X_{p-1} = \vee_{m(p)} S^p$ is a wedge of $m(p)$ p-spheres, then we can calculate the E_1 term as a direct sum $E_1^{p+q} = K^{p+q}(X_p/X_{p-1}) = K^{p+q}(\vee_{m(p)} S^p) = \bigoplus_{m(p)} K^{p+q}(S^p)$. Using the suspension property of generalized cohomology theory, we can rewrite

$$E_1^{p,q} = K^{p+q}(X_p/X_{p-1}) = \bigoplus_{m(p)} K^{p+q}(S^p) = C^p(X, K^q(*))$$

meaning cellular cochains of X with values in K-groups of a point.

4.2. Remark In 9(4.4), we went further to identify the cell boundary morphism with the boundary morphism of the triple (X_{p+1}, X_p, X_{p-1}), that is,

$$d_1 : E_1^{p,q} = K^{p+q}(X_p/X_{p-1}) \longrightarrow K^{p+q+1}(X_{p+1}, X_p) = E_1^{p+1,q}.$$

The result of this calculation is that

$$E_2^{p,q} = H^p(X, K^q(*)).$$

We summarize the main results in the next assertion which holds for any generalized cohomology theory E^* not just K-theory.

4.3. Assertion For a finite CW-complex X there exists a spectral sequence E_r satisfying

(1) $E_1^{p,q} = C^p(X, K^q(*))$, where d_1 is the ordinary cell boundary operator in cohomology,
(2) $E_2^{p,q} = H^p(X, K^q(*))$, and
(3) there is a filtration of $F^p K^*(X)$ such that $E_\infty^{p,q}$ is isomorphic to the associated graded complex, that is,

$$E_\infty^{p,q} = F^p K^{p+q}(X) / F^{p+1} K^{p+q}(X).$$

In the previous assertion (4.3), we have not used any property of K-theory, and in fact, the assertion holds for ordinary cohomology theory in which case the spectral sequence collapses. This is just the content of the analysis in 9(4.4).

4.4. Assertion For ordinary cohomology theory H^*, the spectral sequence in (4.3) collapses with $E_2 = E_\infty$.

In the case of K-theory, we can build Bott periodicity into the form of the spectral sequence.

4.5. Assertion The even–odd grading of $K(X)$ given by Bott periodicity $K(X) = (K^{\text{ev}}(X), K^{\text{od}}(X))$ has a filtration

$$F^p K(X) = \ker(K(X) \longrightarrow K(X/X_{p-1})),$$

with $E_r^{p,q}(X)$ and $E_r^{p,q+2}(X)$ isomorphic and with the odd complementary degrees $E_r^{p,2q+1} = 0$ for $r \geq 1$ and all q. The only differentials are the odd differentials $d_{2s+1} : E_{2s+1}^p \to E_{2s+1}^{p+2s+1}$. The associated graded complex of the filtration on $K(X)$ is of the form $E_\infty^{p,0}$. In particular, the K-theory spectral sequence collapses when $H^i(X, \mathbb{Z})$ is zero for all odd degrees i. As an application we have the following calculation.

4.6. Proposition *The K-theory of the complex projective space $P_m(\mathbb{C})$ is given by* $K^0(P_m(\mathbb{C})) = \mathbb{Z}[x]/x^{m+1}$ *and* $K^1(P_m(\mathbb{C})) = 0$.

For the ring structure, the class $x = L - 1$ must be studied, where L is the canonical line bundle, but as an additive group it is free of rank $m + 1$.

4.7. Remark In Chap 10, Sect. 5, we considered the Chern character ring morphism $K(X) \to H^{**}(X, \mathbb{Q})$. The Chern character induces a morphism from the K-theory spectral sequence to the rational cohomology spectral sequence. The basic result that tensoring with rational numbers $K(X) \otimes \mathbb{Q} \to H^{**}(X, \mathbb{Q})$ gives an isomorphism was shown in 10(5.6). This has the following implication for the K-theory spectral sequence.

4.8. Theorem *In general, the differentials are torsion morphisms. If X is $H^*(X, \mathbb{Z})$ torsion free, then the K-theory spectral sequence collapses.*

5 Formulas for Differentials

In (4.8), we saw that all differentials in the K-theory spectral sequence are torsion operations, it is natural to look for formulas for the differentials which involve cohomology operations with finite field coefficients. Besides the Bockstein operations, we expect the Steenrod operations. For $p = 2$, we used the Steenrod squaring operations in 10(8.2) and 10(8.3) to give another formula for the Stiefel–Whitney classes in 10(8.5).

5.1. Remark In 10(8.6)(1) we had the formula

$$Sq^{2a+1} = Sq^1 Sq^{2a} : H^i(X, \mathbb{F}_2) \longrightarrow H^{i+2a+1}(X, \mathbb{F}_2),$$

where $Sq^1 : H^{i+2a}(X, \mathbb{F}_2) \to H^{i+2a+1}(X, \mathbb{F}_2)$ is the Bockstein morphism associated with the exact sequence of groups

$$0 \longrightarrow \mathbb{Z}/2\mathbb{Z} \longrightarrow \mathbb{Z}/4\mathbb{Z} \longrightarrow \mathbb{Z}/2\mathbb{Z} \longrightarrow 0.$$

In turn, this exact sequence of groups is a quotient of the multiplication by two exact sequence of groups $0 \to \mathbb{Z} \xrightarrow{2} \mathbb{Z} \to \mathbb{Z}/2\mathbb{Z} = \mathbb{F}_2 \to 0$ so there is a factorization of $Sq^1 = r\delta$, where

(1) $\delta : H^j(X, \mathbb{F}_2) \to H^{j+1}(X, \mathbb{Z})$ is the Bockstein associated with multiplication by two exact sequence and
(2) $r : H^q(X, \mathbb{Z}) \to H^q(X, \mathbb{F}_2)$ is the reduction mod 2-induced morphism in cohomology. For an odd degree, we can define an integral-valued Steenrod square

$$Sq^{2a+1} = \delta Sq^{2a} : H^i(X, \mathbb{F}_2) \longrightarrow H^{i+2a+1}(X, \mathbb{Z}).$$

Finally, with the reduction mod 2 morphism, we have an integral operation $Sq^{2a+1} = \delta Sq^{2a} r : H^i(X, \mathbb{Z}) \to H^{i+2a+1}(X, \mathbb{Z})$.

5.2. Remark The first two odd degree cases in the previous discussion are Sq^1 and Sq^3. The integral version of Sq^1 is factorized by $H^i(X, \mathbb{Z}) \xrightarrow{r} H^i(X, \mathbb{F}_2) \xrightarrow{\delta} H^{i+1}(X, \mathbb{Z})$, where r is the reduction mod 2 and δ is the Bockstein. The integral version of Sq^3 is defined by $Sq^3 : H^i(X, \mathbb{Z}) \to H^{i+3}(X, \mathbb{Z})$ as in (5.1)(2). It was realized by Atiyah and Hirzebruch that this is the first nontrivial differential in the complex K-theory spectral sequence. This is always a torsion operation which is killed by 2, and the differentials are torsion morphisms.

5.3. Proposition *In the complex K-theory spectral sequence, the first terms are*

$$E_3^{p,q} = E_2^{p,q} = \begin{cases} H^p(X, \mathbb{Z}) \ for \ q \ even \\ 0 \qquad\qquad for \ q \ odd, \end{cases}$$

and the first possibly nontrivial differential is

$$d_3 = Sq^3 : H^i(X, \mathbb{Z}) = E_3^{p,q} \longrightarrow E_3^{p+3,q-2} = H^{p+3}(X, \mathbb{Z}).$$

6 Calculations for Products of Real Projective Spaces

6.1. Remark The Stiefel–Whitney classes for the real projective spaces are given by formulas

$$w_1(P_n(\mathbb{R})) = n+1 \text{ and } w_2(P_n(\mathbb{R})) = \frac{n(n+1)}{2} \pmod 2.$$

In particular, it is the odd-dimensional real projective spaces which are orientable, that is, $w_1 = 0 \pmod 2$. Of the odd dimensions $n = 4m+1$ and $n = 4m+3$, the $w_2 = 1 \pmod 2$ for $n = 4m+1$ and $w_2 = 0 \pmod 2$ for $n = 4m+3$. In particular, $P_{4m+3}(\mathbb{R})$ is an oriented real manifold which has a spin structure.

6.2. Cohomology of Real Projective Spaces The homology and cohomology with $\mathbb{F}_2$ coefficients is given by

$$H_i(P_n(\mathbb{R}), \mathbb{F}_2) = \mathbb{F}_2 \quad \text{for } 0 \le i \le n$$

and zero otherwise, and

$$H^*(P_n(\mathbb{R}), \mathbb{F}_2) = \mathbb{F}_2[w]/(w)^{n+1}$$

with the cap product algebra structure. With integral coefficients, it depends whether the dimension is even or odd. We give the result of the oriented odd-dimensional case $n = 2m+1$, where we obtain a copy of Z in the top dimension, due to orientability, and

i	0	1	2	3	......	$2m$	$2m+1$
$H^i((P_n(\mathbb{R}), \mathbb{Z})$	Z	0	$\mathbb{Z}/2$	0	......	$\mathbb{Z}/2$	0
$H_i((P_n(\mathbb{R}), \mathbb{Z})$	Z	$\mathbb{Z}/2$	0	$\mathbb{Z}/2$	......	0	Z

6.3. K-Theory Results For an odd-dimensional projective space, we have $K^1(P_n(\mathbb{R}), \mathbb{Z}) = Z$ and $K^0(P_{2m-1}(\mathbb{R}), \mathbb{Z}) = Z \oplus \mathbb{Z}/2^{m-1}$ as additive groups. The ring structure on K^0 is given by

$$K^0(P_{2m-1}(\mathbb{R}), \mathbb{Z}) = \mathbb{Z}[y]/((y+1)^2 - 1, y^m)$$

or by

$$K^0(P_{2m-1}(\mathbb{R}), \mathbb{Z}) = \mathbb{Z}[y]/(y^2 + 2y, y^m) = \mathbb{Z}[x]/((x^2 - 1), (x-1)^m).$$

For the first odd dimensions, we have

$$K^0(P_1(\mathbb{R}), \mathbb{Z}) = \mathbb{Z}, \quad K^0(P_3(\mathbb{R}), \mathbb{Z}) = \mathbb{Z} \oplus \mathbb{Z}/2 = \mathbb{Z}[x]/(x^2 - 1, (x-1)^2)$$

$$K^0(P_5(\mathbb{R}), \mathbb{Z}) = \mathbb{Z} \oplus \mathbb{Z}/2^2 = \mathbb{Z}[x]/(x^2 - 1, (x-1)^3)$$

$$K^0(P_7(\mathbb{R}),\mathbb{Z}) = \mathbb{Z} \oplus \mathbb{Z}/2^3 = \mathbb{Z}[x]/(x^2 - 1, (x-1)^4).$$

7 Twisted K-Theory Spectral Sequence

In twisted K-theory, there is a three-dimensional class which plays a role, and it combines with Sq^3 to give the differential in the twisted K-theory spectral sequence.

7.1. Proposition *For $t \in H^3(X,\mathbb{Z})$, the twisted complex K-theory spectral sequence has initial terms with*

$$E_3^{p,q} = E_2^{p,q} = \begin{cases} H^p(X,\mathbb{Z}) \textit{ for } q \textit{ even} \\ 0 \qquad\qquad \textit{ for } q \textit{ odd}, \end{cases}$$

the first nontrivial differential

$$d_3 = Sq^3 + (\ ) \smile t : H^p(X,\mathbb{Z}) = E_3^{p,q} \longrightarrow E_3^{p+3,q-2} = H^{p+3}(X,\mathbb{Z}).$$

This means that the third differential is deformed by the cup product with the twisting class.

Reference

Rosenberg, J.: Continuous-trace algebras from the bundle-theoretic point of view. J. Austral. Math. Soc. Ser. A **47**(3):368–381 (1989)

Chapter 22
Twisted Equivariant K-Theory and the Verlinde Algebra

As we outlined in the introduction, the role of K-theory in physics started when topological invariants of D-branes in string theory took values in K-groups. When the D-brane has a nontrivial background B-field, the corresponding D-brane invariants take values in the twisted K-groups where the B-field is a three-form defining the twisting in the K-group.

Just as there is a G-equivariant K-theory, there is also a G-equivariant twisted K-theory with a definition in terms of Fredholm operators on a Hilbert space having suitable G-action. One such G-space is $G = \mathrm{Ad}(G)$ with the adjoint action of G on itself, and in this case, it is sometimes possible to determine the twisted G-equivariant K-theory of G.

Throughout this chapter, G is a compact, simply connected, simple real Lie group. In some cases the assertions hold more generally as for the isomorphism

$$R(G) \longrightarrow K_G(*)$$

which is true for a general compact group.

When one considers either the loop group on G or the associated Kac-Moody Lie algebra to $\mathrm{Lie}(G)$, the classical Lie algebra of G, there is a representation theory for a given level or central charge c. The representations of central charge c define a fusion algebra $\mathrm{Ver}_c(G)$, called the Verlinde algebra which is a quotient ring of $R(G)$ for simply connected compact Lie groups G which are simple. The main assertion in this chapter is that this Verlinde fusion algebra is related to twisted K-theory of $\mathrm{Ad}(G)$. The isomorphism

$$R(G) \longrightarrow K_G(*)$$

has a quotient which is an isomorphism giving a commutative square

$$
\begin{array}{ccc}
R(G) & \longrightarrow & K_G(*) \\
\downarrow & & \downarrow \\
\mathrm{Ver}_c(G) & \longrightarrow & K_G^\alpha(\mathrm{Ad}(G)).
\end{array}
$$

D. Husemöller et al.: *Twisted Equivariant K-Theory and the Verlinde Algebra*, Lect. Notes Phys. **726**, 265–274 (2008)
DOI 10.1007/978-3-540-74956-1_23   © Springer-Verlag Berlin Heidelberg 2008

Some of the background material for this chapter is taken from lectures in Bonn of Gerd Faltings. Some of the calculations are taken from Gregory Moore where one finds a physical explanation of the role of the third differential in the Atiyah–Hirzebruch spectral sequence for twisted K-theory which was discussed in the last section of the previous Chap. 23.

1 The Verlinde Algebra as the Quotient of the Representation Ring

There is a Verlinde algebra associated with a finite-dimensional simple complex Lie algebra and a level c. Recall that to a simple and simply connected compact real Lie group G, we can form the complex Lie algebra $\mathfrak{g} = \mathrm{Lie}(G) \otimes_{\mathbb{R}} \mathbb{C}$. This sets up a bijection between isomorphism classes of simple, simply connected compact real Lie groups and the isomorphism classes of finite-dimensional simple complex Lie algebras. The representation rings are also isomorphic under this correspondence. The Verlinde algebra $\mathrm{Ver}_c(\mathfrak{g})$ of representations of the loop algebra $L(\mathfrak{g})$ of level or central charge c with fusion product corresponds to the algebra of positive energy representations of level $c \geq 0$ of the loop Lie group $L(G)$ with the fusion product (Witten (1993)).

1.1. Notation Let $\mathfrak{g}$ be a Lie algebra over the complex numbers, and let A be an associative algebra. The Lie algebra structure on $\mathfrak{g} \otimes A$ is given by the formula

$$[u \otimes a, v \otimes b] = [u,v] \otimes ab \ \ \text{or} \ \ [au,bv] = ab\,[u,v]$$

for $u,v \in \mathfrak{g}$, $a,b \in A$. Two special cases are

$$\mathfrak{g}[t] = \mathfrak{g} \otimes_{\mathbb{C}} \mathbb{C}[t] \ \text{ and } \ \mathfrak{g}[t,t^{-1}] = \mathfrak{g} \otimes_{\mathbb{C}} \mathbb{C}[t,t^{-1}],$$

where the second case $\mathfrak{g}[t,t^{-1}]$ is called the Lie algebra of algebraic loops with values in $\mathfrak{g}$.

The affine Kac–Moody algebra $\tilde{\mathfrak{g}}$ corresponding to a given Lie algebra $\mathfrak{g}$ refers to a basic extension of the Lie algebra $\mathfrak{g}[t,t^{-1}]$ of loops with values in $\mathfrak{g}$, see (1.4). We consider the construction in the following case.

1.2. Simple Lie Algebra Let $\mathfrak{g}$ be a finite-dimensional simple Lie algebra over $\mathbb{C}$. Choose a Cartan subalgebra $\mathfrak{h} \subset \mathfrak{g}$ whose action on $\mathfrak{g}$ gives the root system. Choices of positive root systems correspond to decompositions

$$\mathfrak{g} = \mathfrak{h} \oplus \mathfrak{n}^+ \oplus \mathfrak{n}^-,$$

where $\mathfrak{n}^+$ and $\mathfrak{n}^-$ are the nilpotent subalgebras of $\mathfrak{g}$ with a basis of the positive and negative roots, respectively. Let P be the weight lattice relative to $\mathfrak{h} \subset \mathfrak{g}$, and let P^+ denote the positive weights relative to the Borel subalgebra $\mathfrak{b} = \mathfrak{h} \oplus \mathfrak{n}^+ \subset \mathfrak{g}$.

On the simple Lie algebra $\mathfrak{g}$, we have an invariant bilinear form

$$(\, | \,) : \mathfrak{g} \times \mathfrak{g} \to \mathbb{C}$$

which is unique up to a scalar multiple. It is unique assuming $(\theta|\theta) = 2$ for the longest root θ.

1.3. Representation Ring $R(\mathfrak{g})$ of $\mathfrak{g}$ A basis of the abelian group $R(\mathfrak{g})$ is the set of irreducible representations π_λ up to isomorphism which are parametrized by their highest weight vector $\lambda \in P^+$. The ring structure is given by the tensor product $\pi \otimes \pi'$ of representations. So $[\pi][\pi'] = [\pi \otimes \pi']$, where $[\pi]$ denotes the isomorphism class of the representation π. In all cases, the representations are finite dimensional.

By restriction of the action from $\mathfrak{g}$ to the subalgebra $\mathfrak{h}$, we have an injective ring morphism

$$R(\mathfrak{g}) \to R(\mathfrak{h}) = \mathbb{Z}[t_1, t_1^{-1}, \ldots, t_\ell, t_\ell^{-1}],$$

where $\ell = \dim \mathfrak{h}$ is called the rank of $\mathfrak{g}$. Since $\mathfrak{h}$ is commutative, the irreducible representations of $\mathfrak{h}$ are one dimensional. The Weyl group W of G acts on $\mathfrak{h}$ and $R(\mathfrak{h})$, and the invariants

$$R(\mathfrak{h})^W = \mathrm{image}\,(R(\mathfrak{g}) \to R(\mathfrak{h})) \,.$$

This is made explicit for $\mathfrak{g} = \mathfrak{sl}(2)$ in the next section.

1.4. Affine Kac–Moody Algebra $\tilde{\mathfrak{g}}$ Associated with a Finite-dimensional Simple Lie Algebra $\mathfrak{g}$ We have an exact sequence of Lie algebras over the complex numbers

$$0 \longrightarrow \mathbb{C}K \oplus \mathbb{C}d \longrightarrow \tilde{\mathfrak{g}} \longrightarrow \mathfrak{g}[t, t^{-1}] \longrightarrow 0 \,.$$

The kernel $\mathbb{C}K \oplus \mathbb{C}d$ is an abelian Lie algebra, and the twisting cocycle is the additional term in the formula for the Lie bracket

$$[f(t)u, g(t)v] = f(t)g(t)\,[u, v] - (u|v)\,\mathrm{Res}_{t=0}(f\,dg)K,$$

the twisting of the induced central extension is

$$[d, f(t)u] = t\frac{d}{dt}f(t)u,$$

and

$$[K, f(t)u] = 0 \,; \ [d, K] = 0 \,.$$

1.5. Representations of $\tilde{\mathfrak{g}}$ of Level $c \in \mathbb{N}$ where K acts with Eigenvalue c Form the Verlinde Algebra $\mathrm{Ver}_c(\mathfrak{g})$. This algebra $\mathrm{Ver}_c(\mathfrak{g})$ has an additive basis of irreducible representations on which $K \in L(\mathfrak{g})$ acts by the scalar c. They parametrized by $\lambda \in P^+$ with

$$\langle \lambda, \theta^\vee \rangle \leq c,$$

where $\theta^\vee$ is the maximal coroot. This parameter λ makes a description of the Verlinde algebra as some kind of quotient of $R(\mathfrak{g})$ possible.

1.6. Assertion The Verlinde algebra $\mathrm{Ver}_c(\mathfrak{g})$ is the quotient ring of $\mathrm{R}(\mathfrak{g})$ by an ideal I_c. Additively, $\mathrm{Ver}_c(\mathfrak{g})$ is a finitely generated free abelian group which is a Grothendieck group of central charge c representations of $L(\mathfrak{g})$.

The product on $\mathrm{Ver}_c(\mathfrak{g})$ has an independent definition called the fusion product. In the quotient process, it comes from the tensor product of representations.

1.7. Generators of the Ideal I_c in $\mathrm{R}(g)$ To describe the generators of the ideal I_c, we denote the fundamental weights by Λ_i, where $0 \le i \le \ell = \dim \mathfrak{h}$. For an affine weight of level c,

$$\Lambda = \lambda + c\Lambda_0 + ad$$

with $\lambda \in P^+$, $a \in \mathbb{C}$, the action of an element $w \in W_{\mathrm{aff}}$ in the affine Weyl group is given by the formula

$$w \cdot \Lambda = w \cdot (\lambda + c\Lambda_0 + ad) = w(\lambda, c) + c\Lambda_0 + (a - d_w(\lambda, c))d,$$

where $w(\lambda, c) \in P^+$ and $d_w(\lambda, c) \in \mathbb{Z}$. For a weight λ, the action takes the form $w \cdot \lambda = w \cdot (\lambda + \rho) - \rho$ called shift action, where $\rho = \sum_{i=0}^{\ell} \Lambda_i$.

1.8. Basic Assertion The ideal $I_c = \ker(\mathrm{R}(\mathfrak{g}) \to \mathrm{Ver}_c(\mathfrak{g}))$ is generated by the elements of the form

$$[\lambda] - (-1)^{\ell(w)}[w(\lambda, c)], \ \lambda \in P^+,$$

where $\ell(w) = \mathrm{length}$ of $w \in W_{\mathrm{aff}}$. The basic reference is Sect. 2.2 in Feigin et al. (2002).

2 The Verlinde Algebra for $SU(2)$ and $\mathfrak{sl}(2)$

As an example, we work out the Verlinde algebra $\mathrm{Ver}_c(SU(2)) = \mathrm{Ver}_c(\mathfrak{sl}(2))$ for the compact group $SU(2)$ and the corresponding Lie algebra $\mathrm{Lie}(SU(2)) \cong \mathfrak{sl}(2)$ of two by two complex matrices of trace zero. This algebra which we denote in this section by Ver_c is a quotient of the ring of characters $\mathrm{R}(SU(2))$ of the group $SU(2)$ or also of the Grothendieck ring $\mathrm{R}(\mathfrak{sl}(2))$ of representations of the Lie algebra $\mathfrak{sl}(2)$. The structure is very explicit since the rings $\mathrm{R}(SU(2)) \cong \mathrm{R}(\mathfrak{sl}(2))$ are both isomorphic to the polynomial ring in one variable over $\mathbb{Z}$, and the Verlinde algebra is a quotient algebra which is of finite rank over $\mathbb{Z}$.

2.1. Remark As a ring of characters, $\mathrm{R}(SU(2))$ is determined by the monomorphism $\mathrm{R}(SU(2)) \to \mathrm{R}(U(1))$ where the circle group $T = U(1)$ is the maximal torus. We have $\mathrm{R}(U(1)) = \mathbb{Z}[t, t^{-1}]$ where $t = e^{2\pi i \theta}$, and hence, it is the ring of finite Fourier series. The image of the monomorphism $\mathrm{R}(SU(2)) \to \mathrm{R}(U(1))$ is the subring of the finite Fourier series $f(t)$ with the property that $f(t) = f(\frac{1}{t})$.

2.2. Lemma A finite Fourier series $f(t)$ satisfies $f(t) = f(\frac{1}{t})$ if and only if $f(t) = g(t + \frac{1}{t})$, where $g(x) \in \mathbb{Z}[x]$ is a polynomial.

The proof results from the relation

$$t^{n+1} + t^{-n-1} = (t + t^{-1})(t^n + t^{-n}) - (t^{n-1} + t^{-n+1}),$$

which generates the inductive procedure.

2.3. Proposition *As a subring of* $R(U(1)) = \mathbb{Z}[t, t^{-1}]$, *the ring* $R(SU(2))$ *is the polynomial ring* $\mathbb{Z}[x]$, *where* $x = t + t^{-1}$.

2.4. Remark The irreducible representations of the Lie group $SU(2)$ or of the Lie algebra $\mathfrak{sl}(2)$ are parametrized by natural numbers and are denoted by $V(n), n \in \mathbb{N}$, where $\dim_{\mathbb{C}} V(n) = n + 1$. Under the natural injective ring homomorphism $R(\mathfrak{sl}(2)) \to R(\mathfrak{h}) = \mathbb{Z}[t, t^{-1}]$, the class of $V(n)$ becomes the polynomial

$$[n](t) = t^n + t^{n-2} + \ldots + t^{-n+2} + t^{-n}.$$

In $R(\mathfrak{sl}(2))$, we denote the element representing the irreducible $V(n)$ also by $[n]$.

For example, the first $V(0)$ is the trivial one-dimensional representation, $V(1)$ is the standard two-dimensional matrix representation, and $V(2)$ is the three-dimensional adjoint representation of $\mathfrak{sl}(2)$ on itself. We denote that $V(n) = 0$ for $n < 0$.

2.5. Proposition *For the multiplicative structure on the irreducible representations, we can calculate with the Clebsch–Gordan inductive rule*

$$V(n) \otimes V(m) = V(m+n) \oplus V(n-1) \otimes V(m-1)$$

the formula

$$V(n) \otimes V(m) = V(m+n) \oplus \ldots \oplus V(n-m) \ \text{for} \ n \geq m.$$

For the proof, we consider again the embedding if $R(\mathfrak{sl}(2))$ *as a subring of* $R(\mathfrak{h}) = \mathbb{Z}[t, t^{-1}]$. *The image of the class of* $V(n)$ *is given by* $t^n + t^{n-2} + \ldots + t^{-n+2} + t^{-n}$. *Now,*

$$(t^n + t^{n-2} + \ldots + t^{-n+2} + t^{-n})(t^m + t^{m-2} + \ldots + t^{-m+2} + t^{-m}) =$$
$$t^{n+m} + t^{n+m-2} + \ldots + t^{-n+m+2} + t^{-n+m}$$
$$t^{n+m-2} + t^{n+m-4} \ldots + t^{-n+m} + t^{-n+m-2}$$
$$t^{n+m-4} \ldots + t^{-n+m} + t^{-n+m-2} + t^{-n+m-4}$$
$$\cdots\cdots$$
$$t^{-n+m} + t^{-n+m-2} + \ldots + t^{-n-m+2} + t^{-n-m}$$

which after reordering and summing the terms gives

$$(t^{n+m} + t^{n+m-2} + \ldots + t^{-n-m+2} + t^{-n-m}) + (t^{n-1} + \ldots + t^{-n+1})(t^{m-1} + \ldots + t^{-m+1})$$

which is the desired formula.

2.6. Example The first special cases are $V(n) \otimes V(0) = V(n)$ and $V(n) \otimes V(1) = V(n+1) \oplus V(n-1)$ which are formally the same as the character formula

$$(t+t^{-1})(t^n+t^{n-2}+\ldots+t^{-n})=(t^{n+1}+t^{n-1}+\cdots+t^{-n-1})+(t^{n-1}+t^{n-3}+\cdots+t^{-n+1})$$

used in describing $R(SU(2))$. In particular, $V(2)\otimes V(1)=V(3)\oplus V(1)$ is the adjoint action on the two by two matrices.

2.7. Proposition *The Grothendieck ring* $R(\mathfrak{sl}(2))$ *of finite-dimensional represen-tations is the polynomial ring* $\mathbb{Z}[x]$, *where x is the class of the standard two-dimensional representation* $V(1)$ *of* $\mathfrak{sl}(2)$. *It is the polynomial ring in one variable x over the integers. As an abelian group,* $R(\mathfrak{sl}(2))$ *is of countable rank having a basis given by the isomorphism classes* $[V(n)]$ *of irreducible representations.*

To describe the Verlinde algebra as a quotient ring

$$\mathbb{Z}[t,t^{-1}] \supset R(\mathfrak{sl}(2)) \longrightarrow \mathrm{Ver}_c(\mathfrak{sl}(2))$$

of the representation ring $R(\mathfrak{sl}(2))$, we use the following prescription.

Associated with each $c \in \mathbb{N}$, we have the reflection σ_c of the integers $\mathbb{Z}$ given by $\sigma_c(n) = 2c - n$ around c and fixing c. The conditions of (1.8) for the generators of the ideal I_c in the case of $\mathrm{Ver}_c(\mathfrak{sl}(2))$ take the form

$$[c+1] \in I_c \text{ and } [n] + [2c+2-n] \in I_c \text{ for } 0 < n < c+1.$$

The quotient $\pi : R(\mathfrak{sl}(2)) \to \mathrm{Ver}_c(\mathfrak{sl}(2))$, which depends on the central charge c, is defined by the two conditions:

(1) $\pi(V(c+1)) = 0$ and
(2) $\pi(V(n)+V(2c+2-n)) = 0$ for $0 < n < c+1$.

2.8. Proposition *The ideal I_c in $R(\mathfrak{sl}(2))$ is a principal ideal generated by $[c+1]$ the class of the irreducible representation of dimension $c+2$.*

To see that one relation is sufficient, we have as given $[c+1] \in I_c$ and the relation

$$V(m) \otimes V(1) = V(m+1) \oplus V(m-1)$$

for all $m \in \mathbb{N}$ shows that for $m = c+1$ that we have $[n] + [\sigma_c(n)] \in I_c$ for $n = c$, and by an elementary descending induction also for $0 < n < c+1$. Hence, the kernel is a principal ideal generated by $[c+1]$.

The following theorem can be worked out from the above description, see the article of Faltings (1995), on the Verlinde formula.

2.9. Theorem *The algebra $\mathrm{Ver}_c(\mathfrak{sl}(2))$ is isomorphic to $\mathbb{Z}[x]/(f_c(x))$, where*

$$f_c(x) = \prod_{\zeta^{2c+2}=1, \zeta \neq \pm 1/\text{inverse}} \left(x - (\zeta + \frac{1}{\zeta})\right).$$

In this formula, ζ and $\frac{1}{\zeta}$ are grouped together in one term giving a total of c factors. In particular, the algebra has degree c and tensored with $\mathbb{Q}$ is a product of totally real fields and CM fields. Another form of f_c is

$$f_c(x) = \prod_{m=1}^{c} \left(x - \left(\exp\left(\frac{m}{2c+2} 2\pi i \right) + \exp\left(-\frac{m}{2c+2} 2\pi i \right) \right) \right)$$

$$= \prod_{m=1}^{c} \left(x - 2\cos\left(\frac{m}{2c+2} 2\pi \right) \right).$$

To determine the multiplicative structure of the Verlinde algebras $Ver_c(\mathfrak{sl}(2)) = Ver_c(SU(2))$, we recall the character of the irreducible representation $V(n)$ on $R(U(1))$.

2.10. Remark For $g \in R\left[t, t^{-1}\right]$, where R is a commutative ring, we have used the fact that $g(t) = g(\frac{1}{t})$ if and only if there exists $f(x) \in R[x]$ with

$$g(t) = f\left(t + \frac{1}{t}\right).$$

Also $g(t) = t^m + \ldots + t^{-m}$ if and only if f is monic, $f(x) = x^m +$ lower terms. If $u \in R^*$, then

$$g(u) = 0 \text{ if and only if } f\left(u + \frac{1}{u}\right) = 0.$$

Applying this to $t^c + t^{c-2} + \ldots + t^{-c}$ which appears in the factorization

$$t^{c+1} - t^{c-1} = (t - t^{-1})(t^c + t^{c-2} + \ldots + t^{-c}),$$

we see for $\zeta^{2c+2} = 1$, $\zeta^2 \neq 1$, that

$$0 = g(\zeta), \text{ where } g(t) = t^c + t^{c-2} + \ldots + t^{-c}.$$

Each $\zeta + \frac{1}{\zeta}$ is one of the c different roots of the monic polynomial $f_c(x)$, hence the above formula for $f_c(x)$.

3 The G-Bundles on G with the Adjoint G–Action

To see that $G \times G$ modulo G with the diagonal action is a G-space isomorphic to G with the adjoint action, we make the following remarks.

3.1. Definition The left-diagonal action of G on $G \times G$ is given by

$$G \times (G \times G) \to G \times G \text{ is } s(x,y) = (sx, sy)$$

and the right-diagonal action of G on $G \times G$ is given by

$$(G \times G) \times G \to G \times G \text{ is } (x,y)s = (xs, ys).$$

We will use the orbit notation and write $G(x,y)$ for the set of all $s(x,y) = (sx, sy)$, $s \in G$.

3.2. Definition The adjoint action of G on G is given by

$$\text{Ad} : G \to \text{Aut}(G) \text{ is } \text{Ad}(s)t = sts^{-1}$$

for $s, t \in G$. With this action, we denote the G-space G as $\text{Ad}(G)$.

3.3. Proposition *The map* $\psi' : G \backslash (G \times G) \to G$ *given by the formula* $\psi'(G(x,y)) = x^{-1}y$ *is an isomorphism of spaces which is*

(1) a G-isomorphism of right G-spaces, where $G(x,y)s = G(x, ys)$ for $s \in G$ and $(x,y) \in G \times G$ and
(2) a G-isomorphism of left G-spaces $\psi' : G \backslash (G \times G) \to \text{Ad}(G)$, where $s \cdot G(x,y) = G(xs^{-1}, ys^{-1})$ for $s \in G$.

The map $\psi'' : (G \times G)/G \to G$ *given by the formula* $\psi''((x,y)G) = xy^{-1}$ *is an isomorphism of spaces which is*

(1) a G-isomorphism of left G-spaces, where $s(x,y)G = G(sx,y)$ for $s \in G$ and $(x,y) \in G \times G$ and
(2) a G-isomorphism of left G-spaces $\psi'' : (G \times G)/G \to \text{Ad}(G)$, where $s \cdot (x,y)G = G(sx, sy)$ for $s \in G$.

3.4. Double Diagonal Action Consider the following diagram for a topological group using the analysis of (3.3) with the two-sided diagonal action given by the image of the element $(s, (x,y), t)$ being (sxt, syt).

$$
\begin{array}{ccc}
G \times (G \times G) \times G & \longrightarrow & (G \times G) \times G \\
\downarrow & & \downarrow \\
G \times [(G \times G)/G] & \longrightarrow & (G \times G)/G \\
\scriptstyle{G \times \psi''} \downarrow & & \downarrow \scriptstyle{\psi''} \\
G \times G & \xrightarrow{\ \text{Ad}\ } & G
\end{array}
$$

where $s.((x,y),t) = ((sx,sy),t)$

where $s.((x,y)G) = ((sx,sy)G)$

where $\psi''(s(x,y)G) = (sx)(sy)^{-1}$

$$= s(xy)^{-1}s^{-1} = \text{Ad}(s)(\psi''(x,y)G)$$

We have a principal bundle over G for a left G action, and we introduce this concept in the next definition.

3.5. Definition Let G and Γ be two topological groups. A Γ-equivariant principal G-bundle $p : P \to B$ is a map of Γ-spaces where for each $\gamma \in \Gamma$, the action maps by γ denoted by $\gamma : P \to P$ is a morphism of G-principal bundles over $\gamma : B \to B$.

In terms of elements $z \in P$, $t \in G$, and $\gamma \in \Gamma$, the morphism of G-principal bundles condition takes the form $\gamma(zt) = (\gamma z)t$. The previous construction is an example of G-principal bundle $P = G \times G \to (G \times G)/G = B$ isomorphic to G with a left action of G such that on $B = G$ it is the adjoint action of G on G and on P it is the left diagonal G-action on $G \times G$.

As for fibre bundles with a G-group action, we have the following definition.

3.6. Definition Let G and Γ be two topological groups, and let $p : P \to B$ be a Γ-equivariant principal G-bundle. Let Y be a left $G \times \Gamma$-space. We define a left Γ-bundle structure on the fibre bundle $q : P[Y] = P \times^G Y \to B$ with the quotient action where $\Gamma \times P[Y] \to P[Y]$ is given by the relation $\gamma \cdot (x,y)G = (x, \gamma y)G$. This is well-defined since $(xs, s^{-1}\gamma y)G = (x, \gamma y)G$.

For $b \in B$ and its image $\gamma b \in B$ under $\gamma \in \Gamma$, we have the induced map on the fibres $P[Y]_b = b \times Y \to \gamma b \times Y = P[Y]_{\gamma b}$ which is just the action of $\gamma \in \Gamma$ on Y. In the case of a vector space as fibre, we have the following.

3.7. Remark Let G and Γ be two topological groups, and let $P \to B$ be a Γ-equivariant principal G-bundle. For a Γ-vector space V, the fibre bundle $P[V] \to B$ is a Γ-vector bundle.

3.8. Example We define the adjoint ring morphism $R(G) \to K_G(\mathrm{Ad}(G))$ by assigning to a class $[V] \in R(G)$ the K-class in $K_G(\mathrm{Ad}(G))$ of the G-equivariant vector bundle represented by $(G \times G)[V]$ on $\mathrm{Ad}(G)$.

4 A Version of the Freed–Hopkins–Teleman Theorem

The natural isomorphism $R(G) \longrightarrow K_G(*)$ extends to a commutative triangle

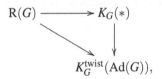

where the vertical arrow is a general Gysin morphism in twisted K-theory associated to the G-map $* \to \mathrm{Ad}(G)$, and twist is a number related to $H^3(G) = H^3_G(\mathrm{Ad}(G)) = \mathbb{Z}$.

4.1. Main Assertion For each twist α on $\mathrm{Ad}(G)$, the above commutative triangle extends to a commutative diagram

$$
\begin{array}{ccc}
R(G) & \longrightarrow & K_G(*) \\
\downarrow & & \downarrow \\
\mathrm{Ver}_c(G) & \longrightarrow & K_G^\alpha(\mathrm{Ad}(G)),
\end{array}
$$

where $\alpha = \alpha(c)$ is c shifted by the dual Coxeter number of G. The lower arrow

$$\mathrm{Ver}_c(G) \longrightarrow K_G^\alpha(\mathrm{Ad}(G))$$

is an isomorphism.

The proof divides into two parts. Firstly, show that $\mathrm{R}(G) \to K_G^\alpha(\mathrm{Ad}(G))$ is surjective, and secondly, establish the identity

$$\ker\left(\mathrm{R}(G) \longrightarrow K_G^\alpha(\mathrm{Ad}(G))\right) = \ker\left(\mathrm{R}(G) \longrightarrow \mathrm{Ver}_c(G)\right),$$

where again α is c shifted by the dual Coxeter number of G. In three preprints of Freed–Hopkins–Teleman (2005), one can find a proof that the lower arrow is an isomorphism. There have been various proofs of aspects of this isomorphism in the electronic archives before 2005. Another reference is the paper of Bunke and Schröder, where the formulation of the theorem as a commutative square is considered.

In Sects. 2 and 3, the case of $G = SU(2)$ and $\mathfrak{g} = \mathfrak{sl}(2)$ are carried out in detail with $\mathrm{Ver}_c(\mathfrak{sl}(2)) = \mathrm{R}(\mathfrak{sl}(2))/I_c$. In general, the space $G\backslash\mathrm{Ad}(G)$ is a simplex of dimension equal to the rank of G. Here, in the special $G = SU(2)$, the space $G\backslash\mathrm{Ad}(G)$ equals the closed interval $[0,\pi]$, where each $t \in [0,\pi]$ corresponds to the conjugacy class of

$$\begin{pmatrix} e^{it} & 0 \\ 0 & e^{-it} \end{pmatrix}.$$

In K-theory, $K_{SU(2)}(*) \longrightarrow K_{SU(2)}(\mathrm{Ad}(SU(2)))$ is just a power series in one parameter divided by one relation. The first calculation in $K_{SU(2)}$-theory is to see that this relation corresponds to a generator of $I_c \subset \mathrm{R}(SU(2))$.

This isomorphism is one of the reasons for being interested in the general theory of twisted K-theory.

References

Feigin, B.L.: Jimbo, M., Kedem, R., Loktev, S., Miwa, T.: Spaces of coinvariants and fusion product I. From equivalence theorem to Kostka polynominals (2002) (arXiv:math.QA/0205324v3)
Freed, D.S., Hopkins, M.S., Teleman, C.: Loop groups and twisted K-theory II. (2005) (arXiv:math.AT/0511232)

Part V
Gerbes and the Three Dimensional Integral Cohomology Classes

In Part IV, we considered algebra bundle over a space B which has a characteristic class in $H^3(B,\mathbb{Z})$. In this part, we introduce other geometric objects giving rise to a three-dimensional integral characteristic cohomology class. For some background, we return to the local coordinate condition (1) in 5(4.3), giving a 1-cocycle,

$$g'_{i,j}(b)g_{j,k}(b)g_{k,i}(b) = 1 \quad \text{for} \quad b \in U_i \cap U_j \cap U_k, \tag{4.1}$$

and extend it to a condition of the form

$$g_{i,j}(b)g_{j,k}(b)g_{k,i}(b) = \lambda_{i,j,k}(b) \quad \text{for} \quad b \in U_i \cap U_j \cap U_k \tag{4.2}$$

where $\lambda_{i,j,k}(b)$ for $b \in U_i \cap U_j \cap U_k$ is a higher level of twisting in terms of the extra data of the 2-cocycle $\lambda_{i,j,k}(b)$ for $b \in U_i \cap U_j \cap U_k$. When this is formulated in more intrinsic terms, it comes under the name of "gerbe," which is an object glued together with by condition (2). If the condition (2) can be reduced to that of (1), then we have again a bundle.

The reason that algebra bundles lead to twisting conditions on change of coordinates in vector bundles is that the functions $g_{i,j}(b)$ relating two trivial algebra bundles over $U_i \cap U_j$ have values in PGL, while for vector bundles, the change of coordinates has values in GL. When the local change of coordinate functions $g_{i,j}(b)$ in PGL satisfying (1) above are lifted to $\alpha_{i,j}(b) \in GL$, the condition (1) takes the twisted form

$$\alpha_{i,j}(b)\alpha_{j,k}(b)\alpha_{k,i}(b) = \lambda_{i,j,k}(b) \quad \text{for} \quad b \in U_i \cap U_j \cap U_k, \tag{4.3}$$

where $\lambda_{i,j,k}(b)$ for $b \in U_i \cap U_j \cap U_k$ has its values in $GL(1) = \ker(GL \to PGL)$. This is an example of the higher level twisting in terms of the extra data of the $GL(1)$ valued 2-cocycle $\lambda_{i,j,k}(b)$ for $b \in U_i \cap U_j \cap U_k$. The characteristic class $\alpha(A) \in H^3(X,\mathbb{Z})$ is the image under the connecting morphism of the class $[\lambda_{i,j,k}] \in H^2(X, GL(1))$ coming from the long exact sequence associated to the exponential exact sequence $e(t) = \exp(2\pi i t)$

$$0 \longrightarrow \mathbb{Z} \longrightarrow \mathbb{C} \longrightarrow \mathbb{C}^* \longrightarrow 1.$$

With gerbes, we have other examples of this lifting to a twisted class as in the condition (3).

In the three chapters of this part, we illustrate this process from (1) to (3) with increasing generality. Finally, arriving at Grothendieck's approach in Chap. 27 with an introduction to stacks and gerbes.

All the topics in this Part V would merit a more extensive development, but we have included this sketch as an introduction.

Chapter 23
Bundle Gerbes

In Part V, we consider constructions, called gerbes, leading to an element of $H^3(X,\mathbb{Z})$, and these cohomology classes classify gerbes in some suitable sense. These constructions are other versions of the same phenomena related to algebra bundles, and they lead also to the general concept of twisted K-theory when properly stabilized.

We begin with a special type of gerbe, called a bundle gerbe. While a vector bundle is rather a direct generalization of vector space over a point, a bundle gerbe is a gerbe construction over the clutching data arising from a general bundle. It is included in the general concept of gerbe, see Chap. 27, but it leads directly to certain examples and illustrates some ideas around the concept of a gerbe.

Bundle gerbes are line bundles L over the fibre product $Y \times_B Y$ of a space (or bundle) Y over B together with a multiplication on L which uses the fibre product structure. While the first Chern class of the line bundle L over B is in $H^2(B,\mathbb{Z})$ and describes the complex line bundle up to isomorphism, the bundle gerbe has a class in $H^3(B,\mathbb{Z})$ which describes the gerbes coming from the bundles on Y over B up to isomorphism.

1 Notation for Gluing of Bundles

1.1. Definition Descent data for a bundle E over $Y = Z_0$ relative to the map $f : Y \to B$, see also 5(4.5), is a two-stage pseudosimplicial space

$$Z_2 = Y \times_B Y \times_B Y \xrightarrow{\ d_0,d_1,d_2\ } Z_1 = Y \times_B Y \xrightarrow{\ d_0,d_1\ } Z_0 = Y$$

together with an isomorphism $\alpha : d_1^*(E) \to d_0^*(E)$ of bundles over $Z_1 = Y \times_B Y$. The maps d_i refer to deleting the ith coordinate, and we require the following compatibility condition for the three induced versions of α on $Z_2 = Y \times_B Y \times_B Y$, that is, the commutativity of the diagram

D. Husemöller et al.: *Bundle Gerbes*, Lect. Notes Phys. **726**, 277–286 (2008)
DOI 10.1007/978-3-540-74956-1_24   © Springer-Verlag Berlin Heidelberg 2008

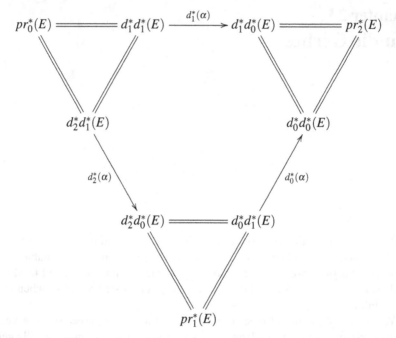

or we have the following formula

$$d_0^*(\alpha) d_2^*(\alpha) = d_1^*(\alpha).$$

Recall we have the following gluing theorem as in 5(4.6).

1.2. Theorem *For an étale map $f : Y \to B$, a bundle E over Y has descent data if and only if E is isomorphic to $f^{-1}(V)$, where V is a bundle over B.*

1.3. Remark We can speak about maps $f : Y \to B$ with descent data for vector bundles E over B, that is, for a given $\alpha : d_1^*(E) \to d_0^*(E)$ an isomorphism of vector bundles over $Z_1 = Y \times_B Y$ satisfying the compatibility conditions, and we have a unique vector bundle E' over B with $f^*(E')$ isomorphic to E.

Now, we consider the elementary conditions related to gluing as motivation for gerbes. These refer to an open covering of the base space together with the twofold and threefold intersections of the open sets.

1.4. Notation Let $Y = \coprod_{U \in \Phi} U$, where Φ is a covering of B by contractible open subsets, and let $\pi : Y \to B$ be the étale map induced by the inclusions $U \subset B$ for each $U \in \Phi$. Then, observe that we have the simplicial maps two-stage simplicial space

$$\coprod_{U,V,W \in \Phi} U \cap V \cap W = Y \times_B Y \times_B Y \xrightarrow{d_0, d_1, d_2} \coprod_{U,V \in \Phi} U \cap V = Y \times_B Y \xrightarrow{d_0, d_1} \coprod_{U \in \Phi} U = Y.$$

1.5. Zero-Order Gluing or Cocycle Condition Consider the cosimplicial object by mapping the simplicial space of (1.4) into a fixed space Z

$$\mathrm{Map}(B,Z) \xrightarrow{\ \delta\ } \mathrm{Map}(Y,Z) \xrightarrow{\ \delta_0,\delta_1\ } \mathrm{Map}(Y\times_B Y,Z) \xrightarrow{\ \delta_0,\delta_1,\delta_2\ } \mathrm{Map}(Y\times_B Y \times_B Y,Z),$$

where $\delta = \mathrm{Map}(\pi,Z)$ is induced by the étale projection $\pi : Y \to B$ and $\delta_i = \mathrm{Map}(d_i,Z)$. Now, the gluing statement is the equivalence between giving a map $f : B \to Z$ and giving a family of maps $(f_U) \in \mathrm{Map}(Y,Z) = \prod_{U\in\Phi} \mathrm{Map}(U,Z)$ satisfying the coboundary gluing condition which takes the form either

$$\delta_0(f_U) = \delta_1(f_U) \quad \text{or} \quad f_U|(U\cap V) = f_V|(U\cap V) \ \text{ for } U,V \in \Phi.$$

1.6. First-Order Gluing or Cocycle Condition Consider the cosimplicial object in (1.5) for $Z = S^1$. A line bundle L on B induces by π a trivial line bundle $\pi^*(L)$ isomorphic to $Y \times \mathbb{C}$ together with an isomorphism $d_1^*(Y \times \mathbb{C}) \to d_0^*(Y \times \mathbb{C})$ which is defined by $g : Y \times_B Y \to S^1$. The compatibility condition as in (1.1) has the form $\delta_1^*(g) = \delta_0^*(g)\delta_2^*(g)$. If we denote by $g_{U,V} = g|(U\cap V)$, then the compatibility becomes the two relations

(a)
$$g_{U,V} = g_{V,U}^{-1} \text{ on } U\cap V$$

and
(b)
$$g_{U,V}\,g_{V,W}\,g_{W,U} = 1 \text{ on } U\cap V\cap W.$$

This data $g : Y \times_X Y \to S^1$ or equivalently the collection of $g_{U,V} : U\cap V \to S^1$ is the descent data for the line bundle on B.

1.7. Second-Order Gluing or Cocycle Condition Consider again the cosimplicial object in (1.5) for $Z = S^1$ with a map $g : Y \times_B Y \to S^1$ or equivalently a family of maps $g_{U,V} : U\cap V \to S^1$. On $U\cap V\cap W$, we form the function $\lambda_{U,V,W} : U\cap V\cap W \to S^1$, or equivalently, $\lambda : Y \times_B Y \times_B Y \to S^1$ given by relation $\lambda_{U,V,W} = g_{U,V}\,g_{V,W}\,g_{W,U}$. We assume that λ has the following properties:

(1) $\lambda_{U,V,W} = \lambda_{U,W,V}^{-1} = \lambda_{W,V,U}^{-1} = \lambda_{V,U,W}^{-1}$ on $U\cap V\cap W$ and
(2) $\delta(\lambda)_{U,V,W,T} = \lambda_{V,W,T}\,\lambda_{U,W,T}^{-1}\,\lambda_{U,V,T}\,\lambda_{U,V,W}^{-1} = 1$ on $U\cap V\cap W\cap T$

for $U,V,W,T \in \Phi$. The function λ satisfying (1) and (2) is equivalent to a line bundle L on $Y \times_B Y$ satisfying the following:

(a) $\tau^*(L)$ and $L^\vee = L^{(-1)\otimes}$ are isomorphic, where τ interchanges the two factors of $Y \times_B Y$ and
(b) the line bundle $d_0^*(L) \otimes d_1^*(L)^{(-1)\otimes} \otimes d_2^*(L)$ is trivial.

2 Definition of Bundle Gerbes

For the definition of bundle gerbes, we use the previous fibre product notation related to gluing of bundles for the projections.

2.1. Fibre Product Three-Stage Simplicial Space We start with a general three stage simplicial space

$$Z_3 \xrightarrow{\ d_0,d_1,d_2,d_3\ } Z_2 \xrightarrow{\ d_0,d_1,d_2\ } Z_1 \xrightarrow{\ d_0,d_1\ } Z_0 .$$

In that case, $Z_3 = Y \times_B Y \times_B Y \times_B Y$, $Z_2 = Y \times_B Y \times_B Y$, and $Z_1 = Y \times_B Y$, $Z_0 = Y$ for a space Y over B, that is, a map $Y \to B$. The morphisms d_i refer to the projection which deletes the ith coordinate $0 \le i \le q$ on Z_q. These morphisms compose with the following relation

$$d_i d_j = d_{j-1} d_i \quad \text{for} \quad 0 \le i < j \le q \text{ on } Z_q .$$

2.2. Definition A bundle gerbe over a space B is a triple consisting of a locally trivial bundle $p : Y \to B$, a line bundle $L \to Y \times_B Y$ over the fibre product of Y with Y over B, and an isomorphism of line bundles

$$\gamma : d_2^*(L) \otimes d_0^*(L) \longrightarrow d_1^*(L)$$

on $Z_2 = Y \times_B Y \times_B Y$ satisfying an associative law on the fourfold product $Z_3 = Y \times_B Y \times_B Y \times_B Y$ which is the following commutative diagram where we use the cosimplicial relation $d_j^* d_i^* = d_i^* d_{j-1}^*$ for $i < j$ freely.

$$
\begin{array}{ccc}
d_3^*(d_2^*(L) \otimes d_0^*(L)) \otimes d_1^* d_0^*(L) & == & d_3^* d_2^*(L) \otimes d_0^*(d_2^*(L) \otimes d_0^*(L)) \\
\Big\downarrow{\scriptstyle d_3^*(\gamma) \otimes d_1^* d_0^*(L)} & & \Big\downarrow{\scriptstyle d_3^* d_2^*(L) \otimes d_0^*(\gamma)} \\
d_3^* d_1^*(L) \otimes d_1^* d_0^*(L) & & d_3^* d_2^*(L) \otimes d_0^* d_1^*(L) \\
\Big\| & & \Big\| \\
d_1^*(d_2^* \otimes d_0^*(L)) & & d_2^*(d_2^*(L) \otimes d_0^*(L)) \\
\Big\downarrow{\scriptstyle d_1^*(\gamma)} & & \Big\downarrow{\scriptstyle d_2^*(\gamma)} \\
d_1^* d_1^*(L) & == & d_2^* d_1^*(L)
\end{array}
$$

2.3. Definition A morphism of bundle gerbes is a triple consisting of a map $f : B' \to B$, an f-bundle morphism $u : Y' \to Y$, and a line bundle morphism $w : L' \to L$ such that over $u \times_{B'} u \times_{B'} u$ we have a commutative diagram

$$
\begin{array}{ccc}
d_2^*(L') \otimes d_0^*(L') & \xrightarrow{\;\;\gamma\;\;} & d_1^*(L') \\
{\scriptstyle d_2^*(w) \otimes d_0^*(w)} \downarrow & & \downarrow {\scriptstyle d_1^*(w)} \\
d_2^*(L) \otimes d_0^*(L) & \xrightarrow{\;\;\gamma\;\;} & d_1^*(L).
\end{array}
$$

If f is the identity on B, then u is a B-morphism of bundle gerbes.

2.4. Remark The composition of morphisms of bundle gerbes is again a morphism of bundle gerbes. We can speak of the category (bun/gerbes) of bundle gerbes and the category (bun/gerbes/B) of bundle gerbes over B.

2.5. Remark The categories (bun/gerbes) and (bun/gerbes/B) have finite products beginning with the product of bundles over B and then the fibrewise tensor product of the line bundles on the product of their base space. For the bundle gerbes $Y' \to B'$ with line bundle L' on $Y' \times_{B'} Y'$ and $Y'' \to B''$ with line bundle L'' on $Y'' \times_{B''} Y''$, the product $Y' \times Y'' \to B' \times B''$ has the line bundle $q'^*(L') \otimes q''^*(L'')$ on $(Y' \times Y'') \times_{B' \times B''} (Y' \times Y'') = (Y' \times_{B'} Y') \times (Y'' \times_{B''} Y'')$, where q' and q'' are the projections to the first and second factors.

2.6. Example For a bundle $p : Y \to B$ and a line bundle J on Y by using the dual $J^\vee$, we form the line bundle $L = d_1^*(J) \otimes d_0^*(J^\vee)$ and define the following bundle gerbe $\delta(J)$ as p and L with the natural morphism given by

$$
\begin{array}{ccc}
\gamma : d_2^*(d_1^*(J) \otimes d_0^*(J^\vee)) \otimes d_0^*(d_1^*(J) \otimes d_0^*(J^\vee)) & \longrightarrow & d_1^*(d_1^*(J) \otimes d_0^*(J^\vee)) \\
\| & & \| \\
d_2^* d_1^*(J) \otimes d_2^* d_0^*(J^\vee) \otimes d_0^* d_1^*(J) \otimes d_0^* d_0^*(J^\vee) & \longrightarrow & d_1^* d_1^*(J) \otimes d_1^* d_0^*(J^\vee) \\
\| & & \| \\
d_2^*(L) \otimes d_0^*(L) & \longrightarrow & d_1^*(L)
\end{array}
$$

Here, we have used relations $d_2^* d_1^* = d_1^* d_1^*$, $d_2^* d_0^* = d_0^* d_1^*$, and $d_0^* d_0^* = d_1^* d_0^*$.

2.7. Definition A bundle gerbe $Y \to B$ with L on $Y \times_B Y$ is trivial provided there exists a line bundle J on Y with L isomorphic to $\delta(J)$.

3 The Gerbe Characteristic Class

3.1. Remark Now, we define the characteristic class of a bundle gerbe. The first Chern class of a line bundle L on B is given by choosing an open covering $U_i, i \in I$

and sections $s_i : U_i \to L|U_i$ which are never zero. Then on the intersection $U_i \cap U_j$, we have relations between the everywhere nonzero sections of the form $s_i = g_{ij} s_j$, where $g_{ij} : U_i \cap U_j \to C^*$. On the intersection $U_i \cap U_j \cap U_k$, the restrictions of the section changing functions satisfy

$$g_{ij} g_{jk} g_{ki} = 1 \quad \text{or} \quad g_{ij} g_{jk} = g_{ik}.$$

The inclusions $U_i \to B$ induce a map $f : Y = \coprod_{i \in I} U_i \to B$, and $f^{-1}(L) = M$ is trivial on Y. The bundle L is the result of the gluing isomorphism $d_1^*(M) \to d_0^*(M)$ on $Y \times_B Y$ which is just multiplication by g_{ij} on $U_i \times_B U_j$ isomorphic to $U_i \cap U_j$. Recall $\{g_{ij}\}$ gives the Chern class $c_1(L)$.

3.2. Definition A bundle gerbe over B with map $p : Y \to B$ and line bundle $L \to Y \times_B Y$ is locally trivial provided there exists an open covering $U_i, i \in I$ of B and sections $s_i : U_i \to Y$ of p such that on $U_{ij} = U_i \cap U_j$ the induced line bundle $(s_i, s_j)^{-1}(L)$ is trivial. The sections $s_i : U_i \to Y$ are called trivializing sections of the gerbe, and they can be collected to $Z = Z_0 = \coprod_{i \in I} U_i$ giving a map $Z \to B$ and a lifting $s : Z \to Y$ over B such that $(s \times_Z s)^{-1}(L) = M$ is a trivial line bundle over $Z \times_B Z$.

3.3. Definition Consider a locally trivial bundle gerbe over B with map $p : Y \to B$, line bundle $L \to Y \times_B Y$ over the fibre product of Y over B, and an isomorphism of line bundles

$$\gamma : d_2^*(L) \otimes d_0^*(L) \longrightarrow d_1^*(L)$$

on $Y \times_B Y \times_B Y$ satisfying the above associative law on the fourfold product $Y \times_B Y \times_B Y \times_B Y$. The local cocycle condition with the trivializing sections $s_i : U_i \to Y$ of the gerbe is given by inducing the isomorphism γ to $s^{-1}(\gamma) : d_2^*(M) \otimes d_0^*(M) \to d_1^*(M)$ on $Z \times_B Z \times_B Z$ given by a function $a : Z \times_B Z \times_B Z \to C^*$ or equivalently a family of complex valued functions $a_{ijk} : U_{ijk} \to C^*$ satisfying the relations

$$g_{ij} g_{jk} g_{ki} = a_{ijk} \quad \text{or} \quad g_{ij} g_{jk} = a_{ijk} g_{ik}.$$

3.4. Remark The cocycle condition on the $a_{ijk} : U_{ijk} \to C^*$ is derived from the associative law condition in the isomorphism γ. The result is a continuous cohomology class or equivalently a sheaf cohomology class $\{a_{ijk}\}$ in $H^2(B, C^*)$.

3.5. Definition The gerbe cohomology class of a locally trivial gerbe is the cohomology class in $H^3(B, \mathbb{Z})$ coming from the class $\{a_{ijk}\} \in H^2(B, C^*)$ under the coboundary operator arising from the short exact sequence

$$0 \longrightarrow \mathbb{Z} \longrightarrow \mathbb{C} \longrightarrow C^* \longrightarrow 1.$$

3.6. Remark The notion of continuous cohomology $H^q(X,A)$ with values in a topological abelian group can be defined by the iterated classifying space construction $B^q(A) = B(B^{q-1}(A))$. This means that $B(A)$ is naturally an abelian topological group if A is an abelian topological group. This can be carried out in the simplicial language which we have not mentioned up to now or by a modification of the Milnor construction by Milgram.

3.7. Notation For a bundle gerbe $L \to Y \times_B Y$, also denoted later by $L/Y \times_B Y$, the corresponding characteristic class is denoted by

$$\alpha(L/Y \times_B Y) \in H^3(B,\mathbb{Z}).$$

The elementary properties of α are the following

3.8. Remark We have $\alpha(L^\vee/Y \times_B Y) = -\alpha(L/Y \times_B Y)$ and for the tensor product $\alpha(L' \otimes L''/Y \times_B Y) = \alpha(L'/Y \times_B Y) + \alpha(L''/Y \times_B Y)$.

With respect to morphism, we have the naturality property directly from the definitions.

3.9. Proposition *Let $u : Y' \to Y''$ be a bundle map over a map of the base space $f : B' \to B''$, and let $w : L' \to L''$ be a $u \times_B u$ map of line bundles. Then, we have*

$$f^*(\alpha(L''/Y'' \times_{B''} Y'')) = \alpha(L'/Y' \times_{B'} Y').$$

4 Stability Properties of Bundle Gerbes

Bundle gerbes have stability properties relative to trivial bundle gerbes just as vector bundles have stability properties relative to trivial bundles. The bundle gerbe characteristic class $\alpha(L/Y \times_B Y) \in H^3(B,\mathbb{Z})$ gives a complete classification as in the case of $\mathscr{B}(H)$-algebra bundles A over X for an infinite-dimensional Hilbert space H, where the characteristic class is denoted by $\alpha(A) \in H^3(B,\mathbb{Z})$, see 20(4.12).

4.1. Proposition *A bundle gerbe $L/Y \times_B Y$ is a trivial bundle gerbe if and only if $\alpha(L/Y \times_B Y) = 0$ in $H^3(B,\mathbb{Z})$.*

For the proof, note that a trivialization of $L/Y \times_B Y$ gives a coboundary for the cocycle $\alpha(L/Y \times_B Y)$ and conversely, a coboundary relation for the cocycle is the descent data for the trivialization of the gerbe.

4.2. Definition Two bundle gerbes $L'/Y' \times_{B'} Y'$ and $(L''/Y'' \times_{B''} Y'')$ are stably isomorphic provided that $L' \otimes T'$ and $L'' \otimes T''$ are isomorphic for trivial bundle gerbes T and T' or equivalently $L' \otimes (L'')^\vee$ is a trivial gerbe.

4.3. Proposition *The function which assigns to a stable isomorphism class $[L/Y \times_B Y]$ of bundle gerbes the characteristic class $\alpha(L/Y \times_B Y) \in H^3(B,\mathbb{Z})$ is a bijection.*

For the proof, we use an extension of the idea of (4.1).

5 Extensions of Principal Bundles Over a Central Extension

In geometry, structure groups of bundles usually are reduced by putting additional structure on the basic geometric construction. This is the case with vector bundles where by introducing a metric the structure groups are reduced from $GL(n,\mathbb{C})$ to the compact $U(n)$. The fact that it is always possible to reduce the structure group from the general linear group to the unitary group is related to the fact that all vector bundles can be endowed with a metric.

In physics, structure groups of bundles are changed by going from the symmetries of a classical system which are representations to the symmetries of the related quantized system which are projective representations. This change in structure group is measured by a characteristic class.

5.1. Notation Let

$$ 1 \longrightarrow \mathbb{C}^* \longrightarrow G' \longrightarrow G \longrightarrow 1 $$

be a central extension with classifying space sequence

$$ \dots \longrightarrow G \longrightarrow B(\mathbb{C}^*) \longrightarrow B(G') \xrightarrow{\gamma} B(G) \xrightarrow{\beta} B(B(\mathbb{C}^*)) = K(\mathbb{Z},3). $$

As mentioned in (3.6), the double classifying space $B(B(\mathbb{C}^*))$ can be defined, and the map β and the map $G \to B(\mathbb{C}^*)$ come from a similar construction.

5.2. Extension of Structure Group Class Let P be a principal G-bundle over B with a classifying map $f_P : B \to B(G)$. We look for a G' principal bundle which factors $P' \to P \to B$, where the first map is $G' \to G$ equivariant. In particular, the quotient map $P'/\mathbb{C}^* \to P$ is defined and G-equivariant. This principal G'-bundle is classified by a map $f_{P'} : B \to BG'$ where $\gamma f_{P'} = f_P$ up to homotopy, see 7(3.3). Since we have a fibre sequence, we have the following assertion:

5.3. Proposition *The principal G-bundle P has a lifting to P' if and only if the three-dimensional cohomology class defines βf_P as 0 in $H^3(B,\mathbb{Z})$.*

We denote the obstruction to lifting by $\alpha(P) = \beta f_P$ in $H^3(B,\mathbb{Z})$.

These considerations resemble the role of w_1 for an orientation of a vector bundle, see 12(2.3).

6 Modules Over Bundle Gerbes and Twisted K-Theory

Starting with a $\mathscr{B}(H)$ algebra bundle, we introduce the related twisted K-theory by using the action of a bundle gerbe on gerbe modules. The stability theory of gerbe modules gives another formulation of twisted K-theory, see 22(3.2).

Recall that a bundle gerbe $L \to Y \times_B Y$ or just $L/Y \times_B Y$ has a structure morphism $\gamma : d_2^*(L) \otimes d_0^*(L) \to d_1^*(L)$ on $Y \times_B Y \times_B Y$ satisfying an associative law on the four-fold product $Y \times_B Y \times_B Y \times_B Y$.

6.1. Definition A module over $L/Y \times_B Y$ is a pair consisting of a vector bundle $E \to Y$ together with a vector bundle isomorphism $\sigma : L \otimes d_0^*(E) \to d_1^*(E)$ satisfying the associative law which is the following commutative diagram where we use the cosimplicial relations $d_j^* d_i^* = d_i^* d_{j-1}^*$ for $i < j$ freely.

$$
\begin{array}{ccc}
(d_2^*(L) \otimes d_0^*(L)) \otimes d_1^* d_0^*(E) & =\!=\!= & d_2^*(L) \otimes d_0^*(L \otimes d_0^*(E)) \\
\scriptstyle{\gamma \otimes d_1^* d_0^*(E)} \downarrow & & \downarrow \scriptstyle{d_2^*(L) \otimes d_0^*(\sigma)} \\
d_1^*(L) \otimes d_1^* d_0^*(E) & & d_2^*(L) \otimes d_0^* d_1^*(E) \\
\| & & \| \\
d_1^*(L \otimes d_0^*(E)) & & d_2^*(L \otimes d_0^*(E)) \\
\scriptstyle{d_1^*(\sigma)} \downarrow & & \downarrow \scriptstyle{d_2^*(\sigma)} \\
d_1^* d_1^*(E) & =\!=\!= & d_2^* d_1^*(E)
\end{array}
$$

The dimension of the module is the dimension of the vector bundle.

A morphism $f : E' \to E''$ of modules over the bundle gerbe $L/Y \times_B Y$ is a vector bundle morphism f over Y such that the following diagram of action morphisms is commutative.

$$
\begin{array}{ccc}
L \otimes d_0^*(E') & \xrightarrow{\ \sigma'\ } & d_1^*(E') \\
\scriptstyle{d_0^*(f)} \downarrow & & \downarrow \scriptstyle{d_1^*(f)} \\
L \otimes d_0^*(E'') & \xrightarrow{\ \sigma''\ } & d_1^*(E'')
\end{array}
$$

The composition of module morphisms over the bundle gerbe $L/Y \times_B Y$ as vector bundle morphisms is again a morphism of modules over the gerbe $L/Y \times_B Y$. Let $\mathrm{Mod}(L/Y \times_B Y)$ denote the category of modules over the gerbe $L/Y \times_B Y$ with the operations of direct sum $E' \oplus E''$ and tensor product $E' \otimes E''$ defined in the obvious manner.

6.2. Notation Let $\mathrm{Iso}(L/Y \times_B Y)$ denote the set of isomorphism classes of elements in $\mathrm{Mod}(L/Y \times_B Y)$ with the semiring structure given by direct sum and tensor product.

6.3. Proposition *If a gerbe $L/Y \times_B Y$ has a module E of dimension n, then $n \cdot \alpha(L/Y \times_B Y) = 0$ in $H^3(B, \mathbb{Z})$.*

Proof. When E is one dimensional, then the action σ defines a trivialization of $L/Y \times_B Y$ of L, and if E is n dimensional, then $L^{n\otimes}/Y \times_B Y$ acts on the nth exterior power $\Lambda^n E$ which is one dimensional, and thus $L^{n\otimes}/Y \times_B Y$ is trivializable so that $0 = \alpha(L^{n\otimes}/Y \times_B Y) = n \cdot \alpha(L/Y \times_B Y)$. This proves the proposition.

6.4. Proposition *A trivialization of $L/Y \times_B Y$ defines a semiring morphism Mod $(L/Y \times_B Y) \to Vect(B)$. Further, two stably equivalent gerbes $(L'/Y' \times_B Y')$ and $(L''/Y'' \times_B Y'')$ have isomorphic semiring $Mod(L'/Y' \times_B Y')$ and $Mod(L''/Y'' \times_B Y'')$.*

Observe that the action of a trivialized gerbe is just descent data for a bundle, and stabilizing relations gives also descent data for comparing the module categories.

6.5. Definition Let B be a compact space. For a given torsion element $\alpha \in H^3(B, \mathbb{Z})$ the twisted K-theory constructed by bundle gerbes is the Grothendieck group $K_\alpha(B)$ of the semigroup $\mathrm{Iso}(L/Y \times_B Y)$ where $\alpha = \alpha(L/Y \times_B Y)$.

For the comparison of this definition of twisted K-theory with 22(3.2), we use the class defined for the extension in (5.1)

$$1 \longrightarrow \mathbb{C}^* \longrightarrow G' \longrightarrow G \longrightarrow 1,$$

where $G' = GL(H)$ and $G = PGL(H)$. An algebra bundle has the form $P[\mathscr{B}(H)]$, where P is a principal $U(H)$-bundle or $\mathscr{B}(H)$-bundle. This has a clear relation to the gerbe, and the sections of the Fredholm subbundle is related to gerbe modules.

We will not go into this further here, but the interested reader can find the details and also the discussion of the nontorsion case in the reference: Bouwknegt et al. (2002). In the references of this paper, one will find articles developing the theory of bundle gerbes including the first paper of Murray.

Reference

Bouwknegt, P., Carey, A.L., Mathai, V., Murray, M.K., Stevenson, D.: Twisted K-theory and K-theory of bundle gerbes. Comm. Math. Phys. **228**: 17–45 (2002)

Chapter 24
Category Objects and Groupoid Gerbes

We introduce the notions of category objects and groupoid objects in a category $\mathscr{C}$ with finite fibre products. A groupoid object is a category object with an inversion structure morphism which sometimes is unique. In this case, a groupoid is a category object with an axiom, otherwise it is a category with additional structure. For this, we begin with n-level pseudosimplicial objects in a category in order to formulate the notion of a category object. These have been used already for a description of local triviality of bundles and the construction of bundles from local data or from descent data. Then, we go further with simplicial objects, geometric realization, the nerve of a category, and the final step in the detour away from gerbes to the definition of algebraic K-theory. All this illustrates the vast influence of these general concepts in mathematics for which there are more and more applications to physics.

Groupoid gerbes are line bundles over the morphism space $G(1)$ in a groupoid $G(*)$, which in case $G = G(1)$ is a group, correspond to central extensions. In the case of bundle gerbes, these groupoid gerbes are line bundles L over the fibre product $Y \times_B Y$ of a space Y over B together with a multiplication which uses the fibre product structure. Moreover, for bundle gerbes, there is a line bundle isomorphism on $Y \times_B Y \times_B Y$ with an associativity condition on $Y \times_B Y \times_B Y \times_B Y$, and for groupoid gerbes, there is a line bundle isomorphism on $G(1) \times_{G(0)} G(1)$ with an associativity condition on

$$G(1) \times_{G(0)} G(1) \times_{G(0)} G(1).$$

The line bundle on $G(1)$ usually is taken to be symmetric with respect to the inverse mapping of the groupoid.

1 Simplicial Objects in a Category

In several places, we have already used the formalism of part of a simplicial object to organize data and axioms especially for the local description of bundles, see 5(4.5) and 25(1.1), for the basic ingredients of a bundle gerbe 25(2.2). We will use these concepts to describe the category objects and groupoid objects in a category $\mathscr{C}$.

D. Husemöller et al.: *Category Objects and Groupoid Gerbes*, Lect. Notes Phys. **726**, 287–301 (2008)
DOI 10.1007/978-3-540-74956-1_25   © Springer-Verlag Berlin Heidelberg 2008

1.1. Definition An n-level pseudosimplicial object in a category $\mathscr{C}$ is a sequence of objects Z_q for $0 \leq q \leq n$ in $\mathscr{C}$ together with morphisms $d_i : Z_q \to Z_{q-1}$ for $0 \leq i \leq q$ in $\mathscr{C}$ such that

$$d_i d_j = d_{j-1} d_i : Z_q \longrightarrow Z_{q-2}$$

for $0 \leq i < j \leq q$. A morphism of n-level pseudosimplicial objects is a sequence of morphisms $f_q : Z_q' \to Z_q''$ for $0 \leq q \leq n$ such that

$$d_i f_n = f_{n-1} d_i \quad \text{for} \quad 1 \leq i \leq n.$$

The reason for the term pseudosimplicial instead of simplicial is that a simplicial object $Z_{\bullet}$ will also have morphisms $s_j : Z_q \to Z_{q+1}$ for $0 \leq j \leq q$ with suitable relations.

We can display every morphism and object easily up to three levels

$$Z_3 \xrightarrow{\ d_0, d_1, d_2, d_3\ } Z_2 \xrightarrow{\ d_0, d_1, d_2\ } Z_1 \xrightarrow{\ d_0, d_1\ } Z_0.$$

and for descent questions for a bundle E over $Y = Z_0$, the terms arise as fibre products relative to a map $f : Y \to B$

$$Z_2 = Y \times_B Y \times_B Y \xrightarrow{\ d_0, d_1, d_2\ } Z_1 = Y \times_B Y \xrightarrow{\ d_0, d_1\ } Z_0 = Y.$$

This extends further with $Z_3 = Y \times_B Y \times_B Y \times_B Y$, $Z_2 = Y \times_B Y \times_B Y$, $Z_1 = Y \times_B Y$, and $Z_0 = Y$, the morphisms d_i refer to the projection which deletes the ith coordinate $0 \leq i \leq q$ on Z_q.

1.2. Definition Associated to a first level or first stage pseudosimplicial object $Z_1 = Z(1) \xrightarrow{d_0, d_1} Z(0) = Z_0$ in a category, we have an object, namely the fibre product denoted by $Z(1) *_{Z_0} Z(1) = Z(1)_{d_0}(\times_{Z(0)})_{d_1} Z(1)$ which has a first and second projection denoted by $pr_1 = d_2 : Z(1) *_{Z_0} Z(1) \to Z(1)$ and $pr_2 = d_0 : Z(1) *_{Z_0} Z(1) \to Z(1)$, respectively. Note that d_2 is the last factor deleted and d_0 is the first factor deleted in the case $Z(1) = Z(0) \times_B Z(0)$ for $f : Z(0) \to B$.

We have used pseudosimplicial objects of order n to organize gluing data, and now we consider the complete concept of a simplicial object, and in the case of simplicial set, we define a filtered space called the geometric realization.

1.3. Definition A simplicial object $X_{\bullet}$ in a category $\mathscr{C}$ is a sequence of objects X_n for $n \geq 0$ in $\mathscr{C}$ together with morphisms in $\mathscr{C}$ defined by

$$d_i(X) = d_i : X_n \longrightarrow X_{n-1} \quad \text{and} \quad s_j(X) = s_j : X_n \longrightarrow X_{n+1}$$

for $0 \leq i, j \leq n$ such that the following relations hold

(1) $\qquad\qquad\qquad\qquad d_i d_j = d_{j-1} d_i \ (i < j)$

(2) $\qquad\qquad\qquad\qquad s_i s_j = s_j s_{i-1} \ (i > j)$

and

(3)
$$d_i s_j = \begin{cases} s_{j-1} d_i & (i < j) \\ 1 & (i = j, j+1) \\ s_j d_{i-1} & (i > j+1). \end{cases}$$

1.4. Definition A morphism $f_\bullet : X_\bullet \to Y_\bullet$ of simplicial objects in a category is a sequence of morphisms $f_n : X_n \to Y_n$ satisfying the relations

$$d_i(Y) f_n = f_{n-1} d_i(X) \quad \text{and} \quad s_i(Y) f_n = f_{n+1} s_i(X)$$

with $0 \le i, j \le n$.

1.5. Remark The composition of two morphisms $f_\bullet : X_\bullet \to Y_\bullet$ and $g_\bullet : Y_\bullet \to Z_\bullet$ defined by the relation $(gf)_n = g_n f_n$ is a morphism $X_\bullet \to Z_\bullet$ of simplicial objects in the category $\mathscr{C}$. With these definitions, we have the category $\Delta(\mathscr{C})$ of simplicial objects in $\mathscr{C}$.

We introduce a topological space associated to a simplicial set, or in other words, we define a functor $R : \Delta(\text{set}) \to (\text{top})$ from the category of simplicial sets to the category of topological spaces, called geometric realization using affine simplexes.

1.6. Definition The affine simplex $A(n)$ is the closed subset of the compact cube $[0,1]^{n+1} \subset \mathbb{R}^{n+1}$ consisting of all points $(t_0, \ldots, t_n) \in [0,1]^{n+1} \subset \mathbb{R}^{n+1}$ with $t_0 + \ldots + t_n = 1$. The coordinates t_i of these points in $A(n)$ are called barycentric coordinates.

1.7. Remark The boundary $\partial A(n)$ of $A(n)$ is related to the cosimplicial operations $\delta_i : A(n-1) \to A(n)$ defined for $0 \le i \le n$ in barycentric coordinates by

$$\delta_i(t_0, \ldots, t_{n-1}) = (t_0, \ldots, t_{i-1}, 0, t_i, \ldots, t_{n-1}).$$

The boundary $\partial A(n) = \bigcup_{i \in [n]} \text{im}(\delta_i)$ is a union of the images of the face morphisms δ_i for $i \in [n] = \{0, 1, \ldots, n\}$.

1.8. Remark The iterates of the cosimplicial operations satisfy $\delta_j \delta_i = \delta_i \delta_{j-1}$: $A(n-2) \to A(n)$ for $0 \le i < j \le n$, where both compositions will take $(t_0, \ldots, t_{n-2})$ to $(t_0, \ldots, t_{i-1}, 0, t_i, \ldots t_{j-1}, 0, t_j, \ldots, t_{n-2})$. The other cosimplicial operations σ_i : $A(n+1) \to A(n)$ for $0 \le j \le n$ are defined by the formulas

$$\sigma_j(t_0, \ldots, t_{n+1}) = (t_0, \ldots, t_{j-1}, t_j + t_{j+1}, t_{j+2}, \ldots, t_{n+1}).$$

1.9. Definition The geometric realization $R(X_\bullet)$ of a simplicial set $X_\bullet$ is the quotient of the disjoint union $\coprod_{0 \le n} X_n \times A(n)$ by the simplicial relations which are generated by

$$(d_i(x), t) \sim (x, \delta_i(t)) \quad \text{and} \quad (s_j(x'), t') \sim (x', \sigma_j(t'))$$

for $x \in X_n$, $t \in A(n-1)$, $x' \in X_{n-1}$, $t' \in A(n+1)$ for $0 \le i, j \le n$.

1.10. Remark The homology and cohomology of the space $R(X_\bullet)$ can be calculated in terms of the simplicial set $X_\bullet$. The geometric realization $R(X_\bullet)$ is a filtered space with $R(X_\bullet)_k$ being the image of $\coprod_{0 \le n \le k} X_n \times A(n)$ in $R(X_\bullet)$, and level

$R(X_\bullet)_k / R(X_\bullet)_{k-1}$ is a wedge of k-spheres each one for each element of $X_k -$ $\coprod_{0 \leq j \leq k} s_j(X_{k-1})$, the set of nondegenerate k-simplexes in $X_\bullet$.

Thus, a simplicial set generates a filtered space whose homology is controlled by the combinatorial topology of the simplicial set.

2 Categories in a Category

This is a short introduction to the concept of category in a form leading to the definition of category objects in a category $\mathscr{X}$.

2.1. Small Categories as Pairs of Sets Let $\mathscr{C}$ be a small category which means that the class of objects $C(0)$ is a set. Form the set $C(1)$ equal to the disjoint union of all $\mathrm{Hom}(X,Y)$ for $X, Y \in C(0)$. These two sets are connected by several functions. Firstly, we have the domain (left) and ring (right) functions $l, r : C(1) \to C(0)$ defined by the requirement that

$$l(\mathrm{Hom}(X,Y)) = \{X\} \quad \text{and} \quad r(\mathrm{Hom}(X,Y)) = \{Y\}$$

on the disjoint union. For $C(1)$, we have $f : l(f) \to r(f)$ is a notation for the morphism f in C.

Secondly, we have an identity morphism for each object of $\mathscr{C}$ which is a function $e : C(0) \to C(1)$ having the property that le and re are the identities on $C(0)$. Here, $e(X) = \mathrm{id}_X : X \to X$ for $e \in \mathrm{Hom}(X,X)$.

Thirdly, we have composition gf of two morphisms f and g but only in case where $r(f) = l(g)$. Hence, composition is not defined in general on the entire product $C(1) \times C(1)$, but it is defined on all subsets of the form $\mathrm{Hom}(X,Y) \times \mathrm{Hom}(Y,Z) \in C(1) \times C(1)$. This subset is called the fibre product of $r : C(1) \to C(0)$ and $l : C(1) \to C(0)$ consisting of pairs $(f,g) \in C(1) \times C(1)$ where $r(f) = l(g)$. The fibre product is denoted by $C(1)_r \times_{C(0)l} C(1)$ with two projections $r, l : C(1)_r \times_{C(0)l} C(1) \to C(0)$ defined by

$$l(f,g) = l(f) \quad \text{and} \quad r(f,g) = r(g).$$

Then, composition is defined by $m : C(1)_r \times_{C(0)l} C(1) \to C(1)$, where $m(f,g) = gf$ satisfying $lm(f,g) = l(f,g) = l(f)$ and $rm(f,g) = r(f,g) = r(g)$. Now, the reader can supply the unit and associativity axioms.

Fourthly, the notion of opposite category $\mathscr{C}^{\mathrm{op}}$ where $f : X \to Y$ in $\mathscr{C}$ becomes $f^{\mathrm{op}} : Y \to X$ in $\mathscr{C}^{\mathrm{op}}$, and $(gf)^{\mathrm{op}} = f^{\mathrm{op}} g^{\mathrm{op}}$ can be described as $\mathscr{C}^{\mathrm{op}} = (C(0), C(1), e, l^{\mathrm{op}} = r, r^{\mathrm{op}} = l, m^{\mathrm{op}} = m\tau)$, where τ is the flip in the fibre product $\tau : \mathscr{C}(1)_r \times_{C(0)l} C(1) \to C(1)_r \times_{C(0)l} C(1)$.

2.2. Remark We rewrite the previous setup where left l is replaced by d_1 deleting the first factor and right r is replaced by d_0 deleting the 0-factor in a pair using the second-level pseudosimplicial notation of the previous section $l = d_1$,

$r = d_0 : C(1) \to C(0)$. Furthermore, we have the fibre product mixing d_0 and d_1 giving $C(1) *_{C(0)} C(1) = C(1)_r \times_{C(0)} {}_l C(1)$ with the projection on the second and first factors $d_0, d_2 : C(1) *_{C(0)} C(1) \to C(1)$, respectively. A category object structure fills in this picture with a morphism $d_1 : C(1) *_{C(0)} C(1) \to C(1)$ extending the structure to a second-level pseudosimplicial object where $C(2) = C(1) *_{C(0)} C(1)$ giving $d_0, d_1, d_2 : C(2) \to C(1)$. In addition, we have a base or identity morphism $e : C(0) \to C(1)$.

2.3. Definition Let $\mathscr{C}$ be a category with finite fibre products. A category object $C(*)$ in $\mathscr{C}$ is a sextuple $(C(0), C(1), l, r, e, m)$ consisting of two objects, $C(0)$ and $C(1)$, and four morphisms

(1) $l, r : C(1) \to C(0)$ called domain (left) and range (right)
(2) $e : C(0) \to C(1)$ a unit morphism, and
(3) $m : C(1)_r \times_{C(0)} {}_l C(1) \to C(1)$ called multiplication or composition satisfying the following axioms:

(Cat1) The compositions le and re are the identities in $C(0)$.
(Cat2) Domain and range are compatible with multiplication

$$
\begin{array}{ccc}
C(1)_r \times_{C(0)} {}_l C(1) & \xrightarrow{\ m\ } & C(1) \\[4pt]
{\scriptstyle pr_1/pr_2}\Big\downarrow & & \Big\downarrow {\scriptstyle l/r} \\[4pt]
(C(1)) & \xrightarrow{\ l/r\ } & C(0)
\end{array}
$$

(Cat3) (associativity) The following diagram is commutative

$$
\begin{array}{ccc}
C(1)_r \times_{C(0)} {}_l C(1)_r \times_{C(0)} {}_l C(1) & \xrightarrow{\ m \times C(1)\ } & C(1)_r \times_{C(0)} {}_l C(1) \\[4pt]
{\scriptstyle C(1) \times m}\Big\downarrow & & \Big\downarrow {\scriptstyle m} \\[4pt]
C(1)_r \times_{C(0)} {}_l C(1) & \xrightarrow{\qquad m \qquad} & C(1)
\end{array}
$$

(Cat4) (unit property of e) $m(C(1), er)$ and $m(le, C(1))$ are each identities on $C(1)$.

2.4. Definition A morphism $u(*) : C'(*) \to C''(*)$ from the category object $C'(*)$ in $\mathscr{C}$ to the category object $C''(*)$ in $\mathscr{C}$ is a pair of morphisms $u(0) : C'(0) \to C''(0)$ and $u(1) : C'(1) \to C''(1)$ commuting with the four structure morphisms of $C'(*)$ and $C''(*)$. The following diagrams have to be commutative

$$C'(0) \xrightarrow{e'} C''(1) \qquad C'(1) \xrightarrow{l'/r'} C''(0)$$

$$u(0) \downarrow \qquad \downarrow u(1) \qquad u(1) \downarrow \qquad \downarrow u(0)$$

$$C''(0) \xrightarrow{e''} C''(1) \qquad C''(1) \xrightarrow{l''/r''} C''(0)$$

and

$$C'(1)_r \times_{C'(0)} {}_l C'(1) \xrightarrow{m'} C'(1)$$

$$u(1) \times_{u(0)} u(1) \downarrow \qquad \qquad u(1) \downarrow$$

$$C''(1)_r \times_{C''(0)} {}_l C''(1) \xrightarrow{m''} C''(1)$$

2.5. Example A category object $C(*)$ in the category of sets (set) is a small category as in (1.1).

2.6. Example A category object $C(*)$ with $C(0)$, the final object in $\mathscr{C}$, is just a monoidal object $C(1)$, object together with a multiplication, in the category $\mathscr{C}$.

2.7. Example The pair of morphisms consisting of the identities $u(0) = C(0)$ and $C(1) = u(1)$ is a morphism $u(*) : C(*) \to C(*)$ of a category object called the identity morphism on $C(*)$. If $u(*) : C'(*) \to C''(*)$ and $v(*) : C''(*) \to C(*)$ are two morphisms of category objects in $\mathscr{C}$, then $(vu)(*) : C'(*) \to C(*)$ defined by $(vu)(0) = v(0)u(0)$ and $(vu)(1) = v(1)u(1)$ is a morphism of category objects in $\mathscr{C}$.

2.8. Definition With the identity morphisms and the composition of morphisms in (2.7), we see that the category objects in $\mathscr{C}$ and the morphisms of category objects form a category called cat($\mathscr{C}$), that is, the category of category objects in $\mathscr{C}$. There is a full subcategory mon($\mathscr{C}$) of cat($\mathscr{C}$) consisting of those categories $C(*)$, where $C(0)$ is the final object in $\mathscr{C}$. This is the category of monoids in the category $\mathscr{C}$.

Now, we return to the point of view of pseudosimplicial objects as a perspective on category objects.

2.9. Remark The category cat(set) of category objects in the category (set) of sets is just the category of small categories where morphisms of categories are functors and composition is composition of functors. Also mon(set) is just the category of monoids. There is an additional structure of equivalence between morphisms as natural transformation of functors, and this leads to the notion of 2-category which we will not go into here.

2.10. Remark Let $C(*) = (C(0), C(1), l, r, e, m)$ be a category object in a category $\mathscr{C}$. In (2.2), we have identified the first two terms in a pseudosimplicial object with $d_0 = r$ and $d_1 = l$ with $s_0 = e : C(0) \to C(1)$ we have a simplicial object. Then, by introducing the special fibre products using r on the left and l on the right, we have $C(2) = C(1) *_{C(0)} C(1)$ and $C(3) = C(1) *_{C(0)} C(1) *_{C(0)} C(1)$ but also $C(3) = C(2) *_{C(0)} C(1) = C(1) *_{C(0)} C(2)$. The outside projections give the simplicial morphism

$$d_0, d_3 : C(3) \longrightarrow C(2) \quad \text{and} \quad d_0, d_2 : C(2) \longrightarrow C(1).$$

The composition in the category is $d_1 : C(2) \to C(1)$. The relation $d_0 d_2 = d_1 d_0$ is just compatibility in the fibre product, and two relations $d_0 d_1 = d_0 d_0$ and $d_1 d_2 = d_1 d_1$ are just the fact that the domain of gf is the domain of f and the range of gf is the range of g. The simplicial morphisms $d_1, d_2 : C(3) \to C(2)$ are defined as $c *_{C(0)} C(1)$ and $C(1) *_{C(0)} c$, where $c = d_1 : C(2) \to C(1)$. There are six relations $d_i d_j = d_{j-1} d_i$ for $i < j$, where

(a) $d_1 d_2 = d_1 d_1$ is associativity
(b) $d_0 d_2 = d_1 d_0$ and $d_1 d_3 = d_2 d_1$ are compositions, and
(c) $d_0 d_1 = d_0 d_0$, $d_2 d_3 = d_2 d_2$, and $d_0 d_3 = d_2 d_0$ are compatibility in the fibre product.

3 The Nerve of the Classifying Space Functor and Definition of Algebraic K-Theory

3.1. Definition The classifying space functor or nerve functor

$$\text{Ner: (cat)} \longrightarrow \Delta(\text{set})$$

is defined by assigning to a small category $\mathscr{X}$ the nerve $\text{Ner}(\mathscr{X})$ in the category of simplicial sets $\Delta(\text{set})$ and assigning to a functor $F : \mathscr{X} \to \mathscr{Y}$ the morphism $\text{Ner}(F) : \text{Ner}(\mathscr{X}) \to \text{Ner}(\mathscr{Y})$ of simplicial sets given by the formula

$$\text{Ner}(F)(X_0, \ldots, X_p; u_1, \ldots, u_p) = (F(X_0), \ldots F(X_p); F(u_1), \ldots, F(u_p)).$$

For functors $F : \mathscr{X} \to \mathscr{Y}$ and $G : \mathscr{Y} \to \mathscr{Z}$, we have $\text{Ner}(GF) = \text{Ner}(G)\text{Ner}(F) : \mathscr{X} \to \mathscr{Z}$ and $\text{Ner} : (\text{cat}) \to \Delta(\text{set})$ is a well-defined functor with the following simplicial operations.

3.2. Definition The nerve $\text{Ner}(\mathscr{X})$ of a small category $(\mathscr{X})$ is the simplicial set, where $\text{Ner}(\mathscr{X})_p$ is set of all sequences $(X_0, \ldots, X_p; u_1, \ldots, u_p)$ where $u_i : X_{i-1} \to X_i$ is a morphism in $\mathscr{X}$ for $1 \le i \le p$. The simplicial operations are defined as follows:

(1) $d_i : \text{Ner}(\mathscr{X})_p \to \text{Ner}(\mathscr{X})_{p-1}$ is defined by deleting the object X_i in the sequence. More precisely, this is given by the formula

$$d_i(X_0, \ldots X_p; u_1, \ldots, u_p) = \begin{cases} (X_1, \ldots, X_p; u_2, \ldots, u_p) & i = 0 \\ (X_0, \ldots X_{i-1}, X_{i+1}, \ldots, X_p; u_1, \ldots, u_{i+1} u_i, \ldots, u_p) & 0 < i < p \\ (X_0, \ldots, X_{p-1}; u_1, \ldots, u_{p-1}) & i = p . \end{cases}$$

(2) $s_j : \text{Ner}(\mathscr{X})_p \to \text{Ner}(\mathscr{X})_{p+1}$ is defined by putting identity on X_j at X_j. More precisely, this is given by the formula

$$s_j(X_0, \ldots, X_p; u_1, \ldots, u_p) = (X_0, \ldots, X_j, X_j, \ldots, X_p; u_1, \ldots, u_j, X_j, u_{j+1} \ldots, u_p),$$

where $X_j : X_j \to X_j$ is used to denote the identity on X_j. That is, the object has the same symbol as the identity morphism on the object.

The 0 simplexes of $\mathrm{Ner}(\mathscr{X})$ are just the objects of $\mathscr{X}$, and the 1 simplexes of $\mathrm{Ner}(\mathscr{X})$ are just the morphisms of $\mathscr{X}$.

The reader is invited to check the simplicial relations.

3.1 Special Low Degree Examples

For the nerve of $\mathscr{X}$, we have the following formulas and relations:

(1) $d_1(X \xrightarrow{f} Y) = X$ and $d_0(X \xrightarrow{f} Y) = Y$,

(2) $d_2(X \xrightarrow{f} Y \xrightarrow{g} Z) = X \xrightarrow{f} Y$, $d_1(X \xrightarrow{f} Y \xrightarrow{g} Z) = X \xrightarrow{gf} Z$,

$d_0(X \xrightarrow{f} Y \xrightarrow{g} Z) = Y \xrightarrow{g} Z$,

The relations are

$$d_0 d_1(X \xrightarrow{f} Y \xrightarrow{g} Z \xrightarrow{h} T) = d_0 d_0(X \xrightarrow{f} Y \xrightarrow{g} Z \xrightarrow{h} T) = Z \xrightarrow{h} T,$$

$$d_1 d_2(X \xrightarrow{f} Y \xrightarrow{g} Z \xrightarrow{h} T) = X \xrightarrow{(hg)f} T = X \xrightarrow{h(gf)} T = d_1 d_1(X \xrightarrow{f} Y \xrightarrow{g} Z \xrightarrow{h} T),$$

that is, associativity, and

$$d_2 d_3(X \xrightarrow{f} Y \xrightarrow{g} Z \xrightarrow{h} T) = d_2 d_2(X \xrightarrow{f} Y \xrightarrow{g} Z \xrightarrow{h} T) = X \xrightarrow{f} Y,$$

$$d_0 d_2(X \xrightarrow{f} Y \xrightarrow{g} Z \xrightarrow{h} T) = d_1 d_0(X \xrightarrow{f} Y \xrightarrow{g} Z \xrightarrow{h} T) = Y \xrightarrow{hg} T,$$

$$d_1 d_3(X \xrightarrow{f} Y \xrightarrow{g} Z \xrightarrow{h} T) = d_2 d_1(X \xrightarrow{f} Y \xrightarrow{g} Z \xrightarrow{h} T) = X \xrightarrow{gf} Z,$$

$$d_0 d_3(X \xrightarrow{f} Y \xrightarrow{g} Z \xrightarrow{h} T) = d_2 d_0(X \xrightarrow{f} Y \xrightarrow{g} Z \xrightarrow{h} T) = Y \xrightarrow{g} Z.$$

An easy check shows that $\mathrm{Ner}(F)$ in (3.1) commutes with d_i and s_j defining a morphism of simplicial sets and that Ner(identity on $\mathscr{X}$) equals the identity on $\mathrm{Ner}(\mathscr{X})$.

Before we go to groupoids and gerbes, we wish to indicate how the higher algebraic K-theory of a ring R was defined by Quillen using the geometric realization of the nerve of a category.

3.3 *Remark* Given a ring R, the category of finitely generated projective modules (vect/R) has the Grothendieck group $K(R)$ of isomorphism classes of objects with the direct sum of modules as semigroup structure, see 4(3.1). Using the subcategory $\mathscr{C}$ of (vect/R) of nonzero finitely generated projective modules and split monomorphisms, Quillen defines the higher K-groups $K_i(R)$ as the homotopy groups $\pi_{i-1}(\mathrm{Ner}(\mathscr{C}))$. The basic properties follow from homotopy property of nerves.

4 Groupoids in a Category

In the case of a groupoid where each f is an isomorphism with inverse $\iota(f) = f^{-1}$, this formula defines a map $\iota : C(1) \to C(1)$ with domain and range interchanged $r(\iota(f)) = l(f)$ and $l(\iota(f)) = r(f)$. There is also the inverse property $m(f, \iota(f)) = e(l(f))$ and $m(\iota(f), f) = e(r(f))$.

4.1. Definition Let $\mathscr{C}$ be a category with fibre products. A groupoid $G(*)$ in $\mathscr{C}$ is a septuple $(G(0), G(1), l, r, e, m, \iota)$ where the first six items form a category object $(G(0), G(1), l, r, e, m)$ and where the inverses morphism $\iota : C(1) \to C(1)$ satisfies in addition to (Cat1–Cat4) the following axioms:

(grpoid1) The following compositions hold $l\iota = r$ and $r\iota = l$.
(grpoid2) The following commutative diagram give the inverse property of ι :
$$G(1) \to G(1)$$

$$
\begin{array}{ccc}
G(1) & \xrightarrow{\;((G(1),\iota)/(\iota,G(1)))\;} & G(1)_r \times_{G(0) \, l} G(1) \\
{\scriptstyle l/r} \downarrow & & \downarrow {\scriptstyle m} \\
G(0) & \xrightarrow{\qquad\qquad e \qquad\qquad} & G(1).
\end{array}
$$

In general, we can think of a groupoid as a category where every morphism is an isomorphism, and the process of associating to an isomorphism its inverse is an isomorphism of the category to its opposite category. Moreover, this isomorphism is an involution when the double opposite is identified with the original category.

4.2. Example In the category of sets, a category object $G(*)$ is a groupoid when every morphism $u \in G(1)$, which is defined $u : l(u) \to r(u)$, is a bijection, and in this case, the morphism ι is the inverse given by $\iota(u) = u^{-1} : r(u) \to l(u)$.

4.3. Remark If ι' and ι'' are two groupoid structures on a category object $(C(0), C(1), l, r, e, m)$, then $\iota' = \iota''$. To see this, we calculate as with groups using the associative law

$$\iota'(u) = m(\iota'(u), m(u, \iota''(u))) = m(m(\iota'(u), u), \iota''(u)) = \iota''(u).$$

This means that a groupoid is not a category with an additional structure, but a category satisfying an axiom, namely ι exists. Also, we have $\iota\iota = G(1)$, the identity on $G(1)$ by the same argument.

4.4. Example A groupoid $G(*)$ with $G(0)$, the final object in $\mathscr{C}$, is just a group object in the category $\mathscr{G}$.

4.5. Definition A morphism $u(*) : G'(*) \to G''(*)$ of groupoids in $\mathscr{C}$ is a morphisms of categories in $\mathscr{C}$.

4.6. Remark A morphism of groupoids has the additional property that $\iota''u(1) = u(1)\iota'$. This is seen as with groups from the relation $m''(\iota''u(1), u(1)) = l''e'' = m''(u(1)\iota', u(1))$. We derive (2.3) by applying this to the identity functor.

4.7. Definition Let grpoid($\mathscr{C}$) denote the full subcategory of cat($\mathscr{C}$) determined by groupoids.

There is a full subcategory grp($\mathscr{C}$) of grpoid($\mathscr{C}$) consisting of those groupoids $G(*)$, where $G(0)$ is the final object. This is the category of groups over the category $\mathscr{C}$.

The category grpoid(set) of groupoids over the category of sets is just the category of small categories with the property that all morphisms are isomorphisms. Also grp(set) is just the category (grp) of groups.

4.8. Example Let G be a group object in a category $\mathscr{C}$ with fibre products. An action of G on an object X of $\mathscr{C}$ is a morphism $\alpha : G \times X \to X$ satisfying two axioms given by commutative diagrams

(1) (associativity)

$$
\begin{array}{ccc}
G \times G \times X & \xrightarrow{\;\;G \times \alpha\;\;} & G \times X \\
{\scriptstyle \mu \times X}\big\downarrow & & \big\downarrow{\scriptstyle \alpha} \\
G \times X & \xrightarrow{\;\;\alpha\;\;} & X
\end{array}
$$

where $\mu : G \times G \to G$ is the product on the group object, and
(2) (unit)

$$
\begin{array}{ccc}
X = \{*\} \times X & \xrightarrow{\;\;e \times X\;\;} & G \times X \\
& {\scriptstyle X}\searrow \quad \swarrow {\scriptstyle \alpha} & \\
& X. &
\end{array}
$$

4.9. Remark The related groupoid $X\langle G\rangle(*)$ is defined by $X\langle G\rangle(0) = X$ and $X\langle G\rangle(1) = G \times X$ with structure morphisms

$$ e = e_G \times X : X\langle G\rangle(0) \longrightarrow X\langle G\rangle(1) = G \times X, $$

$$ l = pr_2 : X\langle G\rangle(1) = G \times X \longrightarrow X = X\langle G\rangle(0), $$

and

$$ r = \alpha : X\langle G\rangle(1) \longrightarrow X = X\langle G\rangle(0). $$

For the composition, we need natural isomorphism

$$ \theta : X\langle G\rangle(1){}_{r}\underset{X\langle G\rangle(0)}{\times}{}_{l}X\langle G\rangle(1) \longrightarrow G \times G \times X $$

given by $pr_1\theta = pr_G pr_1$, $pr_2\theta = pr_G pr_2$ and $pr_3\theta = pr_X pr_1$. Then, composition m is defined by the following diagram

$$\begin{array}{ccc}
X\langle G\rangle(1)_r \times_{X\langle G\rangle(0)^l} X\langle G\rangle(1) & \xrightarrow{\theta} & G \times G \times X \\
\downarrow{\scriptstyle m} & & \downarrow{\scriptstyle \mu \times X} \\
X\langle G\rangle(1) & =\!\!=\!\!=\!\!=\!\!=\!\!= & G \times X.
\end{array}$$

The unit and associativity properties of m come from the unit and associativity properties of $\mu : G \times G \to G$.

4.10. Definition With the above notation, $X\langle G\rangle(*)$ is the groupoid associated to the G action on the object X in $\mathscr{C}$. It is also called the translation category.

5 The Groupoid Associated to a Covering

Let $X = \bigcup_{\alpha \in I} U_\alpha$ be a covering of X where X is a set, a space and U_α are as usually open sets, or a smooth manifold and U_α are open submanifolds.

5.1. Definition Let $(U_\alpha)_{\alpha \in I}$ be a covering of X. The groupoid $U(*)$ associated to the covering is defined as the coproducts $U(0) = \coprod_{\alpha \in I} U_\alpha$ and $U(1) = \coprod_{(\alpha,\beta) \in I \times I} U_\alpha \cap U_\beta$, and the structure morphisms are defined using the coproduct injections

$$q_\alpha : U_\alpha \to \coprod_{\alpha \in I} U_\alpha \quad \text{and} \quad q_{\alpha,\beta} : U_\alpha \cap U_\beta \to \coprod_{(\alpha,\beta) \in I \times I} U_\alpha \cap U_\beta$$

as follows.

(1) The source and target morphisms $s, t : U(1) \to U(0)$ are defined by the following relations

$$s q_{\alpha,\beta} = q_\alpha(\text{inc}_1) \quad \text{and} \quad t q_{\alpha,\beta} = q_\beta(\text{inc}_2),$$

where $\text{inc}_1 : U_\alpha \cap U_\beta \to U_\alpha$ and $\text{inc}_2 : U_\alpha \cap U_\beta \to U_\beta$ are inclusions.
(2) The unit morphism $e : U(0) \to U(1)$ is defined by the following relation $e q_\alpha = q_{\alpha,\alpha} : U_\alpha = U_\alpha \cap U_\alpha$.
(3) The multiplication morphism $m : U(1)_t \times_{U(0)^s} U(1) \to U(1)$ is defined by observing that the fibre product $U(1)_t \times_{U(0)^s} U(1) \to U(1)$ is a coproduct of inclusions

$$q_{\alpha,\beta'} \times q_{\beta'',\gamma} : (U_\alpha \cap U'_\beta)_t \times_{U(0)^s} (U''_\beta \cap U_\gamma) = U_\alpha \cap U_\beta \cap U_\gamma \longrightarrow U(1)_t \times_{U(0)^s} U(1)$$

and using the notation $\text{inc}_{\alpha,\beta,\gamma}$ for the inclusions, where $\beta = \beta' = \beta''$

$$\text{inc}_{\alpha,\beta,\gamma} : (U_\alpha \cap U'_\beta)_t \times_{U(0)^s} (U''_\beta \cap U_\gamma) = U_\alpha \cap U_\beta \cap U_\gamma \longrightarrow U_\alpha \cap U_\gamma.$$

With this description of the fibre product, we see that the groupoid multiplication is defined by the relation

$$m(q_{\alpha,\beta'} \times q_{\beta'',\gamma}) = q_{\alpha,\gamma}(\mathrm{inc}_{\alpha,\beta,\gamma}).$$

(4) The inversion morphism $\iota : U(1) \to U(1)$ is defined by the requirement that $\iota q_{\alpha,\beta} = q_{\beta,\alpha}$ for all $\alpha,\beta \in I$. The axioms for a groupoid are easily checked, and this is left to the reader.

5.2. Remark The calculation of the fibre product in terms of triple intersections

$$q_{\alpha,\beta'} \times q_{\beta'',\gamma} : (U_\alpha \cap U_\beta')_t \underset{U(0)}{\times}{}_s (U_\beta'' \cap U_\gamma) = U_\alpha \cap U_\beta \cap U_\gamma \longrightarrow U(1)_t \underset{U(0)}{\times}{}_s U(1)$$

has the following extension to n factors $U(1)$, and it is a coproduct of $(n-1)$-fold intersections, which is seen by calculating the fibre product of n injections $q_{\alpha(0),\alpha'(1)} \times q_{\alpha''(1),\alpha'(2)} \times \cdots \times q_{\alpha''(n-1),\alpha(n)}$. As with the simple fibre product, this reduces to the case

$$\alpha'(1) = \alpha''(1) = \alpha(1), \ldots, \alpha'(n-1) = \alpha''(n-1) = \alpha(n-1),$$

and the notation is abbreviated to

$$q_{\alpha(0),\ldots,\alpha(n)} : U_{\alpha(0)} \cap \ldots \cap U_{\alpha(n)} \longrightarrow (U(1)_t \underset{U(0)}{\times}{}_s *)^n.$$

6 Gerbes on Groupoids

Now, we return to the simplicial notation for a category which is a groupoid in order to carry over to groupoids the ideas contained in the previous chapter on bundle gerbes.

6.1. Notation Let $G(*) = (G(0), G(1), l, r, e, m, \iota)$ be a groupoid in $\mathscr{C}$ where sextuple $(G(0), G(1), l, r, e, m)$ is a category object and inverse morphism is $\iota : G(1) \to G(1)$. As in (2.2) and (2.9), we identify the first two terms in a simplicial object with $d_0 = r, d_1 = l$, and the unit is $s_0 = e : G(0) \to G(1)$. Again by introducing the special fibre product using r on the left and l on the right, we have $G(2) = G(1) *_{G(0)} G(1)$ and $G(3) = G(1) *_{G(0)} G(1) *_{G(0)} G(1) = G(2) *_{G(0)} G(1) = G(1) *_{G(0)} G(2)$. The outside projections are given by the simplicial morphisms $d_0, d_3 : G(3) \longrightarrow G(2)$ and $d_0, d_2 : G(2) \longrightarrow G(1)$. The composition in the groupoid or category is $d_1 : G(2) \to G(1)$, and we have all the considerations (a), (b), and (c) contained in (2.9) with a new consideration for $\iota : G(1) \to G(1)$, where

(d) $d_0\iota = d_1, d_1\iota = d_0 : G(1) \to G(0)$ and
(e) $\iota d_1 = d_1(\iota \times \iota)\tau : G(2) \to G(1)$, where $\tau : G(2) \to G(2)$ interchanges the two factors of $G(2) = G(1) *_{G(0)} G(1)$ as it is seen from the formulas in (3.3)(2)

$$d_2(X \xrightarrow{f} Y \xrightarrow{g} Z) = X \xrightarrow{f} Y, \quad d_1(X \xrightarrow{f} Y \xrightarrow{g} Z) = X \xrightarrow{gf} Z$$

and

$$d_0(X \xrightarrow{f} Y \xrightarrow{g} Z) = Y \xrightarrow{g} Z.$$

6.2. *Example* Let $Y \to B$ be a bundle. Then $G(0) = Y$ and $G(1) = Y \times_B Y$ has a groupoid structure with composition

$$d_1 : G(2) = G(1) *_{G(0)} G(1) = (Y \times_B Y) \times_B (Y \times_B Y) \longrightarrow Y \times_B Y = G(1)$$

resulting from projection from the middle two factors to B. The inverse $\iota : G(1) = Y \times_B Y \to Y \times_B Y = G(1)$ just switches the two factors. We leave it to reader to check the axioms.

6.3. Definition Let $G(*)$ be a groupoid in the category of spaces. A gerbe on $G(*)$ is a pair of a line bundle L on $G(1)$ together with an isomorphism $\gamma : d_2^*(L) \otimes d_0^*(L) \to d_1^*(L)$ over $G(2)$ satisfying the associative law on $G(3)$ which is the following commutative diagram where we use the cosimplicial relations $d_j^* d_i^* = d_i^* d_{j-1}^*$ for i, j freely.

$$
\begin{array}{ccc}
d_3^*(d_2^*(L) \otimes d_0^*(L)) \otimes d_1^* d_0^*(L) & =\!=\!= & d_3^* d_2^*(L) \otimes d_0^*(d_2^*(L) \otimes d_0^*(L)) \\
{\scriptstyle d_3^*(\gamma) \otimes d_1^* d_0^*(L)} \downarrow & & \downarrow {\scriptstyle d_3^* d_2^*(L) \otimes d_0^*(\gamma)} \\
d_3^* d_1^*(L) \otimes d_1^* d_0^*(L) & & d_3^* d_2^*(L) \otimes d_0^* d_1^*(L) \\
\| & & \| \\
d_1^*(d_2^* \otimes d_0^*(L)) & & d_2^*(d_2^*(L) \otimes d_0^*(L)) \\
{\scriptstyle d_1^*(\gamma)} \downarrow & & \downarrow {\scriptstyle d_2^*(\gamma)} \\
d_1^* d_1^*(L) & =\!=\!=\!=\!=\!=\!= & d_2^* d_1^*(L)
\end{array}
$$

6.4. Definition Let $(u, v) : G'(*) \to G''(*)$ be a morphism of groupoids of spaces, and let L', γ' be a groupoid gerbe over $G'(*)$ and L'', γ'' be a groupoid gerbe over $G''(*)$. A v-morphism of gerbes is a v-morphism $w : L' \to L''$ of line bundles such that over $v \times_u v : G'(2) \to G''(2)$ we have the commutative diagram

$$
\begin{array}{ccc}
d_2^*(L') \otimes d_0^*(L') & \xrightarrow{\gamma'} & d_1^*(L') \\
{\scriptstyle d_2^*(w) \otimes d_0^*(w)} \downarrow & & \downarrow {\scriptstyle d_1^*(w)} \\
d_2^*(L) \otimes d_0^*(L) & \xrightarrow{\gamma} & d_1^*(L)
\end{array}
$$

If (v, u) is the identity on $G(*)$, then w is a B-morphism gerbes over $G(*)$.

6.5. *Remark* The composition of morphisms of groupoid gerbes is again a morphisms of gerbes. We can speak of the category (gerbes) of groupoid gerbes and the category of (gerbes/$G(*)$) over $G(*)$.

6.6. Remark The category (gerbes/$G(*)$) has tensor products given by the tensor products of line bundles. An exterior tensor product exists on the product groupoid $G'(*) \times G''(*)$ for a gerbe L', γ' and gerbe L'', γ''.

6.7. Example For a groupoid $G(*)$ and a line bundle J on $G(0)$ by using the dual $L = d_1^*(J) \otimes d_0^*(J^\vee)$, we define the following groupoid gerbe $\delta(J)$ as γ and L with the natural morphism given by

$$\gamma : d_2^*(d_1^*(J) \otimes d_0^*(J^\vee)) \otimes d_0^*(d_1^*(J) \otimes d_0^*(J^\vee)) \longrightarrow d_1^*(d_1^*(J) \otimes d_0^*(J^\vee))$$

$$\|$$
$$\qquad\qquad\qquad\qquad\qquad\qquad\qquad\qquad\qquad\qquad\qquad\qquad\|$$

$$d_2^* d_1^*(J) \otimes d_2^* d_0^*(J^\vee) \otimes d_0^* d_1^*(J) \otimes d_0^* d_0^*(J^\vee) \longrightarrow d_1^* d_1^*(J) \otimes d_1^* d_0^*(J^\vee)$$

$$\|$$
$$\qquad\qquad\qquad\qquad\qquad\qquad\qquad\qquad\qquad\qquad\qquad\qquad\|$$

$$d_2^*(L) \otimes d_0^*(L) \longrightarrow d_1^*(L).$$

Here, we have used relations $d_2^* d_1^* = d_1^* d_1^*$, $d_2^* d_0^* = d_0^* d_1^*$ and $d_0^* d_0^* = d_1^* d_0^*$ as in 25(2.6).

6.8. Definition A groupoid gerbe L on $G(*)$ is trivial provided there exists a line bundle J on $G(0)$ with L isomorphic to $\delta(J)$.

7 The Groupoid Gerbe Characteristic Class

The characteristic class for a groupoid gerbe will lie in the Čech hypercohomology of the groupoid which is defined using the Čech cochain functor $C^*(X, M)$ on a space X with values in a module or sheaf M.

7.1. Definition Let $G(*)$ be a category of spaces. The Čech cosimplicial object associated with $G(*)$ with values in M is

$$C^*(G(*), M) : C^*(G(0), M) \underset{\delta_1}{\overset{\delta_0}{\rightrightarrows}} C^*(G(1), M) \ldots \underset{\delta_n}{\overset{\delta_0}{\rightrightarrows}} C^*(G(n), M) \underset{\delta_{n+1}}{\overset{\delta_0}{\rightrightarrows}} \cdots ,$$

where $\delta_i = C^*(d_i, M)$.

7.2. Definition The Čech double cochain complex associated to a category of spaces $G(*)$ is the double complex where the first coboundary is

$$\delta = \sum_{i=1}^{n} (-1)^i \delta_i : C^*(G(n-1), M) \longrightarrow C^*(G(n), M)$$

and the second is the Čech coboundary. Let $C^*(G(*), M)$ denote this double complex. The Čech cohomology of category of spaces $G(*)$ with values in M denoted by $\mathbb{H}^*(G(*), M)$ is the cohomology of the single complex associated with the double complex $C^*(G(*), M)$.

7.3. Remark Let $G(*)$ be a groupoid gerbe with line bundle L on $G(1)$ together with an isomorphism $\gamma : d_2^*(L) \otimes d_0^*(L) \to d_1^*(L)$ over $G(2)$ satisfying the associative law on $G(3)$. The Čech cocycle representing the first Chern class $c_{(1)}(L)$ of L is a cocycle in $C^2(G(1), \mathbb{Z})$ of the associated single complex of the double complex $C^*(G(*), \mathbb{Z})$.

7.4. Definition With previous notation in (7.3), the characteristic class of gerbe $(G(*), L)$ denoted by $\alpha(G(*), L)$ is the cohomology class in $\mathbb{H}^*(G(*)), \mathbb{Z})$ represented by $c_{(1)}(L)$.

7.5. Remark If the groupoid gerbe $(G(*), L)$ is trivial, then $c_{(1)}(L)$ is trivial in the first direction, and the characteristic class $\alpha(G(*), L) = 0$.

Chapter 25
Stacks and Gerbes

A basic structure in mathematics for the study of a space X is to give to each open set an object $A(U)$ in a category $\mathscr{C}$ together with restriction morphisms $r_{V,U} : A(U) \to A(V)$ for $V \subset U$ satisfying $r_{U,U}$ is the identity and the composition property

$$r_{W,U} = r_{W,V}\, r_{V,U} \quad \text{for} \quad W \subset V \subset U.$$

Such a structure is called a presheaf with values in $\mathscr{C}$.

Again, there is a gluing condition which is realized by a universal construction. To formulate this, we use adjoint functors which are introduced in Sect. 2. An example of such a functor is the sheaf associated to a presheaf.

In this chapter, we consider another approach to gerbes by the more general concept of stack. A stack over a space X is a configuration of categories over X associated to open sets of X with various gluing properties. Instead of gluing trivial bundles to obtain a general bundle, we consider categories $\mathscr{F}_U$ indexed by the open subsets of a space X together with restriction functors $r_{V,U} : \mathscr{F}_U \to \mathscr{F}_V$ for $V \subset U$. Normally, we expect to have the composition property $r_{W,U} = r_{W,V}\, r_{V,U}$ for $W \subset V \subset U$ of a presheaf, but now we can only assume that a third element of structure is given, namely an isomorphism between the functors $r_{W,U}$ and $r_{W,V}\, r_{V,U}$. When these data satisfy a suitable coherence relation, this triple of data is called a category over X. This concept was introduced by Grothendieck with the term fibred category over X. It is a generalization of the presheaf of categories where the key feature is that transitivity of restriction is just defined up to isomorphism.

Although the entire collection of categories, denoted by $\mathscr{F}(U)$ or $\mathscr{F}_U$, over open sets do not form a presheaf, but only a presheaf up to an isomorphism in the transitivity relation, it is possible to extract from the category over X various presheaves on X. The condition that these presheaves are sheaves of sets is the first step in formulating the definition of a stack. We have seen that descent conditions play a basic role in many parts of bundle theory, and we formulate descent data concepts in a category over a space. The condition that a category over a space be a stack is formulated in terms of descent data in a category being realized as objects in the category.

D. Husemöller et al.: *Stacks and Gerbes*, Lect. Notes Phys. **726**, 303–322 (2008)
DOI 10.1007/978-3-540-74956-1_26   © Springer-Verlag Berlin Heidelberg 2008

1 Presheaves and Sheaves with Values in Category

In Chap. 5, principal bundles are introduced without reference to local charts, but in 5(4.5), we describe what form a local description would take using an étale map $q : U \to B$, where U is related to an open covering of B. Then in terms of this description with the concept of descent data, we are able to reverse the process and go from descent data back to a principal bundle. Now, we return to these concepts which are also basic in sheaf theory with a brief introduction to sheaf theory as a preliminary to the discussion of stacks and their related gerbes.

1.1. Definition Let X be a topological space, and let $\mathscr{C}$ be a category. A presheaf P on X with values in $\mathscr{C}$ assigns to each open set U of X an object $P(U)$ in $\mathscr{C}$ and to each inclusion $V \subset U$ of open sets in X a morphism $r_{V,U} : P(U) \to P(V)$ in $\mathscr{C}$ such that the following two axioms hold:

(a) $r_{U,U} = $ identity on $P(U)$ and
(b) $r_{W,U} = r_{W,V} r_{V,U} : P(U) \to P(W)$ for open sets $W \subset V \subset U$.

1.2. Remark For a topological space X, we have the category $\mathrm{Op}(X)$ coming from the ordered set of open subsets of X. There is at most one morphism $V \to U$, and the morphism set $\mathrm{Hom}_{\mathrm{Op}(X)}(V,U)$ is the inclusion $V \subset U$ otherwise is empty. With this category, we see that the presheaf P with values in the category $\mathscr{C}$ is just a functor $P : \mathrm{Op}(X)^{\mathrm{op}} \to \mathscr{C}$ on the opposite category, that is, a contravariant functor.

A sheaf is defined as a presheaf where gluing is always possible, or in other words, descent data is always realized. For the definition we use products in the category.

1.3. Definition A presheaf F on X with values in category $\mathscr{C}$ is a sheaf provided that for all open coverings $\mathscr{U}$ the following sequence is exact

$$F(U) \longrightarrow \prod_{U' \in \mathscr{U}} F(U') \rightrightarrows \prod_{(U',U'') \in \mathscr{U}^2} F(U' \cap U'').$$

Here, the first morphism in $\mathscr{C}$ is defined from the restriction $F(U) \to F(U')$, and the second pair of morphisms is defined from two restrictions

$$F(U') \to F(U' \cap U'') \text{ and } F(U'') \to F(U' \cap U'')$$

in all cases commuting with the projections from the product.

We remark that the exactness says that the object $F(U)$ is completely determined from its restrictions to all objects $F(U')$ in the open covering $U = \coprod_{U' \in \mathscr{U}} U'$ and if the family of objects $F(U')$ in $\mathscr{C}$ have the property that their restrictions to overlapping open sets in $F(U' \cap U'')$ are the same, then we can glue the objects together to an object in $F(U)$ over the open set U covered by the $U' \in \mathscr{U}$.

1.4. Remark But this is not exactly the gluing that takes place for bundles where we use an isomorphism between the two restrictions of $a' \in F(U') \to F(U' \cap U'')$ and of $a'' \in F(U'') \to F(U' \cap U'')$ in $F(U' \cap U'')$ over the overlapping set $U' \cap U''$. The restrictions are not assumed to be equal, but they come with a given isomorphism as the part of the data.

1.5. Definition Let P' and P'' be two presheaves on X with values in the category $\mathscr{C}$. A morphism $f : P' \rightarrow P''$ is a family of morphisms $f_U : P'(U) \rightarrow P''(U)$ in $\mathscr{C}$ indexed by the open sets of X commuting with restriction, that is, the following diagram is commutative for $V \in U$

$$
\begin{array}{ccc}
P'(U) & \xrightarrow{\ f_U\ } & P''(U) \\
{\scriptstyle r'_{V,U}}\big\downarrow & & \big\downarrow{\scriptstyle r''_{V,U}} \\
P'(V) & \xrightarrow{\ f_V\ } & P''(V).
\end{array}
$$

A morphism of sheaves is a morphism of underlying presheaves.

Composition of morphisms of presheaves is defined indexwise on the open sets of X.

1.6. Notation. Let $\mathrm{preSh}(X,\mathscr{C})$ denote the category of presheaves on X with values in $\mathscr{C}$ and morphisms of presheaves. Let $\mathrm{Sh}(X,\mathscr{C})$ be the subcategory of sheaves with values in $\mathscr{C}$. It is a full subcategory in the sense that for two sheaves F' and F'', the morphism sets $\mathrm{Hom}_{\mathrm{Sh}(X,\mathscr{C})}(F',F'') = \mathrm{Hom}_{\mathrm{preSh}(X,\mathscr{C})}(F',F'')$. We denote by $J : \mathrm{Sh}(X,\mathscr{C}) \rightarrow \mathrm{preSh}(X,\mathscr{C})$ the inclusion functor for the subcategory of sheaves in the category of presheaves.

1.7. Key Construction A fundamental process is to turn a presheaf into a sheaf $s(P)$ to obtain a morphism of presheaves $\beta : P \rightarrow J(s(P))$ in a universal manner, that is, the morphism β of functors of $\mathrm{preSh}(X,\mathscr{C}) \rightarrow \mathrm{preSh}(X,\mathscr{C})$ has a universal property. We explain this universal mapping property of β in terms of adjoint functors in the next assertion, and we give an introduction to adjoint functors in Sect. 2. We should emphasize that if J has such an adjoint functor s, called sheafification, then β exists and satisfies the universal property. Moreover, s is unique up to isomorphism of functors when it is the adjoint functor to J. In particular, sheafification is part of this very basic idea which is central for so many constructions in mathematics.

1.8. Theorem *For a space X and a category $\mathscr{C}$ with colimits, the inclusion functor $J : \mathrm{Sh}(X,\mathscr{C}) \rightarrow \mathrm{preSh}(X,\mathscr{C})$ has a left adjoint functor $s : \mathrm{preSh}(X,\mathscr{C}) \rightarrow \mathrm{Sh}(X,\mathscr{C})$. (It is called sheafification.)*

1.9. Construction of s for the Category $\mathscr{C} = $ (Set) of Sets Let P be a presheaf of sets on a space X. The stalk P_x of P at $x \in X$ is the quotient of the union of all sets $P(U)$ with $x \in U$ under the relation $a' \in P(U')$ is equivalent to $a'' \in P(U'')$ provided there exists an open set U with $x \in U \subset U' \cap U''$ such that $r_{U,U'}(a') = r_{U,U''}(a'')$. This is an equivalence relation. The equivalence class of $b \in P(V)$ is denoted by $[b,x]$, and it is called the germ of $b \in P(V)$ for $x \in V$. Let $\mathrm{\acute{e}t}(P) = \coprod_{x \in X} P_x$ be the space of all germs with projection $\pi : \mathrm{\acute{e}t}(P) \rightarrow X$ defined by $\pi(P_x) = x$. The basis of the open sets is given by all sets of the form $\langle a \rangle$ consisting of all $[a, x']$ for all $x' \in U$ and $a \in P(U)$. Then, π is an étale map. With $\beta : P \to Js(P)$ where $a \in P(U)$ is carried to the section $\sigma(x) = [a, x]$, we get the identification $s(P)(U) = \Gamma(U, \text{ét}(P))$.

2 Generalities on Adjoint Functors

This section is a detour from the aim of the chapter, but it is a basic tool used for the constructions of the chapter.

Abelianization is defined by a universal property relative to the category of abelian objects. The theory of adjoint functors, which we sketch now, is the formal development of this idea of a universal property, and this theory also gives a method for constructing equivalences between categories. We approach the subject by considering morphisms between the identity functor and composite of two functors. For an object X in a category, we frequently use the symbol X also for the identity morphism $X \to X$ along with 1_X, and the same for a category $\mathscr{X}$, the identity functor on $\mathscr{X}$ is frequently denoted by $\mathscr{X}$. Let (set) denote the category of sets.

2.1. Remark Let $\mathscr{X}$ and $\mathscr{Y}$ be two categories and $T : \mathscr{X} \to \mathscr{Y}$ and $S : \mathscr{Y} \to \mathscr{X}$ two functors. Morphisms of functors $\mathscr{X} \to \mathscr{X}$ of the form $\alpha : ST \to \mathscr{X}$ are in bijective correspondence with morphisms

$$a : \text{Hom}_{\mathscr{Y}}(Y, T(X)) \longrightarrow \text{Hom}_{\mathscr{X}}(S(Y), X)$$

of functors $\mathscr{Y}^{\text{op}} \times \mathscr{X} \to$ (set), where

(1) the morphism α defines a by the relation $a(f) = \alpha(X)S(f)$ and
(2) the morphism a defines α by the relation

$$a(1_{T(X)}) = \alpha(X) : ST(X) \to X.$$

Morphisms of functors $\mathscr{Y} \to \mathscr{Y}$ of the form $\beta : \mathscr{Y} \to TS$ are in bijective correspondence with morphisms

$$b : \text{Hom}_{\mathscr{X}}(S(Y), X) \longrightarrow \text{Hom}_{\mathscr{Y}}(Y, T(X))$$

of functors $\mathscr{Y}^{\text{op}} \times \mathscr{X} \to$ (set), where

(1) the morphism β defines b by the relation $b(g) = T(g)\beta(Y)$ and
(2) the morphism b defines β by the relation

$$b(1_{S(Y)}) = \beta(Y) : Y \to TS(Y).$$

2.2. Remark For $f : Y \to T(X)$, we calculate

$$b(a(f)) = T(a(f))\beta(Y) = T(\alpha(X))[TS(f)\beta(Y)] = [T(\alpha(X))\beta(T(X))]f$$

using $TS(f)\beta(Y) = \beta(T(X))f$, where β is a morphism of functors, and for $g : S(Y) \to X$, we calculate

$$a(b(g)) = \alpha(X)S(b(g)) = [\alpha(X)ST(g)]S(\beta(Y)) = g[\alpha(S(Y))S(\beta(Y))]$$

using $\alpha(X)ST(g) = g\alpha(S(Y))$, where α is a morphism of functors. Thus, we have that $b(a(f)) = f$ for all $f : Y \to T(X)$ if and only if

$$T(\alpha(X))\beta(T(X)) = 1_{T(X)} \,,$$

and we have $a(b(g)) = g$ for all $g : S(Y) \to X$ if and only if

$$\alpha(S(X))S(\beta(Y)) = 1_{S(Y)} \,.$$

2.3. Definition An adjoint pair of functors is a pair of functors $T : \mathscr{X} \to \mathscr{Y}$ and $S : \mathscr{Y} \to \mathscr{X}$ together with an isomorphism of functors of X in $\mathscr{X}$ and Y in $\mathscr{Y}$

$$b : \mathrm{Hom}_{\mathscr{X}}(S(Y), X) \longrightarrow \mathrm{Hom}_{\mathscr{Y}}(Y, T(X)),$$

or equivalently, the inverse isomorphism

$$a : \mathrm{Hom}_{\mathscr{Y}}(Y, T(X)) \longrightarrow \mathrm{Hom}_{\mathscr{X}}(S(Y), X).$$

The functor S is called the left adjoint of T and T is the right adjoint of S, and this is denoted by $S \dashv T$.

2.4. Remark In the view of (1.1) and (1.2), an adjoint pair of functors can be defined as $T : \mathscr{X} \to \mathscr{Y}$ and $S : \mathscr{Y} \to \mathscr{X}$ together with two morphisms of functors

$$\beta : \mathscr{Y} \longrightarrow TS \quad \text{and} \quad \alpha : ST \longrightarrow \mathscr{X}$$

satisfying

$$T(\alpha(X))\beta(T(X)) = 1_{T(X)} \quad \text{and} \quad \alpha(S(X))S(\beta(Y)) = 1_{S(Y)}.$$

2.5. Remark If $S : \mathscr{Y} \to \mathscr{X}$ is the left adjoint of $T : \mathscr{X} \to \mathscr{Y}$, then for the dual categories $S : \mathscr{Y}^{\mathrm{op}} \to \mathscr{X}^{\mathrm{op}}$ is the right adjoint of $T : \mathscr{X}^{\mathrm{op}} \to \mathscr{Y}^{\mathrm{op}}$.

The relation between adjoint functors and morphism with a universal property is contained in the next property.

2.6. Universal Property for $T : \mathscr{X} \to \mathscr{Y}$. For each object Y in $\mathscr{Y}$, we assume we are given an object $s(Y)$ in $\mathscr{X}$ and a morphism $\beta(Y) : Y \to T(s(Y))$ such that for all $f : Y \to T(X)$ there exists a unique morphism $g : s(Y) \to X$ in $\mathscr{X}$ with $T(g)\beta(Y) = f$. Then, there exists a unique functor $S : \mathscr{Y} \to \mathscr{X}$ with $S(Y) = s(Y)$ and $\beta : 1_{\mathscr{Y}} \to TS$, a morphism of functors with $b(g) = T(g)\beta(Y)$ defining a bijection

$$b : \mathrm{Hom}_{\mathscr{X}}(S(Y), X) \longrightarrow \mathrm{Hom}_{\mathscr{Y}}(Y, T(X),.$$

The pair $(s(Y), \beta(Y))$ for each Y in $\mathscr{Y}$ defines an adjoint pair $S \dashv T$.

Proof. Let $v : Y' \to Y''$ be a morphism in $\mathscr{Y}$. We define $S(v) : s(Y') \to s(Y'')$ by considering the composite

$$\beta(Y'')v : Y' \longrightarrow T(s(Y''))$$

and applying the universal property giving a unique morphism called $S(v) : s(Y') \to s(Y'')$ such that $T(S(v))\beta(Y'') = \beta(Y'')v$. Denote $S(Y) = s(Y)$, and then one checks that S is a functor with $S \dashv T$. This proves the proposition.

2.7. Universal Property for $S : \mathscr{Y} \to \mathscr{X}$. For each object X in $\mathscr{X}$, we assume there exists an object $t(X)$ in $\mathscr{Y}$ and a morphism $\alpha(X) : S(t(X)) \to X$ such that for all $g : S(Y) \to X$ there exists a unique morphism $f : Y \to t(X)$ such that $\alpha(X)S(f) = g$. Then, there exists a unique functor $T : \mathscr{X} \to \mathscr{Y}$ such that for each object X in $\mathscr{X}$ the object $T(X) = t(X)$ and $\alpha : ST \to 1_{\mathscr{X}}$, a morphism of functors with $a(f) = \alpha(X)S(f)$ defining a bijection

$$a : \mathrm{Hom}_{\mathscr{Y}}(Y, T(X)) \longrightarrow \mathrm{Hom}_{\mathscr{X}}(S(Y), X).$$

The pair $(t(X), \alpha(X))$ for each X in $\mathscr{X}$ defines an adjoint pair $S \dashv T$.

Proof. To define T on morphisms, we use the universal property. If $u : X \to X'$ is a morphism in $\mathscr{X}$, then there exist a unique morphism $T(u) : T(X) \to T(X')$ such that $\alpha(X')(S(T(u)) = u\alpha(X)$ as morphisms $ST(X) \to X'$. The reader can check that this defines a functor T, and the rest follows from the fact that the universal property asserts that this is a bijection. This proves the proposition.

Also, we deduce (2.7) immediately applying (2.6) to the dual category (2.5).

2.8. Examples of Adjoint Functors There are two general classes of examples of adjoint pairs of functors. The first kind is related to a universal construction and the second kind to a function of two variables $f(x,y)$ being viewed as a function $f(x)(y) = f(x,y)$ of x for each value of the second variable y.

Examples related to universal constructions:

(U1) Let (set) denote the category of sets and functions, and let (gr) denote the category of groups. The stripping of the group law from a group is a functor $T : (\mathrm{gr}) \to (\mathrm{set})$, and it has a left adjoint functor $S : (\mathrm{set}) \to (\mathrm{gr})$, denoted by $S \dashv T$. For a set E, the group $S(E)$ is the free group on the set E.

(U2) Let (k) denote the category of k-modules, where k is a commutative ring (with unit). The stripping of the module structure from a module is a functor $T : (k) \to (\mathrm{set})$, and it has a left adjoint functor $S : (\mathrm{set}) \to (k)$. For a set E, the module $S(E)$, sometimes denoted $k[E]$, is the free k-module on the set E. A special case is $(\mathbb{Z}) = (\mathrm{ab})$ the category of abelian groups.

(U3) Let $J : (\text{ab}) \to (\text{gr})$ be the natural inclusion functor of the category of abelian groups into the category of groups. The functor J has a left adjoint functor $\langle(\text{gr})\rangle : (\text{gr}) \to (\text{ab})$ called abelianization $\langle(\text{ab})\rangle(G) = G/(G,G) = G^{\text{ab}}$, the quotient of G by the commutator subgroup.H

Examples related to two ways of viewing a function of two variables:

(T1) For the three sets X, W, and Y and $\text{Hom}(X \times W, Y)$ the set of functions $f : X \times W \to Y$, we have a natural bijection

$$\theta : \text{Hom}(X \times W, Y) \longrightarrow \text{Hom}(X, \text{Hom}(W, Y))$$

given by $(\theta(f)(x)(w) = f(x,w)$ for $x \in X$, $w \in W$. We can interpret this as an adjoint functor $S \dashv T$, where for fixed W the functor $S(X) = X \times W$ and $T(Y) = \text{Hom}(W, Y)$.

(T2) For three separated spaces X, W, and Y and $\text{Hom}(X \times W, Y)$ the set of all continuous functions $f : X \times W \to Y$ with the compact open topology and W locally compact, we have a natural bijection $\theta : \text{Hom}(X \times W, Y) \to \text{Hom}(X, \text{Hom}(W, Y))$ given by $(\theta(f)(x)(w) = f(x,w)$ for $x \in X$, $w \in W$. We can interpret this as an adjoint functor $S \dashv T$, where for fixed W, the functor $S(X) = X \times W$ and $T(Y) = \text{Hom}(W, Y)$.

There is a version of (T2) for the category of pointed separated spaces where $\text{Map}_*(X, Y)$ is the space of base-point-preserving maps with the compact open topology.

(T3) The functor $S(X) = X \wedge W = X \times W / X \vee W$ is the smash product, and the adjunction takes the form of a bijection

$$\theta : \text{Map}_*(X \wedge W, Y) \longrightarrow \text{Map}_*(X, \text{Map}_*(W, Y)),$$

where again $(\theta(f))(x)(w) = f(x \wedge w)$. For $W = S^1$, this specializes to the suspension $S(X) = X \wedge S^1$ adjunction with loop space $\Omega(Y) = \text{Map}_*(S^1, Y)$ on Y. The adjunction takes the form of a bijection

$$\theta : \text{Map}_*(S(X), Y) \longrightarrow \text{Map}_*(X, \Omega(Y)),$$

where again $(\theta(f)(x)(s) = f(x \wedge s)$ for $s \in S^1$.

(T4) For three k-modules L, M, and N, there is a natural isomorphism of k-modules

$$\theta : \text{Hom}(L \otimes M, N) \longrightarrow \text{Hom}(L, \text{Hom}(M, N)),$$

where $\theta(f)(x)(y) = f(x \otimes y)$. For a fixed module M, the functors $S(L) = L \times M$ is the left adjoint of $T(N) = \text{Hom}(M, N)$ where both $S, T : (k) \to (k)$.

3 Categories Over Spaces (Fibred Categories)

3.1. Definition Let X be a space. A category $\mathscr{F}$ over X is the following data:

(a) For each open subset U of X, we are given a category $\mathscr{F}_U$ (also denoted $\mathscr{F}(U)$),

(b) For each inclusion of open sets $V \subset U$, we are given a functor $i^* = i_{V,U} : \mathscr{F}_U \to \mathscr{F}_V$, and

(c) For each double inclusion of open sets $W \subset V \subset U$, we are given an isomorphism of functors $\theta = \theta_{W,V,U} : i_{W,U} \to i_{W,V} i_{V,U}$ satisfying compatibility the relation for triple inclusions: for $W \subset V \subset U \subset T$ the following diagram is commutative

$$
\begin{array}{ccc}
i_{W,T} & \xrightarrow{\ \theta_{W,U,T}\ } & i_{W,U} i_{U,T} \\
{\scriptstyle \theta_{W,V,T}} \downarrow & & \downarrow {\scriptstyle \theta_{W,V,U} i_{U,T}} \\
i_{W,V} i_{V,T} & \xrightarrow[\ i_{W,V}\theta_{V,U,T}\ ]{} & i_{W,V} i_{V,U} i_{U,T}.
\end{array}
$$

Here, we have used the simple notation where for a morphism of functors $\theta : T' \to T''$ defined by $T', T'' : \mathscr{X} \to \mathscr{Y}$ and functors $F : \mathscr{X}' \to \mathscr{X}, G : \mathscr{Y} \to \mathscr{Y}'$ we have a morphism $G\theta F : GT'F \to GT''F$ of functors defined by $\mathscr{X}' \to \mathscr{Y}'$ given by substitution.

The isomorphisms θ make precise the notion that transitivity of restriction holds only up to natural equivalence, and they are also part of the structure. When the isomorphisms θ are identities, then $\mathscr{F}$ is a presheaf of categories.

3.2. Definition Let X be a space, and let $\mathscr{F}$ and $\mathscr{G}$ be two categories over X. A morphism $\phi : \mathscr{F} \to \mathscr{G}$ of categories over X is a functor $\phi(U) = \phi_U : \mathscr{F}_U \to \mathscr{G}_U$ for each open set U and for each open inclusion of open sets $V \subset U$ an isomorphism

$$\eta_{V,U} : \phi_V i_{V,U} \longrightarrow i_{V,U} \phi_U$$

which make the following pentagon commutative with respect to the isomorphisms in the restriction condition for $W \subset V \subset U$

$$
\begin{array}{ccc}
\phi_W i_{W,U} & \xrightarrow{\ \ \ \eta_{W,U}\ \ \ } & i_{W,U} \phi_U \\
{\scriptstyle \phi_W \theta_{W,V,U}} \downarrow & & \downarrow {\scriptstyle \theta_{W,V,U}\phi_U} \\
\phi_W i_{W,V} i_{V,U} & \xrightarrow[\eta_{W,V}i_{V,U}]{} i_{W,V}\phi_V i_{V,U} \xrightarrow[i_{W,V}\eta_{V,U}]{} & i_{W,V} i_{V,U} \phi_U
\end{array}
$$

3.3. Remark The composition of two morphisms $\phi : \mathscr{F} \to \mathscr{G}$ and $\psi : \mathscr{G} \to \mathscr{H}$ with related isomorphisms $\eta_{V,U} : \phi_V i_{V,U} \to i_{V,U}\phi_U$ and $\zeta_{V,U} : \psi_V i_{V,U} \to i_{V,U}\psi_U$, respectively, is defined by the simple composition of functors $(\psi\phi)_U = \psi_U \phi_U$ with related isomorphism $\xi_{V,U} : \psi_V \phi_V i_{V,U} \to i_{V,U}\psi_U \phi_U$ equal to the composite

$$\xi_{V,U} = (\zeta_{V,U}\phi_U)(\psi_V \eta_{V,U})$$

of two isomorphisms $\psi_V \phi_V i_{V,U} \to \psi_V i_{V,U}\phi_U \to i_{V,U}\psi_U \phi_U$. To check associativity and unit of composition of morphisms is immediate.

In particular, we have the category of categories (of fibre categories) over a space X. In the next definition, we show how local concepts come into the discussion of categories over a space.

3.4. Definition A morphism $\phi : \mathscr{F} \to \mathscr{G}$ of categories over X is locally full provided for each open T in X, each object b in $\mathscr{G}(T)$, and each $x \in T$ there exist an open neighborhood U of x and a in $\mathscr{F}(U)$ such that $\phi_U(a)$ is isomorphic to $i_{U,T}(b) \in \mathscr{G}(U)$. A morphism ϕ is a weak equivalence provided $\phi(T)$ is fully faithful for each open set T in X and ϕ is locally full.

To speak about morphisms which are equivalences, we need the notion of a fibred morphism and a fibred isomorphism of morphisms $\mathscr{F} \to \mathscr{G}$. This takes us into the domain of 2-categories. We come back to these considerations later.

3.5. Example Let X be a space, and let G be a topological group. For each open set $U \subset X$, let $\mathscr{P}_G(U)$ denote the category of principal G-bundles over U with the explicit restriction $\mathscr{P}_G(U) \to \mathscr{P}_G(V)$ given by $\mathrm{res}(P) = V \times_U P$. The composition of restrictions is the natural isomorphism $W \times_U P \to W \times_V V \times_U P$ between fibre products.

4 Prestacks Over a Space

A category $\mathscr{F}$ over a space, even though it is not a presheaf, gives arise to presheaves associated to morphism sets.

4.1. Proposition *Let $\mathscr{F}$ be a category over the space X. If T is an open set, and if a', a'' are objects in the category $\mathscr{F}(T)$, then the function which assigns to each open set $U \subset T$ the set $\mathrm{Hom}_{\mathscr{F}(U)}(i_{U,T}(a'), i_{U,T}(a''))$ is a presheaf on T. This presheaf will be denoted by $\underline{\mathrm{Hom}}_{\mathscr{F}}(a', a'')$.*

Proof. For the inclusion $i : V \to U$ in T, we must define the restriction function of the presheaf

$$\underline{\mathrm{Hom}}_{\mathscr{F}}(a', a'')(i) : \underline{\mathrm{Hom}}_{\mathscr{F}}(a', a'')(U) \longrightarrow \underline{\mathrm{Hom}}_{\mathscr{F}}(a', a'')(V).$$

For $u \in \underline{\mathrm{Hom}}_{\mathscr{F}}(a', a'')(U) = \mathrm{Hom}_{\mathscr{F}(U)}(i_{U,T}(a'), i_{U,T}(a''))$ we form

$$i_{V,U}(u) : i_{V,U} i_{U,T}(a') \longrightarrow i_{V,U} i_{U,T}(a'') \quad \text{in } \mathscr{F}(V).$$

Then $\underline{\mathrm{Hom}}_{\mathscr{F}}(a', a'')(i)(u) \in \mathrm{Hom}_{\mathscr{F}(V)}(i_{V,T}(a'), i_{V,T}(a''))$ is the composite

$$v = \underline{\mathrm{Hom}}_{\mathscr{F}}(a', a'')(i)(u) = (\theta_{V,U,T}(a''))^{-1} [i_{V,U}(u)] \theta_{V,U,T}(a'),$$

or in other words, we have the following commutative diagram in $\mathscr{F}(V)$

$$
\begin{array}{ccc}
i_{V,U}i_{U,T}(a') & \xrightarrow{\;\;i_{V,U}(u)\;\;} & i_{V,U}i_{U,T}(a'') \\
{\scriptstyle \theta_{V,U,T}}\big\uparrow & & \big\uparrow{\scriptstyle \theta_{V,U,T}} \\
i_{V,T}(a') & \cdots\cdots\cdots\xrightarrow{\;\;v\;\;}\cdots\cdots\cdots & i_{V,T}(a'').
\end{array}
$$

For $W \subset V \subset U \subset T$ and inclusions $j : W \to V$ and $i : V \to U$, we must check the transitivity of the restriction functions, that is,

$$
\underline{\mathrm{Hom}}_{\mathscr{F}}(a',a'')(j)\underline{\mathrm{Hom}}_{\mathscr{F}}(a',a'')(i) = \underline{\mathrm{Hom}}_{\mathscr{F}}(a',a'')(ij).
$$

For this, we start by applying the functor $i_{W,V}$ to the previous diagram

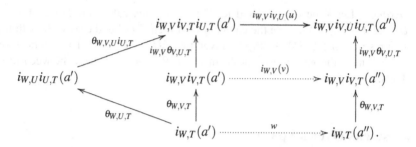

With this notation, $w = \underline{\mathrm{Hom}}_{\mathscr{F}}(a',a'')(j)(v)$, $u = \underline{\mathrm{Hom}}_{\mathscr{F}}(a',a'')(i)(u)$, and $w = \underline{\mathrm{Hom}}_{\mathscr{F}}(a',a'')(ij)(u)$ by the outer commutative square.

Although the category over a space is not a presheaf in general with the proof of this proposition, we see that for each open set T of X and for each pair of objects a' and a'' in $\mathscr{F}$, there is a morphism presheaf $\underline{\mathrm{Hom}}_{\mathscr{F}}(a',a'')$ on the space T.

4.2. Definition Let $\phi : \mathscr{F} \to \mathscr{G}$ be a morphism of categories over X. If T is an open set and if a', a'' are objects in the category $\mathscr{F}(T)$, then induced functions $\phi_{a',a''}(U)$ where $U \subset T$ are defined by the following commutative diagram

$$
\begin{array}{ccc}
\underline{\mathrm{Hom}}_{\mathscr{F}}(a',a'')(U) & \xrightarrow{\;\;\phi_{a',a''}(U)\;\;} & \underline{\mathrm{Hom}}_{\mathscr{G}}(\phi_T(a'),\phi_T(a''))(U) \\
\big\| & & \big\| \\
\mathrm{Hom}_{\mathscr{F}(U)}(i_{U,T}(a'),i_{U,T}(a'')) & & \big\| \\
{\scriptstyle \phi_U}\big\downarrow & & \big\| \\
\mathrm{Hom}_{\mathscr{G}(U)}(\phi_U i_{U,T}(a'),\phi_U i_{U,T}(a'')) & \xrightarrow{\;\;\mathrm{Ad}(\eta)\;\;} & \mathrm{Hom}_{\mathscr{G}(U)}(i_{U,T}\phi_T(a'),i_{U,T}\phi_T(a''))
\end{array}
$$

Here, the function $\mathrm{Ad}(\eta)(f)$ is the conjugation defined by requiring the following diagram to be commutative

$$\phi_U(i_{U,T}(a')) \xrightarrow{\quad f \quad} \phi_U(i_{U,T}(a''))$$

$$\eta_{U,T} \downarrow \qquad\qquad\qquad \downarrow \eta_{U,T}$$

$$i_{U,T}(\phi_T(a')) \xrightarrow{\ \mathrm{Ad}(\eta)(f)\ } i_{U,T}(\phi_T(a'')).$$

4.3. Proposition *Let* $\phi : \mathscr{F} \to \mathscr{G}$ *be a morphism of categories over* X, *let* T *be an open set, and let* a', a'' *be objects in the category* $\mathscr{F}(T)$. *Then, the induced functions*

$$\phi_{a',a''}(U) : \underline{\mathrm{Hom}}_{\mathscr{F}}(a',a'')(U) \longrightarrow \underline{\mathrm{Hom}}_{\mathscr{G}}(\phi_T(a'),\phi_T(a''))(U),$$

where U *is open in* T *is a morphism of presheaves of sets on* T.

Proof. We have to check the commutativity relation

$$\phi_{a',a''}(V)\underline{\mathrm{Hom}}_{\mathscr{F}}(a',a'')(i) = \underline{\mathrm{Hom}}_{\mathscr{G}}(\phi_T(a'),\phi_T(a''))(i)\phi_{a',a''}(U).$$

For $u \in \underline{\mathrm{Hom}}_{\mathscr{F}}(a',a'')(U) = \mathrm{Hom}_{\mathscr{F}(U)}(i_{U,T}(a'),i_{U,T}(a''))$, we form

$$i_{U,V}(u) : i_{V,U}i_{U,T}(a') \longrightarrow i_{V,U}i_{U,T}(a'') \text{ in } \mathscr{F}(V).$$

Then $v \in \underline{\mathrm{Hom}}_{\mathscr{F}}(a',a'')(i)(u) = \mathrm{Hom}_{\mathscr{F}(V)}(i_{V,T}(a'),i_{V,T}(a''))$ is the morphism in the following commutative diagram in $\mathscr{F}(V)$

$$i_{V,U}i_{U,T}(a') \xrightarrow{\quad i_{U,V}(u) \quad} i_{V,U}i_{U,T}(a'')$$

$$\theta_{V,U,T} \uparrow \qquad\qquad\qquad \uparrow \theta_{V,U,T}$$

$$i_{V,T}(a') \cdots\cdots\xrightarrow{\ v\ }\cdots\cdots i_{V,T}(a'').$$

Now applying the morphism $\phi_{a',a''}(V)$ to v gives

$$\underline{\mathrm{Hom}}_{\mathscr{F}}(a',a'')(V) \xrightarrow{\quad \phi_{a',a''}(V) \quad} \underline{\mathrm{Hom}}_{\mathscr{G}}(\phi_T(a'),\phi_T(a''))(V)$$

$$\|$$

$$\mathrm{Hom}_{\mathscr{F}(V)}(i_{V,T}(a'),i_{V,T}(a''))$$

$$\phi_V \downarrow \qquad\qquad\qquad \|$$

$$\mathrm{Hom}_{\mathscr{G}(V)}(\phi_V i_{V,T}(a'),\phi_V i_{V,T}(a'')) \xrightarrow{\ \mathrm{Ad}(\eta)\ } \mathrm{Hom}_{\mathscr{G}(V)}(i_{V,T}\phi_T(a'),i_{V,T}\phi_T(a'')).$$

Putting the diagrams together, we have following commutative diagram for $i : V \to U$

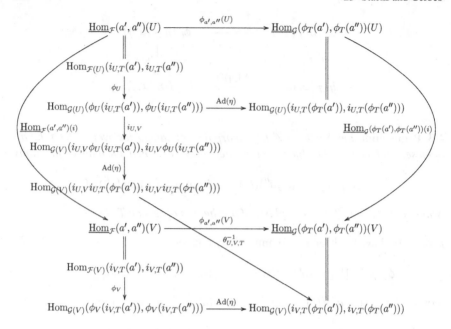

With this diagram, we establish the commutativity of the restriction morphisms with the morphism induced by ϕ.

In terms of these presheaves we can introduce the next concept.

4.4. Definition A category $\mathscr{F}$ over a space X is a prestack provided for each open T in X and each pair of objects a', a'' in $\mathscr{F}(T)$ the presheaf $\underline{\mathrm{Hom}}_{\mathscr{F}}(a', a'')$ on T is a sheaf.

In the category of categories $\mathscr{T}$ over a space X, we have the full subcategory of prestacks.

4.5. Remark For a space X, the category of sheaves $\mathrm{Sh}(X)$ is a subcategory of the category $\mathrm{Op}(X)^{\vee}$ of sets on X with inclusion $j : \mathrm{Sh}(X) \to \mathrm{Op}(X)^{\vee}$. There is a functor $\sigma : \mathrm{Op}(X)^{\vee} \to \mathrm{Sh}(X)$ which associates to every presheaf $\mathscr{P}$ on X, a universal sheaf $\sigma(\mathscr{P})$ on X with a morphism $\beta : \mathscr{P} \to j\sigma(\mathscr{P})$ having the universal property that the function $b(\phi) = j\beta(\phi)$ defines a bijection

$$b : \mathrm{Hom}_{\mathrm{Sh}(X)}(\sigma(\mathscr{P}), \mathscr{E}) \longrightarrow \mathrm{Hom}_{\mathrm{Op}(X)^{\vee}}(\mathscr{P}, j(\mathscr{E})).$$

We say that σ is the left adjoint functor to the inclusion functor $j : \mathrm{Sh}(X) \to \mathrm{Op}(X)^{\vee}$, and from the universal property which incorporated in the bijection b, sheafification σ is unique up to isomorphism of functors.

4.6. Notation Let X be a space. Let (Cat/X) denote the category of categories over X, and let (preSt/X) denote the full subcategory of prestacks over X.

4.7. Proposition *The inclusion functor*

$$j : (\mathrm{preSt}/X) \longrightarrow (\mathrm{Cat}/X)$$

has a left adjoint functor $\sigma : (\mathrm{Cat}/X) \to (\mathrm{preSt}/X)$, *where* σ *is related to the sheafification* σ *by the requirements that* $\sigma\mathscr{F}(U)$ *has the same objects as* $\mathscr{F}(U)$ *and the morphism set is given by cross sections of the sheafification of the presheaves* $\underline{\mathrm{Hom}}_{\mathscr{F}}(a', a'')$

$$\mathrm{Hom}_{\sigma\mathscr{F}(U)}(a', a'') = \Gamma(U, \sigma\underline{\mathrm{Hom}}_{\mathscr{F}}(a', a'')(U)).$$

Proof. In the statement, we have defined σ as identity on objects and as sheafification on the morphism sets. We have to see that the composition operation is well defined for the sheafification of the presheaves $\underline{\mathrm{Hom}}_{\mathscr{F}}(a', a'')$. For a morphism of presheaves $\mathscr{P}' \times \mathscr{P}'' \to \mathscr{P}$, we have the following composite morphism of sheafification $\sigma(\mathscr{P}') \times \sigma(\mathscr{P}'') \to \sigma(\mathscr{P}' \times \mathscr{P}'') \to \sigma(\mathscr{P})$ which is used to construct the composition in $\sigma(\mathscr{F})$. The universal property follows from the universal property in (4.5). The proposition now follows easily.

5 Descent Data

In the previous section, we saw that descent data in a prestack were present on the morphism sets through the condition that $\underline{\mathrm{Hom}}_{\mathscr{F}}(a', a'')$ is a sheaf. Now, we consider a descent condition on the objects which is used to define a stack as a prestack. We use the notation $a|V = i_{V,U}(a)$ for an object in $\mathscr{F}_U$ and $V \subset U$.

5.1. Definition Let $\mathscr{F}$ be a category over a space X, and let $\mathscr{U} = \{U_i\}_{i \in I}$ be an open covering of U in X. A descent datum is two families $\{a_i, \theta_{i,j}\}_{i,j \in I}$, where a_i is an object in $\mathscr{F}(U_i)$ and $\theta_{i,j} : a_j|U_i \cap U_j \to a_i|U_i \cap U_j$ is an isomorphism in $\mathscr{F}(U_i \cap U_j)$ satisfying the cocycle condition

$$\theta_{ii} = \mathrm{id} \quad \text{and} \quad \theta_{i,k} = \theta_{i,j}\theta_{j,k} \in \mathscr{F}(U_i \cap U_j \cap U_k).$$

5.2. Example Each object a in $\mathscr{F}_U$ determines a descent datum $\{a_i, \theta_{i,j}\}_{i,j \in I}$ for every covering $\mathscr{U} = \{U_i\}_{i \in I}$ of U, where $a_i = i_{U_i,U}(a)$ and $\theta_{i,j}$ is the composition of an isomorphism and its inverse between $i_{U_{i,j},U}(a)$ and the two restrictions $i_{U_{i,j},U_i}(a_i)$ and $i_{U_{i,j},U_j}(a_j)$. Let $R(a, \mathscr{U})$ denote this descent datum associated to a.

5.3. Definition Let $\{a_i', \theta_{i,j}'\}_{i,j \in I}$ and $\{a_i'', \theta_{i,j}''\}_{i,j \in I}$ be two descent data for a covering $\mathscr{U} = \{U_i\}_{i \in I}$. A morphism is an indexed family $f_i : a_i' \to a_i''$ in $\mathscr{F}_i$ of U such that for $i, j \in I$ the following diagram is commutative

$$
\begin{array}{ccc}
a_j'|U_{i,j} & \xrightarrow{f_j} & a_j''|U_{i,j} \\
\theta_{i,j}' \downarrow & & \downarrow \theta_{i,j}'' \\
a_i'|U_{i,j} & \xrightarrow{f_i} & a_i''|U_{i,j}.
\end{array}
$$

Indexwise composition of morphisms is again a morphism. The resulting category of all descent data for a covering $\mathscr{U} = \{U_i\}_{i \in I}$ of U is denoted by $\Delta(\mathscr{U}, \mathscr{F})$.

5.4. Remark The restriction in (5.2)

$$R(\ , \mathscr{U}) : \mathscr{F}_U \longrightarrow \Delta(\mathscr{U}, \mathscr{F})$$

is a functor for each open covering $\mathscr{U} = \{U_i\}_{i \in I}$ of U. We call $R(a, \mathscr{U})$ the descent datum of a on $\mathscr{U}$.

5.5. Definition A prestack $\mathscr{F}$ over X is a stack provided for each open set U, and for each open covering $\mathscr{U} = \{U_i\}_{i \in I}$ of U, the restriction functor $\mathscr{F}_U \to \Delta(\mathscr{U}, \mathscr{F})$ is an equivalence of categories. Let (St/X) denote the full subcategory of (preSt/X) consisting of stacks.

5.6. Remark Let $\mathscr{F}$ be a category over X. The restriction functor

$$\mathscr{F}_U \to \Delta(\mathscr{U}, \mathscr{F})$$

is fully faithful for all open sets U if and only if $\mathscr{F}$ is a prestack. This means that a category $\mathscr{F}$ over X is a stack if and only if it is a prestack and every descent datum on any open covering $\mathscr{U}$ of U is isomorphic to a restriction $R(a, \mathscr{U})$ for some object a in $\mathscr{F}_U$.

5.7. Remark If $\phi : \mathscr{F}' \to \mathscr{F}''$ is a morphism of categories over X, then for each open covering $\mathscr{U}$ the function

$$\Delta(\phi) : \Delta(\mathscr{U}, \mathscr{F}') \longrightarrow \Delta(\mathscr{U}, \mathscr{F}'')$$

defined by $\Delta(\phi)(\{a_i', \theta_{i,j}'\}_{i,j \in I}) = \{\phi(a_i'), \phi(\theta_{i,j}')\}_{i,j \in I}$ is a functor between two categories of descent data.

5.8. Example For a topological space, the function which assigns to each open set U the category Vect(U) is a stack and to each open set U the category Sh($U, \mathscr{C}$) of sheaves with values in $\mathscr{C}$ is also a stack.

6 The Stack Associated to a Prestack

In (4.7), we showed how the inclusion functor from prestacks to categories $j :$ (preSt/X) $\to$ (Cat/X) over X has a universal property, that is, it is a left adjoint functor. To show that the analog is true for the inclusion functor $j :$ (St/X) $\to$ (preSt/X), we study how a descent datum changes under the change of coverings. As with sheaves, a colimit of descent data will construct the universal stack determined by a prestack. For this, we need notions where functors are equivalences and where morphisms of categories are equivalences.

6.1. Definition A functor $F : \mathscr{C}' \to \mathscr{C}''$ is faithful (resp. full, fully faithful) provided $F : \mathrm{Hom}_{\mathscr{C}'}(X,Y) \to \mathrm{Hom}_{\mathscr{C}''}(F(X),F(Y))$ is injective (resp. surjective, bijective) for all objects X in $\mathscr{C}$ and all objects Y in $\mathscr{C}'$.

Faithful, full, and fully faithful properties preserved by composition of functors.

6.2. Definition A functor $F : \mathscr{C}' \to \mathscr{C}''$ is an equivalence provided F is fully faithful and surjective up to isomorphism, that is, for each object Z in $\mathscr{C}''$, there exists an object X in $\mathscr{C}'$ with $F(X)$ and Z isomorphic.

6.3. Remark A functor $F : \mathscr{C}' \to \mathscr{C}''$ is an equivalence if and only if there exists a functor $G : \mathscr{C}'' \to \mathscr{C}'$ such that GF is isomorphic to identity on $\mathscr{C}'$ and FG is isomorphic to the identity on $\mathscr{C}''$. It is the case that G is both right and left adjoint of F, and for this reason, the first step in studying whether or not a functor F is an equivalence is to look for a left or right adjoint functor of F.

In (3.4), we introduced the notion of locally full morphisms of categories over X. Now, we consider equivalences of categories over a space.

6.4. Definition A morphism $\phi : \mathscr{F} \to \mathscr{G}$ of categories over X is an equivalence provided $\phi(U) : \mathscr{F}_U \to \mathscr{G}_U$ is an equivalence of categories for each open set U in X. A morphism $\phi : \mathscr{F} \to \mathscr{G}$ of categories over X is a weak equivalence provided $\phi(U)$ is fully faithful for each open set U in X and ϕ is locally full.

6.5. Remark Let $\phi : \mathscr{F} \to \mathscr{G}$ be an equivalence of categories over X. Then, $\mathscr{F}$ is a prestack (resp. stack) over X if and only if $\mathscr{G}$ is a prestack (resp. stack).

6.6. Remark If $\phi : \mathscr{F} \to \mathscr{G}$ is a weak equivalence of prestacks over X and if $\mathscr{F}$ is a stack, then ϕ is an equivalence. In other words, the local surjectivity gives arise to surjectivity map up to isomorphism by descent construction.

Similarly, we have the factorization property.

6.7. Remark Let $\phi : \mathscr{F}' \to \mathscr{F}''$ is a weak equivalence of prestacks over X and $\psi' : \mathscr{F}' \to \mathscr{G}$ be a morphism of prestacks over X. If $\mathscr{G}$ is a stack, then there exists a morphism $\psi'' : \mathscr{F}'' \to \mathscr{G}$ of prestacks over X such that $\psi''\phi$ and ψ' are naturally equivalent morphisms.

With sheaves, the inclusion $J : \mathrm{Sh}(X,\mathscr{C}) \to \mathrm{preSh}(X,\mathscr{C})$ has a left adjoint, but with the inclusion $\mathrm{Stack}(X) \to \mathrm{preSt}(X)$, we must use the following definition which is based on the assertion (6.7).

6.8. Definition For a prestack $\mathscr{F}$ over X, the associated stack is a pair consisting of a stack $s(\mathscr{F})$ together with a weak equivalence $\phi : \mathscr{F} \to s(\mathscr{F})$.

6.9. Theorem *For a prestack $\mathscr{F}$ over X, the associated stack exists and is unique up to a strong equivalence.*

The uniqueness is contained in (6.7), and the existence is a colimit construction as with the associated sheaf to a presheaf. The limit is over all coverings related to the descent data where for a sheaf of sets the related colimit is given by forming the stacks of the presheaf.

7 Gerbes as Stacks of Groupoids

We have considered gerbes as bundle gerbes and groupoid gerbes. Now, we return to the concept of a gerbe in terms of groupoids over a space X.

7.1. Definition A gerbe $\mathscr{G}$ on X is a stack over X such that each $\mathscr{G}(U)$ is a groupoid and the union of all open sets U such that $\mathscr{G}(U)$ is nonempty is X. It satisfies the following local connectivity or transitivity condition: For two objects a', a'' in any $\mathscr{G}(U)$ and x in U, there exists an open set V with $x \in V \subset U$ and a morphism $u : i_{V,U}(a') \to i_{V,U}(a'')$.

The basic example extends the idea of principal bundle. Instead of a group we start with a sheaf G of groups over X.

7.2. Definition A principal G-sheaf P over X is a sheaf of sets with an action $P \times G \to P$ over X of sheaves such that X is the union of all U with $P(U) = \Gamma(U,P)$ nonempty and $G(U)$ is free and transitive on $P(U)$ for each open subset U of X. A morphism $P' \to P''$ of principal G-sheaves is a morphism of sheaves commuting with the action G.

Another word for principal G-sheaf is torsor (it is a rather silly word).

7.3. Remark In terms of the étale spaces over X, the action is defined by $s(P) \times s(G) \to s(P)$ over X, where $s(P) \to X$ is surjective and projection and action $s(P) \times s(G) \to s(P)$ is a homeomorphism.

7.4. Remark The trivial principal G-sheaf is G itself, and a principal G-sheaf P is isomorphic to G if and only if $P(X) = \Gamma(X,P)$ is nonempty. Then, any element defines an isomorphism $G \to P$ by the action on the cross section of P over X.

7.5. Example Let G be a sheaf over X. Let $(\mathrm{Bun}_G(U))$ denote the category of principal $G(U)$ sheaves over U. Then, $(\mathrm{Bun}_G(U))$ is a gerbe over X.

There are two cohomological classifications by $H^1(X,G)$ for isomorphism classes of principal G-sheaves for a sheaf of groups and by $H^2(X,L)$ for gerbes $\mathscr{G}$ over X with given band L, see Sect. 9.

8 Cohomological Classification of Principal G-Sheaves

The cohomological classification of sheaves, bundles, and gerbes depends on the automorphisms of trivial bundles.

8.1. Remark The automorphism group of the trivial principal G-bundle G is just the group of cross sections $\Gamma(X,G)$ where to a cross section σ we associate the automorphism $u_\sigma : G \to G$ given by $u_\sigma(U) : G(U) \to G(U)$ given by $u_\sigma(\tau) = \tau(\sigma|U)$.

8.2. Principal G-Sheaves Trivial Over an Open Covering Let G be a group sheaf over the space X, and let $\mathscr{U} = \{U_i\}_{i \in I}$ be an open covering of X. For a principal

G-sheaf P where $P|U_i$ is trivial, we choose a section $\sigma_i \in \Gamma(U_i, P)$ defining the trivialization. By (8.1) over $U_{i,j} = U_i \cap U_j$, we have a section $g_{i,j} \in \Gamma(U_{i,j}, G)$ such that on $U_{i,j}$, we have $(\sigma_i|U_{i,j})g_{i,j} = (\sigma_j|U_{i,j})$.

(1) *Cocycle Condition*: Over the triple intersection $U_{i,j,k} = U_i \cap U_j \cap U_k$, we have the cocycle relation

$$g_{i,j}g_{j,k} = g_{i,k}.$$

(2) *Change of Section Leads to a Coboundary Relation*: For a second set of sections σ_i' we have new $g_{i,j}'$ such that on $U_{i,j}$, we have $(\sigma_i'|U_{i,j})g_{i,j}' = (\sigma_j'|U_{i,j})$. The two sets of sections are related again by elements $b_i \in \Gamma(U_i, G)$, where $\sigma_i' = \sigma_i b_i$. Hence, we have the cobounding relation

$$b_i g_{i,j}' = g_{i,j} b_j \quad \text{or} \quad g_{i,j}' = b_i^{-1} g_{i,j} b_j \quad \text{on } U_{i,j}$$

which follows from the following substitutions

$$\sigma_i g_{i,j} b_j = \sigma_j b_j = \sigma_j' = \sigma_i' g_{i,j}' = \sigma_i b_i g_{i,j}'.$$

8.3. Definition Let G be a sheaf of groups on a space X, and let $\mathcal{U} = (U_i)_{i \in I}$ be an open covering of X. The first cohomology set $H^1(\mathcal{U}, G)$ of the covering $\mathcal{U}$ of X with values in the sheaf G is the set of cocycles $(g_{i,j})_{i,j \in I}$ with values in G where $g_{i,j} \in \Gamma(U_{i,j}, G)$ modulo the cobounding equivalence $g_{i,j}' = b_i^{-1} g_{i,j} b_j$, where sections $b_i \in \Gamma(U_i, G)$ for $i \in I$.

8.4. Remark Let $(g_{i,j})_{i,j \in I}$ be a cocycle with values in G relative to a covering $\mathcal{U} = \{U_i\}_{i \in I}$. Then, there is a principal G-sheaf P, where $P|U_i$ is trivial for each $i \in I$ and the trivializations are related by the isomorphism defined by $g_{i,j}$ on $U_{i,j}$. Moreover, two cocycles $(g_{i,j})_{i,j \in I}$ and $(g_{i,j}')_{i,j \in I}$ define isomorphic principal G-sheaves if and only if they are cobounding. This gives the following proposition.

8.5. Proposition *The cohomology set $H^1(\mathcal{U}, G)$ classifies principal G-sheaves which are trivial on all $U_i \in \mathcal{U}$ up to isomorphism.*

To remove the dependence on the covering $\mathcal{U}$ in (8.5), we introduce a certain colimit.

8.6. Definition Let G be a sheaf of groups on a space X. The first Čech cohomology set $H^1(X, G)$ is the colimit of all $H^1(\mathcal{U}, G)$ over the open coverings $\mathcal{U}$ of X. A morphism between coverings $\mathcal{U} = \{U_i\}_{i \in I} \to \mathcal{B} = \{V_i\}_{i \in I}$ is defined by a function $\theta : J \to I$, where $U_{\theta(j)} \supset V_j$ for all $j \in J$ in order to define the colimit.

In terms of the first Čech cohomology set, we have the classification of all principal G-sheaves because each principal G-sheaf is trivial for some open covering of X.

8.7. Proposition *The cohomology set $H^1(X, G)$ classifies principal G-sheaves up to isomorphisms.*

9 Cohomological Classification of Bands Associated with a Gerbe

The cohomological classification of bands in a gerbe depends on various sheaves associated to the gerbe.

9.1. Sheaves on Open Sets in a Covering Let G be a gerbe over X. Then, for each object in $G(T)$ over an open $T \subset X$, the automorphisms

$$\underline{\mathrm{Aut}}_G(a,a) = \underline{\mathrm{Hom}}_G(a,a)$$

is a sheaf on T. Let $X = \bigcup_{i \in I} U_i$ be an open covering of X, and let a_i be an object in $G(U_i)$. Then, $\underline{\mathrm{Aut}}_G(a_i, a_i)$ is a sheaf of groups on U_i.

9.2. Relating Automorphism Sheaves on Over Lapping $U_i \cap U_j$ open Sets With the notations of (9.1) and for an open covering

$$U_{i,j} = U_i \cap U_j = \bigcup_{k \in I_{(i,j)}} U_{i,j}(k),$$

we have an isomorphism

$$w_{i,j}(k) : \mathrm{res}(a_i) \longrightarrow \mathrm{res}(a_j)$$

in $G(U_{i,j}(k))$.

9.3. Band Datum for G and the Coverings $U_i \cap U_j = \bigcup_{k \in I_{(i,j)}} U_{i,j}(k)$ With previous isomorphism

$$w_{i,j}(k) : \mathrm{res}(a_i) \longrightarrow \mathrm{res}(a_j)$$

in $G(U_{i,j}(k))$ we can conjugate to obtain a morphism

$$\lambda_{i,j}(k) = \mathrm{Ad}(w_{i,j}(k)) : \mathrm{Aut}(a_i)|U_{i,j}(k) \longrightarrow \mathrm{Aut}(a_j)|U_{i,j}(k)$$

which is an isomorphism of sheaves of groups.

9.4. Equivalence Class of Covering Band Datum for a Gerbe G Two choices in (9.3) of

$$w_{i,j}(k) : \mathrm{res}(a_i) \longrightarrow \mathrm{res}(a_j)$$

give morphisms

$$\lambda_{i,j}(k) = \mathrm{Ad}(w_{i,j}(k))$$

differing by inner automorphisms. Hence, the band data is

$$\{\underline{\mathrm{Aut}}(a_i),\ \lambda_{i,j}(k) \in \underline{\mathrm{Out}}(a_i, a_j)\}$$

which describes the gerbe.

9.5. Outer Automorphism Sheaves For a group sheaf G over X, we have $\underline{\mathrm{Aut}}(G)$, the automorphism sheaf and the exact sequence of sheaves

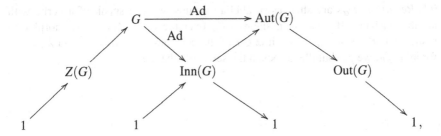

where $\mathrm{Ad}(g)h = ghg^{-1}$ on the stacks and sections. For two sheaves of groups G' and G'', we have the generalization to the isomorphism sheaf

$$X \supset T \longmapsto \mathrm{Iso}(G'|T, G''|T),$$

where $G''|T$ acts by conjugation Ad on the isomorphisms. The quotient

$$X \supset T \longmapsto \mathrm{Iso}(G'|T, G''|T)/\mathrm{Ad}G''|T$$

is a presheaf and the associated sheaf is $\underline{\mathrm{Out}}(G', G'')$. An element $\phi \in \Gamma(T, \underline{\mathrm{Out}}(G', G''))$ is an outer isomorphism $G'|T \to G''|T$.

Let G be a gerbe on X so that over each open set we have a groupoid $G(U)$. Gerbes are analyzed using the notion of the band data.

9.6. Definition A band data for an open covering $X = \bigcup_\alpha U_\alpha$ and a gerbe is a family of sheaves of groups K_α over U_α, a family $\lambda_{\alpha,\beta} : K_\beta|U_\alpha \cap U_\beta \to K_\alpha|U_\alpha \cap U_\beta$ of isomorphisms, and sections $g_{\alpha,\beta,\gamma} \in \Gamma(U_\alpha \cap U_\beta \cap U_\gamma, K_\alpha)$ satisfying

$$\lambda_{\alpha,\beta}\lambda_{\beta,\gamma} = g_{\alpha,\beta,\gamma}\lambda_{\alpha,\gamma},$$

$$g_{\alpha,\beta,\gamma}g_{\alpha,\gamma,\delta} = \lambda_{\alpha,\beta}(g_{\beta,\gamma,\delta})g_{\alpha,\beta,\delta},$$

and

$$\lambda_{\alpha,\alpha} = 1, \quad g_{\alpha,\alpha,\gamma} = g_{\alpha,\gamma,\gamma} = 1.$$

9.7. Definition The cocycle data $(\lambda_{\alpha,\beta}, g_{\alpha,\beta,\gamma})$ is equivalent to the cocycle data $(\lambda'_{\alpha,\beta}, g'_{\alpha,\beta,\gamma})$ provided that there exists sections $\kappa_{\alpha,\beta} \in \Gamma(U_\alpha \cap U_\beta, K_\alpha)$ such that

$$\lambda'_{\alpha,\beta} = \kappa_{\alpha,\beta}\lambda_{\alpha,\beta}$$

and

$$g'_{\alpha,\beta,\gamma} = \kappa_{\alpha,\beta}\lambda_{\alpha,\beta}(\kappa_{\beta,\gamma})g_{\alpha,\beta,\gamma}\kappa_{\alpha,\gamma}^{-1}.$$

The set of equivalence classes of cocycle data for $X = \bigcup_\alpha U_\alpha$ and K_α is denoted

$$\check{H}^2(\{U_\alpha\}_\alpha, K_\alpha).$$

These are the nonabelian Čech classes.

9.8. Remark If K_αs are sheaves of abelian groups, we can speak of a gerbe with an abelian band. If, moreover, sheaves K_α over the opens U_α are isomorphic to restrictions $K_\alpha \sim K|U_\alpha$ for each $\alpha \in I$ of the constant sheaf $K = S^1$ and $\lambda_{\alpha,\beta} = 1$ for all $U_{\alpha,\beta}$, we get bundle gerbes in the sense of Chap. 25.

Bibliography

Adams, J.F., Stable homotopy and generalised homology. Chicago Lectures in Math, University of Chicago (1974)

Atiyah, M.F., On the K-theory of compact Lie groups. Topology **4**: 95–99 (1965)

Atiyah, M.F., K-theory and reality. Quart. J. Math. Oxford Ser. **17(2)**: 367–386 (1966)

Atiyah, M.F., Topological quantum field theory. Publ. Math. Inst. Hautes Etudes Sci. **68**: 175–186 (1989)

Atiyah, M.F., K-theory past and present. Sitzungsberichte der Berliner Mathematischen Gesellschaft. 411–417 (arXiv:math.KT/001221) (2001)

Atiyah, M.F., Bott, R., Shapiro, A.: Clifford modules. Topology 3(1): 3–38 (1964)

Atiyah M.F., Segal, G.B.: Twisted K-theory. Ukrainian Math. Bull. **1** (arXiv:math.KT/0407054) (2004)

Atiyah M.F., Singer, I.M.: Index theory for skew-adjoint Fredholm operators. Publ. Inst. Hautes Études Sci. **37**: 5–26 (1969)

Beilinson, A., Drinfeld, V.: Chiral Algebras. American Mathematical Society Colloquium Publications, Vol. 51, Providence, RI (2004)

Blackadar, B.: K-Theory for Operator Algebras. Mathematical Sciences Research Institute Publications, 5th ed. Springer Verlag, New York (1986)

Borel, A., Serre, J.-P.: Le théoreme de Riemann-Roch (d'après Grothendieck). Bull. Soc. Math. France, **86**: 97–136 (1958)

Borel, A.: Seminar on transformation groups. Ann. of Math. Stud. **46** (1960)

Borel, A.: Sous-Groupes commutatifs et torsion des groupes de Lie compacts connexes. Tôhoku Math. J. **13**: 216–240 (1961)

Bott, R.: Homogeneous vector bundles. Ann. Math. **66**: 203–248 (1957)

Bott, R.: The space of loops on a Lie group. Mich. Math. J. **5**: 35–61 (1958)

Bott, R.: Some remarks on the periodicity theorems. Colloque de Topologie, Lille (1959)

Brown, L.G., Douglas, R.G., Fillmore, P.A.: Extensions of C*-algebras and K-homology. Ann. of Math. **105**: 265–324 (1977)

Brylinski, J.-L.: Loop Spaces, Characteristic Classes and Geometric Quantization. Birkhäuser, Boston (1993)

Carey, A.L., Wang, B.L.: Thom isomorphism and push-forward map in twisted K-theory (2005) (arXiv:math.KT/0507414)

Connes, A.: Noncommutative Geometry. Academic Press (1994)

Dixmier, J.: C*-algebras. North-Holland, Amsterdam and New York (1977)

Dold, A.: Partitions of unity in the theory of fibrations. Ann. of Math. **78**: 223–255 (1963)

Donovan, P., Karoubi, M.: Graded Brauer groups and K-theory with local coefficients. Publ. Math. I.H.É.S. **38**: 5–25 (1970)

Duistermaat, J.J., Kolk, J.A.C.: Lie Groups. Springer Verlag, Berlin (2000)

Eilenberg, S., Steenrod, N.E.: Foundations of Algebraic Topology. Princeton University Press Princeton, NJ (1952)

Feigin, B.L., Frenkel, E.V.: Affine Kac-Moody algebras and semi-infinite flag manifolds. Comm. Math. Phys. **128**: 161–189 (1990)

Freed, D.S.: The Verlinde algebra is twisted equivariant K-theory. Turkish. J. Math. **25**: 159–167 (2001) (arXiv:math.RT/0101038)

Freed, D.S.: Twisted K-theory and loop groups. Proc. Int. Congress of Mathematicians. **3**: 419–430 (2002)

Freed, D.S., Hopkins, M.S., Teleman, C.: Loop groups and twisted K-theory III. (2003a) (arXiv: math.AT/0312155)

Freed, D.S., Hopkins, M.S., Teleman, C.: Twisted K-theory and loop group representations I. (2003) (arXiv:math.AT/0312155)

Frenkel, I.B., Garland, H., Zuckerman, G.J.: Semi-infinite cohomology and string theory. Proc. Nat. Acad. Sci. U.S.A. **83**: 8442–8446 (1986)

Garland, H., Lepowsky, J.: Lie algebra homology and the Macdonald-Kac formulas. Invent. Math. **34**: 37–76 (1976)

Hirzebruch, F.: Some problems on differentiable and complex manifolds. Ann. of Math. **60**: 213–236 (1954)

James, I.M., Segal, G.B.: On equivariant homotopy type. Topology **17**: 267–272 (1987)

Kac, V.G.: Infinite Dimensional Lie Algebras 3rd ed. Cambridge University Press, Cambridge (1990)

Kirillov, A.A.: Lectures on the orbit method. Graduate Studies in Mathematics, Vol. 64 American Mathematical Society, (2004)

Kostant, B.: Lie algebra cohomology and the generalized Borel-Weil theorem. Ann. of Math. **74**: 329–387 (1961)

Kostant, B.: A cubic Dirac operator and the emergence of Euler number mulitplets of representations for equal rank subgroups. Duke Math. J. **100**: 447–501 (1999)

Kostant, B.: A generalization of the Bott-Borel-Weil theorem and Euler number multiplets of representations. Lett. Math. Phys. **52**(1): 61–78 (2000)

Kostant, B., Sternberg, S.: Symplectic reduction, BRS cohomology, and infinite-dimensional Clifford algebras. *Ann. Phys.* **176**: 49–113 (1987)

Kumar, S.: Demazure character formula in arbitrary Kac-Moody setting. *Invent. Math.* **89**: 395–423 (1987)

Landweber, G.D.: Multiplets of representations and Kostant's Dirac operator for equal rank loop groups. Duke Math. J. **110**: 121–160 (2001)

Laredo, J.V.: Positive energy representations of the loop groups of non-simply connected Lie groups. Comm. Math. Phys. **207**: 307–339 (1999)

Lawson, H.B., Michelsson, M.L.: Spin geometry. Princeton Mathematical Series, Vol. 38. Princeton University Press, Princeton, NJ (1989)

Madsen, I., Rosenberg, J.: The universal coefficient theorem for equivariant K-theory of real and complex C*-algebras. In: Kaminker J., Millet K., Schochet C (eds.) Index Theory of Elliptic Operators, Foliations, and Operator Algebras, [Contemp. Math. **70** Amer. Math. Soc., Providence, RI, 145–173] (1998)

Mickelsson, J.: Twisted K-theory invariants. Lett. in Math. Phys. **71**: 109–121 (2005) (arXiv:math. AT/0401130)

Mickelsson, J.: Gerbes, (twisted) K-theory, and the super-symmetric WZW model. Lect. Math. Theor. Phys. **5**: 93–107 (2004)

Moore, G., Segal, G.: D-branes and K-theory in topological field theory (2006) (arXiv:hep-th/0609042)

Murphy, G.J.: C*-algebras and operator theory. Academic Press (1990)

Pressley, A., Segal, G.: Loop Groups. Oxford University Press, New York (1986)

Quillen, D.G.: Homotopical Algebra. Springer Lecture Notes, Vol. 43 (1967)

Quillen, D.G.: Rational homotopy theory, Ann. Math. **90**: 205–295 (1969)

Quillen, D.G.: Superconnections and the Chern character. Topology. **24**: 89–95 (1985)

Rosenberg, J.: Homological invariants of extensions of C*-algebras. In: Kadison R.V.(ed.) Operator Algebras and Applications [Proc. Sympos. Pure Math. **38**, Part 1. Amer. Math. Soc., Providence, RI, 35–75] (1982)

Schottenloher, M.: *A mathematical introduction to conformal field theory*. Lect. Notes Phys. **m43**. Springer-Verlag, Heidelberg (1997) (2nd edition to appear)

Segal, G.B.: Classifying spaces and spectral sequences. Publ. Math. I.H.E.S. Paris, **34**: 105–112 (1968)

Segal, G.B.: Fredholm complexes. Quart. J. Math. **21**: 385–402 (1970)

Segal, G.B.: Topological structures in string theory. Phil. Trans. R. Soc. Ser. A. **359**: 1389–1398 (2001)

Sorger, C.: La formule de Verlinde. Séminaire Bourbaki, 1994/95. Astérisque **237**: 87–114 (1996)

Steenrod, N.E.: A convenient category of topological spaces. Michigan Math. J. **14**: 133–152 (1967)

Sullivan, D.: Infinitesimal computations in topology. Inst. Hautes Études Sci. Publ. Math. **47**: 269–331 (1977)

Taubes, C.H.: Notes on the Dirac operator on loop space. Unpublished manuscript (1989)

Teleman, C.: Borel-Weil-Bott theory on the moduli stack of G-bundles over a curve. Invent. Math. **134**: 1–57 (1998)

Teleman, C.: K-theory of the moduli of principal bundles on a surface and deformations of the Verlinde algebra. Topology, Geometry and Quantum Field Theory: Proceedings of the 2002 Oxford Symposium in Honour of the 60th Birthday of Graeme Segal. CUP (2004)

Thom, R.: Espacés fibrès en sphères et carrés de Steenrod. Ann. Sci. Ecole. Norm. Sup. **69** (1952), 109–182

Tu, J.-L. and Xu, P.: Chern character for twisted K-theory of orbifolds. (2005) (arXiv: math. KT/0505267)

Wendt, R.: A character formula for certain representations of loop groups based on non-simply connected Lie groups. Math. Z. **247**: 549–580 (2004)

Witten, E.: The Verlinde Algebra and the Cohomology of the Grassmannian. Gemoetry, topology, & physics conf. Proc. Lecture Notes Gerom. Topology, IV, 357–422 Int. Press, Cambridge, MA (1995) (arXiv: hep-th/9312104)

Wood, R.: Banach algebras and Bott periodicity. Topology **4**(6): 371–389 (1965)

Index of Notations

See also the *Notation for Examples of Categories*

$_{(G,\,A)}\mathrm{Mod}$	The category of (G,A)-modules and morphisms consisting of A-linear G-equivariant maps, where A is a G-algebra	18(1.5)
$K_G^0(X)$	G-equivariant K-theory	18(1.8)
$K_G(A)$	The Grothendieck group providing the G-equivariant K-theory of a G-algebra A	18(1.9)
$G \ltimes H$	The cross product of groups G, H related to an action of G on H	18(2.1)
$C_c(G, A)$	The $*$-algebra of compactly supported continuous functions $G \to A$ on a topological group G with values in a Banach algebra A	18(2.3)
$C(X,A)$	The algebra of all continuous functions $f : X \to A$ which vanish at ∞	18(3.4)
f_*	Induced morphism $\mathrm{Ext}(X) \to \mathrm{Ext}(Y)$	18(4.9)
$\mathrm{Ext}(B, A)$	The set of isomorphism classes of extensions of B by A for two given locally convex algebras	18(5.2)
$C(B), S(B)$	Continuous cone resp. suspension of a locally convex algebra B	18(6.1)
$C^\vee(A), S^\vee(A)$	Dual cone $C^\vee(A)$ resp. dual suspension $S^\vee(A)$ of a locally convex algebra	18(6.10)
$kk_*(A, B)$	The biinvariant functor	18(8.1)
$\mathrm{Vect}^n(X)$	The set of isomorphism classes of n-dimensional vector bundles over X, see also 21(6.1)	20(1.2), 15(1.2)
$\mathrm{Alg}^n(X)$	The set of isomorphism classes of n by n matrix algebra bundles over X	20(1.4)
$\mathrm{Vect}^H(X)$	The set of isomorphism classes of vector bundles over X with fibre the Hilbert space H	20(2.6)
$\mathrm{Alg}^H(X)$	The set of isomorphism classes of infinite-dimensional algebra bundles modeled on $\mathscr{B}(H)$ or $\mathscr{K}(H)$ for a separable infinite-dimensional Hilbert space H	20(2.6)
$_nH^3(X, \mathbb{Z})$	The subgroup of n-torsion points in $H^3(X,\mathbb{Z})$	20(3.8), 20
$KP(X)$		21(4.3)
$\mathrm{Vect}^n(X)$	The set of isomorphism classes of $\mathscr{C}_X$-modules of sheaves locally isomorphic to $\mathscr{C}_X^n$	21(6.1)

Notation for Examples of Categories

Categories are typically denoted by suitable abbreviations enclosed in brackets. Here are some examples.

(set)	Category of sets
(set)$_*$	Category of pointed sets
(gr)	Category of groups
(ab)	Category of abelian groups
(rg)	Category of rings
(c\rg)	Category of commutative rings
(k)	Category of modules over k, where k is a commutative ring
(vect/k)	Category of finitely generated projective modules over k, where k is a commutative ring
(alg/k)	Category of unital algebras over k
(c\alg/k)	Category of commutative unital algebras over k
(R\mod)	Category of left R-modules for a ring R
(mod/R)	Category of right R-modules
(R\mod/R)	Category of R-bimodules
(top)	Category of topological spaces
(top)$_*$	Category of pointed topological spaces
(htp)	Homotopy category of topological spaces
(htp)$_*$	Homotopy category of pointed topological spaces
(cpt)	Category of compact (Hausdorff) spaces
(cpt)$_*$	Category of pointed compact (Hausdorff) spaces
(bun)	Category of bundles
(bun/B)	Category of bundles over B
(bun/gerbes)	Category of bundle gerbes
(bun/gerbes/B)	Category of bundle gerbes over B
(cat)	Category of small categories
Δ(set)	Category of simplicial sets

Index